History of the Research and Development of Chinese Marine Research Vessels

Han Long · Qu Ningning · Cao Daqiu

History of the Research and Development of Chinese Marine Research Vessels

上海交通大學出版社
SHANGHAI JIAO TONG UNIVERSITY PRESS

Springer

Han Long
Marine Design and Research Institute
of China
Shanghai, China

Qu Ningning
Marine Design and Research Institute
of China
Shanghai, China

Cao Daqiu
Marine Design and Research Institute
of China
Shanghai, China

ISBN 978-981-96-8413-7 ISBN 978-981-96-8414-4 (eBook)
https://doi.org/10.1007/978-981-96-8414-4

Jointly published with Shanghai Jiao Tong University Press

This Springer imprint is published by the registered company Springer Nature Singapore Pte Ltd.
The registered company address is: 152 Beach Road, #21-01/04 Gateway East, Singapore 189721, Singapore

Foreword

The *History of Chinese Ships Research and Development* series is a compilation of the development history of Chinese ships, with a particular focus on civilian ships, working ships, and ocean development equipment. It showcases the remarkable achievements and progress made in the research and development of these ships and relevant equipment since the founding of New China.

China has a long history of shipbuilding, boasting the first in the world to invent the Zhou (small boat), Shan (shampan), Fang (boat), and Ge (barge). As early as the Tang Dynasty (618–907), China had seen significant achievements in shipbuilding technology, with the emergence of watertight compartment, rudder balance, and perforated rudder. In terms of ship type, China had already developed various types of seagoing vessels by the Song and Yuan dynasties, among which were the Fuzhou Ship, Junk Ship, and Canton Ship from the Jiangnan coastal region, known as the three major ship types of ancient China. By the Ming Dynasty (1368–1644), the voyages led by Zheng He further demonstrated China's remarkable shipbuilding achievements, as they constructed 64 large ships within a mere 14 months. In modern times, although China was known to invest efforts in emulating the West and even produced ironclad ships and 10,000-ton ships, its shipbuilding industry inevitably succumbed to a decline and lagged behind Western powers. Ultimately, the nation endured a loss of national dignity and suffered national humiliation under the assaults of the Western powers with their advanced naval technologies. Fortunately, China's shipbuilding industry ushered in a resurgence with the birth of New China. In May 1949, shortly after the liberation of Shanghai, the Shanghai Military Control Committee established the East China Shipbuilding Committee. In September the same year, the Shipbuilding Industry Office under the Ministry of Heavy Industry of the People's Republic of China was inaugurated to administrate the nation's shipbuilding industry. This Office functioned to coordinate the development of the shipbuilding industry, as it pooled shipbuilding talents, expanded, renovated, and constructed new shipyards, as well as restructured and established shipbuilding institutions nationwide. With design and construction soaring ahead, this ancient shipbuilding power, after nearly 500 years of dormancy, eventually awakened! This book series provides elaborations on China's ship research and development history starting from the birth of New China, especially since the reform and opening-up. By giving focus on 10 types of civilian ships, including oil tankers, liquefied gas carriers, working ships, and research vessels, it reveals the spirit of arduous struggles, perseverance, innovation, and pursuit for excellence of New China's "shipbuilders", as they endeavored to rejuvenate China as an ancient shipbuilding power.

Over the 70 years of China's ship research and development history, the development of various types of ships, although not identical, can be roughly divided into three stages:

Foundation Consolidation and Steady Growth (1949–1977). At this stage, the central authorities prioritized the development of transportation industry, providing a broad space for the development of shipbuilding industry. This is a stage when the country was poor and backward, with all industries in need of revitalization. Nevertheless, China still placed great importance on the development of the shipbuilding industry. After several years of integration and development, by 1965, an overall organic system of ship research institutions had taken shape. Just to take the No. 7 Academy of China State Shipbuilding Corporation (China Ship Research and Development Academy) as an example, it alone had over ten research institutes with overall design and special equipment capabilities, covering all aspects of ship design. Furthermore, the country also expanded and built new large and medium-sized shipyards at both the central and local levels, adding facilities and improving technologies. In addition, a huge effort was invested to restore and develop water transportation, and develop various types of civilian ships that were urgently needed for national economic development, thus minimizing the purchase of ships. As a result, a significant number of ships that aligned with the development trends of the world's marine shipping industry at that time were successfully designed and built, which included medium-sized coastal oil tankers, passenger-cargo vessels, luxury ships for cruise on the Yangtze River, channel dredgers, port lifting ships, scientific survey vessel "Shijian", self-elevating drilling platform "Bohai 1", and hovercrafts. Most notably, the rollout of the 10,000-ton "Dongfeng" oceangoing cargo ship, which was independently designed and built by China and ranked top among New China's ten famous ships, marked the end of China's inability to design and build such large cargo ships and meantime heralded a new era in the country's shipbuilding history.

Rapid Development Driven by Reform and Opening-up (1978–2010). Prior to 1978, due to the technological blockade imposed by Western industrial powers, China had limited exchanges in ship science and technology with the outside world. This resulted in the country's heavy reliance on imports for various types of large transport ships, dredging equipment, and ocean development ships. After 1978, benefiting from China's reform and opening-up, the Chinese shipbuilding industry galloped forward like a spirited horse. In 1982, the 27,000-ton bulk carrier "Great Wall" was built. It marked a significant milestone as the first export ship designed and built according to international conventions, regulations, and foreign classification society standards. Since then, China's design and building of various types of working ships and ocean development equipment began to bridge with the world's ship technology development. Through introduction, assimilation, and innovation, the nation's ship research and design technology ushered in leapfrog development, and the shipbuilding capabilities of major shipyards were significantly enhanced. By the late 20th century, China had emerged as a major shipbuilding power in the world. Not only did it end its reliance on imported ships, but it also exported a large quantity of ships to numerous countries. During this period, various ship types witnessed significant development: Container ships emerged from non-existence, with their export ranging from the 700 TEU full-container ships to the 4,700-ton multipurpose container ships; Multiple types of ships were designed and built, including the 50,000-ton large hatch multipurpose bulk carriers, 150,000-ton double-hull Suezmax crude oil carriers, semi-refrigerated and semi-pressurized 16,500-cubic-meter liquefied petroleum gas (LPG) tankers, cable-laying ships, medium-sized dredgers, and strait train ferries; Research vessels made headway into the Antarctic region; In order to support the development of offshore oil and gas resources, China developed the design and construction capabilities for the entire industry chain, covering geophysical vessels and drilling platforms of self-elevating type, semi-submersible type, and submersible type, as well as semi-submersible production platforms, and floating production storage and offloading (FPSO) vessels.

Leapfrog Development Through Independent Innovation (2011–present). Since the new century, particularly since the 18th National Congress of the Communist Party of China (CPC), the CPC Central Committee, with Comrade Xi Jinping at its core, has accurately grasped the overall trend of development in the new era from a strategic perspective of realizing the great rejuvenation of the Chinese nation. It made a crucial strategic decision to build China into a strong maritime nation, guiding the shipbuilding industry to forge forward.

During this period, China witnessed a shipbuilding speed that was rarely seen in the global shipbuilding history. In the course of rapid development, China's shipbuilding industry has conquered multiple key technologies, leading to a substantial improvement in its design and building capacities. As a result, batches of world-class sophisticated ships were born in China. Prominently, the scientific research equipment also achieved leapfrog development: The 3,000-meter deep-water semi-submersible drilling platform "Haiyang Shiyou 981" officially started drilling in the South China Sea, signifying a substantial step forward of China's offshore petroleum industry in deep-water exploration; The first 12-cable geophysical exploration vessel in Asia, "Haiyang Shiyou 720", and the world's first 3,000-meter deep-water engineering survey vessel, "Haiyang Shiyou 708", were delivered and put into use, marking the gradual formation of China's "joint fleet" for deep-water operations; The domestically designed and autonomously integrated manned submersible "Jiaolong" set a new Chinese record by diving to a depth of 7,062 meters in the Mariana Trench, making China the fifth country in the world to possess deep-sea manned diving capabilities; In July 2019, China's first independently built polar scientific research icebreaker, "Xuelong 2", was successfully delivered. Compared to "Xuelong", "Xuelong 2", with a "wider and sturdier body", features stronger icebreaking and research capabilities. This signifies a new breakthrough in the on-site safeguarding and supporting capabilities of China's Antarctic and Arctic research bases.

The 70-year history of shipbuilding research and development showcases the continuous development of China's shipbuilding industry, from weakness to strength. It demonstrates the advantages of the socialist system with Chinese characteristics.

This history bears witness to the constant improvement and innovation in China's ship design and shipbuilding capabilities. It illustrates the nation's transformation from a chaser to a shoulder-to-shoulder runner, and ultimately to a leader in the field of ship research and development.

This history is also a true witness to the vast number of ship researchers and builders in China overcoming hardships, pioneering innovations, reaching new heights, and making contributions. It tells the story of ship entrepreneurs who have remained steadfast in their original aspirations, stayed true to their missions, and worked tirelessly towards their dreams.

Technology serves as the foundation of a strong nation, and innovation as the essence of the nation's progress. As General Secretary Xi Jinping pointed out at the Academician Conference of the Chinese Academy of Sciences and Chinese Academy of Engineering and the 10th Representative Assembly of the China Association for Science and Technology on May 28, 2021, "The world is undergoing profound changes unseen in a century, and the international environment is complex. The global economy is in a downturn, and the reshaping of global industrial and supply chains is underway. Instability and uncertainty have significantly increased." He further noted, "Technological innovation has become the main battlefield of international strategic game, and the competition for technological dominance is unprecedentedly fierce." Against this backdrop, the shipbuilding industry is undoubtedly facing new opportunities and challenges for development.

Looking back at history aims not only to summarize experience and inspire future endeavors, but also to create a better future. As we embark on a new journey of building a modern socialist country in all respects, the trumpet has been sounded to call for us marching towards the second centenary goal. Let us take this history as a mirror, courageously innovating, persistently striving, and consistently making more and greater contributions to building China into a strong maritime nation and realizing the Chinese Dream of national rejuvenation!

Academician of Chinese Academy of Engineering

曾恒一

Preface

The ocean harbors incredibly abundant resources, including biological resources, fossil fuel resources, mineral resources, sand resources, chemical resources, power resources, and medicinal resources. While the ocean provides rich material resources for human survival and development, it also gives rise to natural disasters, such as storms, tsunamis, waves, as well as phenomena closely related to global climate variations, such as El Nino, posing significant threats to human security.

Humanity has been closely connected with the ocean. Until the 1940s, humans' utilization of the ocean was limited to marine biological resources, and investigations focused primarily on marine organisms and general weather observations. The ability to predict and prevent disasters caused by the ocean was still lacking. After World War II, the world's population dramatically increased, leading to an imbalance of ecology with the depletion of land resources and the deterioration of global environment caused by economic development, which prompted humanity to shift their attention to the ocean as a means to solve these challenges. It became increasingly apparent that a comprehensive understanding of the ocean was necessary, whether it be for the development of marine resources, the growth of the marine economy, or the protection of marine rights. Thus, to conduct comprehensive and in depth scientific investigations of the ocean became necessary to provide reliable basis for the rational development of marine resources. Research vessels, serving as important platforms carrying marine research personnel, survey equipment, and operation facilities for ocean exploration and research, are key equipment for humans' exploration and management of the ocean. As human progress in ocean development, research vessels have continuously improved in terms of their size, equipment, and technological levels. Their frequent presence in vast seas, and even polar waters, is a significant reflection of a country's comprehensive strength.

The development of China's research vessels share similarities with those of international counterparts, while also incorporating its own characteristics. The State has regarded ocean development and management as key strategic initiatives for building a prosperous and powerful country. Planning for these initiatives began at the inception of the People's Republic of China. While cultivating marine researchers and establishing marine research institutions, design and building of research vessels were also carried out based on economic development and national needs. The Chinese "shipbuilders", upholding their original aspirations and missions, have been forging ahead with heavy responsibilities carried on their shoulders. With these spirits, they designed and built various types of oceanographic survey vessels, starting from small and medium-sized ones in the 1950s. Particularly in the 21st century, China's research vessels ushered in rapid development. Large and medium-sized research vessels such as "Kexue", "Tan Kah Kee", "Dayang", "Dongfanghong 3", "Shiyan 6", "Zhongshan University", and "Xuelong 2" were successively built and put into use, signifying a substantial progress in the field of research vessel development, with some key technologies reaching the forefront of the global industry.

This book is compiled to showcase the development history and glorious achievements of China's research vessels, tell the stories alongside their development, and inspire research personnel to remain true to their original aspirations, forge ahead with innovation, and achieve even greater accomplishments.

The book consists of eight chapters. Chapter 1 provides an overview of the critical role that research vessels play in advancing human understanding of the oceans. Chapter 2 outlines the key technologies and research and development focuses of research vessels. Chapters 3–5 detail the research and development history and classical ships of China's research vessels. Specifically, there are three stages: the "overcoming challenges and building the foundation" stage from 1949 to 1978, the "rapid development and heading towards the deep blue" stage from 1979 to 2000, and the "innovating and striding towards the forefront" stage from 2001 to 2020. Chapter 6 focuses on China's offshore research vessels. Chapter 7 tells the "Chinese stories" of Chinese ship research and development personnel. Chapter 8 provides an outlook on the characteristics and development trends of future research vessels.

It is hoped that readers could gain an understanding of the remarkable achievements of China's research vessels in their journey exploring towards the deep seas and oceans. Furthermore, it is also aimed to inspire readers to further enhance their ability to explore, develop, utilize, and protect the oceans, while pursuing the great rejuvenation of the Chinese nation and the maritime Chinese Dream.

Shanghai, China

Han Long
Qu Ningning
Cao Daqiu

History of Chinese Ships Research and Development

Editorial Board

Expert Committee

(continued)

(continued)

	Zhou Guoping	Wu Gang	Wang Cailian	Zhang Haiying	Zhang Guangen	Han Ming
	Zhong Weidong	Yu Zaihong	Zhang Taiji	Ding Yong	Ding Weikang	

Editorial Office

Editor-in-Chief	Zhang Yi					
Editor	Ding Yong	Yu Zaihong	Wei Qiang	Wang Bingxiang	Sun Jiapeng	Tian Xin
	Shi Gongqian	Qu Ningning	Liu Jihua	Liu Bingsui	Mou Chaogang	Mou Leipin
	Li Gangqiang	Li Jia	Li Yintao	Li Xiaofeng	Zhang Zhijun	Lin Jie
	Lu Chen	Gui Manhai	Gu Haijun	Kuang Yan	Wu Ying	Wu Yixin
	Qiu Weiqiang	Zhang Fuming	Zhang Taiji	Zhang Haiying	Chen Ying	Zhang Zongke
	Zhang Qixin	Ming Tong	Shang Baoguo	Shan Tiebing	Lu Sheng	Yu Yun
	Yao Liang	Guo Yanliang	He Huiqiong	Duan Xueqiong	Zhou Lanxin	Cao Daqiu
	Cao Caiyi	Yu Minyi	Han Long	Tang Yao	Yang Tian	Tao Xinhua
	Guo Manzhou	Huang Xiaoyan	Liang Dongwei	Qin Qi	Wei Yuefeng	

Contents

Chapter 1
The Ocean and Marine Research Vessels

1 Humanity and the Ocean

The ocean is the source of all life and plays a crucial role in human survival and development. However, 95% of ocean resources have not been properly developed, utilized, and protected. To know more about the ocean, humans have been tirelessly exploring the mysteries of it.

Human research on the ocean relies heavily on various instruments and equipment, with research vessels being a specialized type used for studying the ocean. These vessels serve as mobile laboratories, enabling scientists to investigate and study marine hydrology, geology, geomorphology, meteorology, and marine biology as well as to explore underwater resources.

Marine research is an important manifestation of a nation's scientific and technological prowess. Throughout modern history, every emerging global power has relied on advanced marine science and technology. Since the beginning of the twenty-first century, competition in the maritime domain has intensified globally. To explore and study the ocean with various instruments and equipment of diversified functions, research vessels are necessary for realizing their applications. Therefore, the design and building of advanced research vessels are of great strategic significance.

1.1 Oceans and Seas

The vast majority of the Earth's surface is covered by seawater, with the ocean accounting for approximately 71% of the total surface area. The ocean contains 1.37 billion cubic kilometers of water, approximately 97% of the total water volume on Earth. It not only surpasses the land in terms of its vast area, but its depth is also unparalleled by the height of the continents. With an average depth of around 3,800 m, 75% of the ocean exceeds 3,000 m. In contrast, the average elevation of lands stands

L. Han et al., *History of the Research and Development of Chinese Marine Research Vessels*, https://doi.org/10.1007/978-981-96-8414-4_1

at a mere 840 m, with areas below 1,000 m constituting 71% of the total land area. To date, 95% of the ocean floor remains unexplored by humans.

The ocean, at the center of the marine expanse, stretches far from the continents, accounting for approximately 89% of the total area, which makes it the body of the vast expanse. The average depth of the ocean is generally over 3,000 m, with the Mariana Trench being the deepest point at over 10,000 m. Generally unaffected by land in hydrological elements and features, such as salinity, temperature, water color, and transparency, the ocean maintains relatively stable temperature, salinity, low impurities, and high transparency. Each ocean has its own independent tidal system, ocean currents, and atmospheric circulation system. In terms of the ocean floor, it predominantly comprises basins and ridges.

The world's oceans are divided into four major parts: the Pacific Ocean, the Atlantic Ocean, the Indian Ocean, and the Arctic Ocean. Due to its distinct circulation system and unique water mass structure, the waters near Antarctica are commonly referred to as the Southern Ocean or the Antarctic Ocean in oceanography.

The sea, located at the margins of the ocean, is considered a subsidiary part of the ocean. There are more than 50 seas on Earth, accounting for only 11% of the total ocean area. The seas generally have shallower depths, with average depths ranging from a few meters to 3,000 m. The seas, being in close proximity to the continents, are significantly influenced by land, rivers, climate, and seasons, and the hydrological elements of seawater, such as temperature, salinity, water color, and transparency, and they are also heavily influenced by land and exhibit noticeable cyclic variations. Compared to the oceans, the seas do not possess their own independent tides and oceanic current systems.

Based on their locations, seas can be categorized into marginal seas, inland seas, and intercontinental seas. Marginal seas lie adjacent to continents, separated from the open ocean by islands. Examples include the South China Sea, the East China Sea, and the Sea of Japan, which are considered as marginal seas of the Pacific Ocean. Inland seas reside within landmasses, significantly influenced by continents. Examples include the Baltic Sea and the Bohai Sea in China. Intercontinental seas refer to those situated between continents, characterized by substantial depths and expanses. Examples include the Mediterranean Sea and the Caribbean Sea.

The ocean floor can be differentiated based on their proximity to continents, comprising of the coastal zone, the continental margin, and the floor area. The coastal zone serves as a boundary between land and ocean and is an area closely tied to human life. Approximately two-thirds of the global population reside in the coastal zone with a total length of around 440, 000 km. The continental margin is the transitional zone between the continents and the ocean basins, occupying approximately 22% of the total ocean area. It can be classified into two types: passive margins and active margins. Passive margins experience infrequent crustal activity, with less volcanic activity and earthquakes, such as those surrounding the Atlantic Ocean, Indian Ocean, and Arctic Ocean. Active margins, on the other hand, represent the most vigorous tectonic activity zones on Earth, concentrated along the eastern and western sides of the Pacific Ocean, housing approximately 80% of the world's active volcanoes. The ocean floor refers to the areas beyond the continental margins and constitutes the main body of the ocean. It mainly comprises oceanic ridges and ocean basins. Mid-ocean ridges, also known as central ocean ridges, form the largest global mountain systems on Earth. With a general depth of 2,000 m to 3,000 m underwater, some ridge peaks emerge above the water surface, forming islands. These ridges cover approximately 33% of the total ocean floor area. The ocean basin refers the area lying between the mid-ocean ridge footslopes and the continental margins, with depths typically ranging from 4,500 m to 5,500 m, encompassing deep-sea plains, seamounts, and abyssal hills.

1.2 Ocean Resources

The ocean is considered a treasure-house of resources for humanity. Ocean resources can be broadly categorized into six major types: biological resources, mineral resources, chemical resources, power resources, space resources, and medicinal resources.

1. Biological Resources

The ocean is home to a wide range of living creatures at all depths, providing a vast habitat for biodiversity. Approximately 85% of the world's aquatic products come from the ocean, with fish accounting for over 80% of the total. The ocean also contains rich algal resources.

Among the 33 animal phyla on Earth, nearly all of them can be found in the marine environment, including 15 unique phyla. In contrast, terrestrial environments harbor only 18 phyla, with only one unique phylum.

The ocean is estimated to house around 200,000 species, including 20,000 plant species and 180,000 animal species. Despite the remarkable biological resources, the present development and utilization of the ocean is limited to only around 500 species, primarily in fisheries. The regions currently used for marine fishing and aquaculture activities account for merely 10% of the total ocean area. With advancements in science and technology, humans will further develop the abundant marine biological resources.

2. Mineral Resources

The ocean is rich in diverse mineral resources, including petroleum, natural gas, coal, iron ore, bauxite, manganese nodules, copper ore, quartz rock, and more. Based on their different environmental characteristics and distributions within the ocean, they can be classified into nearshore sand minerals, submarine oil, phosphate and glauconite, manganese nodules and cobalt-rich crusts, submarine hydrothermal minerals, and natural gas hydrates.

Nearshore sand minerals, characterized by easy mining and low investment costs, are the earliest marine mineral resources developed by humans. For example, it is well-known that 90% of diamonds are produced from coastal mining areas.

Submarine oil reserves account for one-third of the world's recoverable petroleum reserves. The Persian Gulf stands out as the richest marine oil region, harboring approximately half of the total proven reserves of marine oil.

Manganese nodules and cobalt-rich crusts hold tremendous reserves, considered to be the most promising deep-sea mineral resources. The reserves of manganese, cobalt, copper, and nickel are respectively 180 times, 1,450 times, 22 times, and 150 times greater than those on land.

Submarine hydrothermal minerals are mineral resources rich in various metallic elements such as copper, lead, zinc, gold, silver, manganese, and iron. They are considered as one of the most promising seabed resources to be developed in the twenty-first century.

There is another valuable resource known as natural gas hydrates, which are crystalline solids formed by the combination of hydrocarbon gases and water molecules. They are also referred to as solid methane or combustible ice. Although they were discovered relatively late, natural gas hydrates hold vast prospects for development and utilization. It is estimated that China's natural gas resources amount to approximately 14.1 trillion cubic meters, and the proven reserves exceed 300 billion cubic meters. The areas that could potentially harbor natural gas hydrates account for approximately 10% of the total ocean area. Natural gas hydrates are expected to become a new type of energy resource for human utilization.

3. Chemical Resources

The ocean contains a multitude of utilizable chemical resources in its waters, organisms, and submarine sediments.

Seawater is a solution of various salt substances. In addition to large amounts of table salt and freshwater, various raw chemical materials such as sodium carbonate, caustic soda and hydrochloric acid, agricultural fertilizers such as potassium-magnesium fertilizers, sodium-magnesium fertilizers and magnesium-hydrogen fertilizers, refractory materials such as magnesium oxide and magnesia bricks, as well as rare and valuable metals such as magnesium, sodium, potassium, uranium, thorium and strontium can also be obtained from the seawater.

4. Power Resources

The movement of seawater can generate wave energy, ocean current energy, and the temperature and salinity variations can produce temperature and salinity gradient energy. Although these energy sources have disadvantages such as uneven distribution, instability, and low energy density, they have large reserves and are renewable. Current research and utilization of marine power resources include tidal power generation, wave power generation, thermoelectric power generation, ocean current power generation, and power generation based on seawater concentration difference, as well as energy utilization from seawater pressure difference.

5. Space Resources

The ocean can be used as a space for transportation, production, storage, military activities, residence, and recreational activities. Specifically, they include marine transportation, marine cities, marine engineering, coastal industrial sites, sea airdromes, marine ranches, marine warehouses and bases, maritime sports, tourism, and recreational activities. Space resources are one of the earliest resource types discovered and utilized by humans.

6. Medicinal Resources

The ocean reserves numerous highly efficient pharmacologically active substances, and "seeking medicines from the ocean" is a consensus among humans. With advancements in marine pharmacology and marine biology, more and more marine organisms are expected to be used by humans for medical treatment and physical fitness.

2 History of Human Exploration of the Ocean

2.1 Early Period

Humans have a long history of utilizing the ocean, as they live on the land surrounded by the vast seas. Human ancestors have long learned to fish from the ocean, produce salt from seawater, and rely on maritime routes for transportation and trade. Prior to the advent of aviation and railways, sailing was the sole means of transportation and travel across continents. As humans exploited and tapped into the ocean, they gradually engaged in investigating and exploring it. However, the time span of recorded and real oceanographic survey is not very long. As recorded, in December 1872, the British navy began an oceanographic survey with "HMS Challenger", which lasted for three years and five months. Taking it as the first multipurpose scientific research of the ocean in human history, then the history of modern oceanographic survey by means of ships extends back only approximately 150 years.

Following the "HMS Challenger" survey, many countries such as Germany, Russia, Norway, Denmark, Sweden, the Netherlands, Italy, and the US then dispatched oceanographic survey vessels for global or regional exploratory surveys. However, due to limited research platforms and oceanographic equipment of that time, most of these surveys primarily focused on biological studies.

After World War I, with the resurgence of oceanography, and oceanographic research activities, along with the advent of advanced detection instruments such as echo sounders, the oceanographic survey experienced a stepwise transition. It gradually shifted from "exploratory navigation survey" focused on biological studies to "specialized survey of specific sea areas" that carried the study of the ocean's physical and chemical properties, as well as the geology and topography of the seabed.

The accomplishments of the global oceanographic survey by the "HMS Challenger" significantly enriched humanity's understanding of the ocean and piqued people's interest in its mysteries. The following oceanographic surveys, prompted by the "HMS Challenger" expedition, expanded their focus beyond the surface of the ocean to the depths below. They started to investigate changes in the physical, chemical, biological, and geological properties of the ocean, such as ocean currents and temperature, which laid the foundation for the establishment and development of modern marine geophysics, marine chemistry, marine biology, and marine geology. Remarkably, the global oceanographic survey by the "HMS Challenger" stands as a milestone in the history of marine science, ushering in a new era of macroscopic study of the ocean and a further exploration of its natural laws.

2.2 *Post-World War II*

In the aftermath of World War II (WWII), oceanographic surveys experienced significant advancements. However, most survey vessels were repurposed from older ships. It was not until the late 1950s that specifically designed and built oceanographic survey vessels emerged. To date, the development of oceanographic survey vessels has roughly gone through two major periods. The first period, spanning from the early twentieth century to the 1980s, saw the gradual emergence of modern and efficient oceanographic survey vessels with the advent of electronic computers and various advanced oceanographic equipment. The second period, from the 1980s to the present, witnessed the proliferation of the electric propulsion system and dynamic positioning system for ships, as well as the improvement of various specialized research equipment and computer networking technology, among other technologies. This led to the birth of advanced oceanographic survey vessels or multipurpose research vessels, both domestically and internationally. These vessels employ various survey methods, categorized as follows:

1. Single-vessel Survey

The early twentieth century saw the development of oceanography primarily focused on biological surveys. During WWII, the need for underwater acoustics technology for war purposes led to the advancement of marine science theories. Prior to the 1950s, oceanographic survey vessels mainly operated by means of single-vessel traverse. These researches were limited in number, scope, duration and means, totaling just over 300 expeditions conducted by various countries in specific regions for short periods with outdated techniques.

Based on the data collected from these over 300 single-vessel surveys, oceanography experts discovered the constancy of the relative content of seawater's main constituents. They measured the weight fractions of seawater's chlorinity, salinity, density, and various elemental contents. In terms of marine geology, initial knowledge regarding seabed topography and sediment distribution was acquired. The classification of larger marine organisms was undertaken, and research on the relationship between organisms and their environment was conducted. In terms of marine geophysics, significant progress was made in the study of tides, waves, and ocean currents. Contour maps of global ocean currents were developed, and the corresponding fundamental theory of global ocean circulation—the wind-driven drift theory—was proposed.

2. Multi-vessel Joint Survey

Due to the increasing demands of oceanographic survey, single-vessel surveys were no longer sufficient. In the 1950s, multi-vessel joint surveys emerged, coexisting with single-vessel surveys that are still in use today.

For example, the China-US Tropical Ocean–Atmosphere Interaction Joint Survey in Western Pacific conducted from 1986 to 1990 involved the participation of China's "Xiangyanghong 05" and "Xiangyanghong 14" survey vessels. Similarly, the China-Japan Kuroshio Cooperation Research conducted from 1986 to 1992 involved the participation of 14 oceanographic survey vessels, including China's "Xiangyanghong 05" and "Shijian" survey vessels, alongside Japan's "Shoyo Maru", "Takuyo Maru", and "Kaiyo Maru", among others. This joint survey resulted in groundbreaking insights into the sources, paths, and hydrological structures of warm currents in regions including the Taiwan Strait and the Tsushima Strait.

In order to understand the influence of the tropical Western Pacific Warm Pool (WPWP) on global climate change through ocean–atmosphere coupling, a continuous four-month offshore survey was conducted from November 1992 to February 1993. Nineteen countries and regions participated this survey by various means. The survey involved 3 satellite systems, 7 aircraft, 14 survey vessels, 31 ground-based sounding stations, and 34 moored buoys, forming a stereoscopic observation network for observing. China's participation engaged the former State Oceanic Administration, the Chinese Academy of Sciences, the State Education Commission, and the China Meteorological Administration, and the survey vessels of "Xiangyanghong 05", "Kexue 1", and "Shiyan 3" were deployed for the observation and survey throughout the whole process.

After over a century of development, oceanographic survey, as a specialized research means, has gradually formed its own system. Present-day oceanographic surveys have developed comprehensive capabilities for conducting large-scale and section surveys, as well as continuous monitoring and auxiliary observations in hydrology, geophysics, chemistry, geology, topography, biology, meteorology, acoustics, and other relevant fields within specific marine regions. Although various research and detection means exist, such as satellite and aerial remote sensing, automatic fixed buoy observation, automatic drift station observation, unmanned aerial and marine vehicle observation, and the combination of the above, ship observation remains the most fundamental, adaptable, and cost-effective means for data collection. Other means can be utilized to complement ship observation.

3 Definition and Classification of Research Vessels

3.1 Definition of Research Vessels

Research vessels, short for marine research vessels, are ships engaged in tasks such as surveys on marine geophysics, chemistry, geology, topography, meteorology, hydrology, archaeology, and biological resources.

In China, these types of ships were referred to as oceanographic survey vessels from the 1950s to the end of the twentieth century. Since the twenty-first century, the Chinese Academy of Sciences and higher education institutions have changed the name from oceanographic survey vessels to marine research vessels, and a number of real vessels have been built. To reflect this change, this book refers to these ships as "research vessels (marine research vessels)", but when referring to early vessels, which were originally called "survey vessels", this book still uses the term "survey vessels (oceanographic survey vessels)".

For the classification of research vessels, in addition to retaining the relevant contents in the national standard GT/T 7391-2002, new ship types, such as the mother ship of submersible, have also been added.

3.2 Classification of Research Vessels

In terms of types, modern research vessels can be categorized as monohull, catamaran, and special types. In terms of operational areas, they can be classified into global, oceanic, regional, and coastal types. In terms of purpose, there are multipurpose research vessels, specialized research vessels, and special research vessels. Specialized research vessels include geophysical survey vessels, acoustic survey vessels, fishery resource survey vessels, geological survey vessels, meteorological observation vessels, channel measuring vessels, and environmental monitoring vessels, among others. Special research vessels include ocean drilling survey vessels, polar research icebreakers, submersible mother ships, and marine archaeological vessels, among others.

1. Multipurpose Oceanographic Research Vessels

As a major category of research vessels, modern multipurpose research vessels have the integrated research capabilities involving multiple disciplines, functionalities, and technological approaches. They provide platforms for scientific research in various disciplines, such as marine hydrology, meteorology, geophysics, geology, geomorphology, biology, chemistry, acoustics, and marine fishery resources.

Multipurpose research vessels are the most numerous of this category. Due to their ability to obtain integrated information from multiple disciplines in a single voyage, they make the greatest contribution to the development of various disciplines in oceanography and have therefore experienced the most rapid growth.

Fig. 1 The multipurpose oceanographic research vessel “Kexue” built in recent years

Figure 1 shows the multipurpose oceanographic research vessel built in China in recent years, the “Kexue” vessel.

2. Specialized Research Vessels

Specialized research vessels refer to vessels that are specifically used for investigating and conducting research in a particular discipline or several closely related disciplines within the field of marine science.

(1) Geophysical Exploration Vessel

Geophysics is a key discipline in geoscience that employs quantitative physical methods and detection techniques such as seismic elastic waves, geomagnetism, geoelectricity, geothermy, gravity, and radioactive energy to study the Earth and explore mineral resources concealed deep within the ground. Geophysical exploration vessels (geophysical vessels) are vessels utilized for conducting surveys and research on mineral resources, such as marine oil and natural gas. By virtue of the superior sound propagation capability of seawater and the varying speed of sound propagation in different substances, it is possible to identify the presence of liquid, gaseous, and solid mineral resources through studying the characteristics of sound wave reflections from various forms of materials in the deep sea.

Using compressed air guns, geophysical research vessels store high-pressure air generated by air compressors in the guns. By releasing the air instantaneously, low-frequency, high-energy sound waves are generated. These sound waves encounter the rock layers on the seabed, creating reflection waves. These waves are then received by the onboard receiving device through the cable, and after computer processing, reflection profiles are obtained. This enables the determination of the distribution of oil

Fig. 2 A geophysical exploration vessel during detection operation

and gas resources, thus facilitating the mapping of offshore oil and gas fields, which is crucial for the planning of future oil and gas drilling and extraction operations.

The capability of geophysical exploration vessels generally depends on the quantity and length of their receiving cables. The more and longer the receiving cables, the higher the detection capability. These vessels can accommodate up to 20 cables, with individual cable lengths exceeding tens of thousands of meters. During towing operations at sea, these cables are spread out in a fan shape at the stern of the vessel, marked by buoys as illustrated in Fig. 2. To avoid entangling with the cables, other vessels are required to maintain a safe distance. Figure 3 shows an example of geophysical exploration vessel—"Haiyang Dizhi 8".

(2) Fishery Resource Survey Vessel

Fishery resource survey vessels are specialized ships used for conducting investigations and research on aquatic resources, fishing grounds, marine environments, fishing gear, and fishing methods. They are primarily used to investigate the development of potential fishing grounds, the occurrence and duration of fishing seasons, and provide guidance for fishery production. These vessels focus on studying the types, distribution density, changes, estimated resource quantities, and recent catchability of fish resources. They are also used to conduct experiments on storage, preservation, and processing methods for various fishing gear, fishing methods, and catch, providing forward-looking resource detection for various fishing vessels. Figure 4 shows the "Songhang" vessel, a fishery resource survey vessel designed and built independently by China.

Fishery resource survey vessels are an integral part of modern fishing vessels. As fishery has expanded from coastal and offshore areas to deep-sea and open ocean regions, and with continuous development and utilization of resources, fishery resource survey vessels have evolved from small tonnage to large tonnage. Their

Fig. 3 Geophysical exploration vessel "Haiyang Dizhi 8"

Fig. 4 Fishery resource survey vessel "Songhang"

functionalities have also expanded from having fishing equipment of single fishing method to incorporating multiple fishing methods. In the 1980s, even small fishery resource survey vessels generally possessed two or more means of surveying.

3. Special Research Vessels

Special research vessels are a type of research vessel specifically designed for a particular scientific research task.

(1) Submersible Mother Ship

An important type of vessel in submersible mother ships is the manned submersible mother ship. It gained its name primarily for two reasons. Firstly, in terms of physical connection, when the submersible is lowered, there is a "umbilical cable" between the submersible and the mother ship that provides the submersible with securing, power, and signal support. Secondly, in terms of maintenance and supply, the preparations, maintenance, and upkeep tasks before and after releasing the submersible, as well as various monitoring and support tasks during the release process, are all carried out on the mother ship, which means the mother ship provides all-around logistical support services for the manned submersible.

The most crucial system on the manned submersible mother ship is the launch and recovery system for the manned submersible. This system generally consists of three parts: A-frame (see Fig. 5) at the stern, transfer track vehicle, and towing winch.

The most crucial system on the manned submersible mother ship is the launch and recovery system for the manned submersible. This system generally consists of three parts: A-frame (see Fig. 5) at the stern, transfer track vehicle, and towing winch.

The A-frame installed at the stern consists of two supporting columns and a crossbeam. The crossbeam is equipped with a lifting winch with constant tension and anti-roll/pitch functions and a gimbal. This allows the manned submersible to enter and exit the water stably, unaffected by the mother ship's sway caused by waves.

The transfer track vehicle is used to transfer the manned submersible between the lifting point and the storage area at the stern. This vehicle can also be raised and lowered, allowing the manned submersible to be lifted for maintenance and upkeep on its bottom in the storehouse.

After the manned submersible surfaces, it is connected to the towing winch by the crew in a small boat. The towing winch then drags the manned submersible from the water surface to the retraction position below the A-frame at the stern of the ship. During the launch and recovery process, the towing winch maintains a constant tension on the manned submersible to minimize pitching and swaying of the submersible.

China's present advanced manned submersible—"Jiaolong" was originally supported by the modified multipurpose survey vessel "Xiangyanghong 09", which served as the mother ship (see Fig. 6). In November 2019, China's newly developed dedicated mother ship for manned submersible "Shenhai 1" was delivered for use. The "Jiaolong" bid farewell to "Xiangyanghong 09" and welcomed its own dedicated mother ship (see Fig. 7).

Fig. 5 A-frame at the stern of the survey vessel "Xiangyanghong 09"

Fig. 6 Early mother ship of submersible "Jiaolong"—survey vessel "Xiangyanghong 09"

(2) Polar Research Vessel

Polar research vessels are specialized research vessels designed for conducting oceanographic surveys and investigations in the South Pole and North Pole regions. These vessels are generally larger in tonnage compared to other research vessels,

Fig. 7 New mother ship of submersible "Jiaolong"—"Shenhai 1"

with a typical full-load displacement of around 10,000 tons. The primary mission of these vessels is to support scientific surveys in polar waters. Due to the remoteness of polar research stations from their home countries, these vessels also possess certain logistical capabilities for transporting supplies. Figure 8 shows China's first polar icebreaker, the "Xuelong" research icebreaker.

Fig. 8 China's first polar icebreaker—research icebreaker "Xuelong"

The polar regions harbor a wealth of potential resources. For instance, the Antarctic glaciers contain valuable climate information, making them excellent samples for studying climate change. Antarctic krill is one of the highest-protein organisms in the world, with a resource potential of one billion tons and an exploitable capacity of 100 million tons. The unique extreme environments of the polar regions also make them important gene resource reservoirs. Although the development and utilization of gene resources from extreme marine organisms is still at an early stage internationally, it has already generated billions of dollars in value. Therefore, polar research is regarded as an essential national strategy for securing the initiative for future development and occupying new commanding heights.

Polar research vessels must possess strong icebreaking capabilities, which come from their specially designed icebreaker bow shape. The icebreaker bow is retractable with a large inclination angle between the ship's surface and the water. Its dagger-like slim and pointed shape allows it to cut through thick ice. Many icebreakers also have icebreaking capabilities at the stern, allowing them to break ice in reverse without the need for maneuvering or turning, which is known as bidirectional icebreaking.

As the underwater parts of the vessels are frequently subject to the impact from ice, the spaces between the hull frames of icebreakers are relatively small. For instance, China's latest-designed and constructed polar icebreaker, the "Xuelong 2", has a frame spacing of only about 350 mm, half of what is typical for general ships. Moreover, the underwater shell plates of icebreakers are thicker than those of general ships. For instance, the ice knife area at the bow of "Xuelong 2" has a plate thickness of 100 mm, while the thickest plate on a 21,000 TEU[1] ultra-large container ship is merely 85 mm. Icebreakers also employ special steel capable of withstanding low temperatures. Furthermore, special ice strengthening measures are taken at exposed seawater inlets, propellers, and other critical areas.

China's newly constructed polar research icebreaker, the "Xuelong 2" (see Fig. 9), possesses bidirectional icebreaking capabilities, which allows it to continuously break through 1.5-m-thick ice plus 0.2-m-thick snow at a speed of 2–3 knots.[2] It is equipped with an array of scientific exploration capabilities, including marine and atmospheric environment observations and experimentation, marine geological and geophysical measurements, as well as marine biological and ecological surveys and analysis.

(3) Ocean Drilling Vessel

Ocean drilling vessels are capable of carrying out drilling operations in the deep sea at depths of several kilometers, obtaining samples from beneath the seabed. These vessels rely on three key systems and equipment: dynamic positioning system, drill hole reentry system, and heave compensation device. Due to their technological complexity and high operation and maintenance costs, there are currently only two

[1] ①TEU, a unit of measurement for container shipping capacity, based on a standard 20-foot container.

[2] ②1 knot = 1.852 km/h.

Fig. 9 China's newly constructed polar research icebreaker "Xuelong 2"

ocean drilling vessels in the world, namely the US's "JOIDES Resolution" and Japan's "Chikyu".

To counteract the vessel's planar directional movements caused by winds, waves and currents under high sea conditions, their dynamic positioning capability ranks among the most demanding of all capabilities. Moreover, in order to counteract the up and down motion, these vessels are usually equipped with heave compensation devices, which can mitigate the adverse effects of the vessel's pitching and rolling on the drill pipe. The drill hole reentry system is the technological core of ocean drilling vessels, as exemplified by Japan's ocean drilling vessel "Chikyu", illustrated in Fig. 10.

4 Overview on the Development of Foreign Research Vessels

During the period from 1492 to 1504, Columbus made four transatlantic voyages, reaching South America and discovering the New World. Portuguese explorer Ferdinand Magellan completed the first circumnavigation of the globe from 1519 to 1522. These voyages, while expanding and enriching our knowledge of ocean geography, also conducted some scientific investigations related to ocean currents and winds. However, these were not professional oceanographic surveys. It was not until the second half of the nineteenth century that the "HMS Challenger" truly inaugurated the era of marine research.

Fig. 10 Japan's ocean drilling vessel "Chikyu"

The "HMS Challenger" boasted the world's first dedicated oceanographic survey vessel. Converted from an old warship, this vessel was measured 68 m in length, with a displacement of 2,300 tons and a crew capacity of 243. Propelled by a combination of sails and steam engines, "HMS Challenger" served as a three-masted steam-powered sailing ship equipped with state-of-the-art survey instruments and a chemistry laboratory, as shown in Fig. 11. From December 7, 1872, to May 26, 1876, The Royal Society organized an expedition using this vessel, which lasted 713 days (three years and five months) and covered a distance of 68,890 nautical miles[3] in the Atlantic, Pacific, and Indian Oceans. Led by Sir Charles Wyville Thomson and with an involvement of six scientists, this expedition marked the first multipurpose oceanographic research around the globe in human history, setting the stage for systematic and goal-oriented modern oceanographic research. The successful refitting and deployment of the "HMS Challenger" became a milestone in the history of marine navigation and marine science. It served as a model and example for the study of oceanography, ushering in a new era of scientific research and exploration of the ocean's natural laws at a macro level. In this investigation, reversing thermometers were first used to measure deep-sea water temperatures and its seasonal variation. A large number of marine plant and animal specimens, seawater samples, and seabed sediment samples were collected. Manganese nodules, containing over 30 elements such as manganese, copper, cobalt, and nickel, were found in deep-sea mud and

[3] ①1 nautical mile = 1,852 m.

red clay. 715 new genera and 4,717 new species of marine organisms were discovered. The constancy of the ratios of major seawater constituents was confirmed, and the first world map of oceanic sediment distribution was compiled. In addition, the geomagnetic and bathymetric conditions of the survey area were measured. The data and samples obtained from these surveys were analyzed and studied by 76 scientists for 23 years, resulting in a 50-volume, 29,500-page report. Their research greatly enriched human understanding of the ocean and laid the foundation for the establishment and development of marine physics, marine chemistry, marine biology, and marine geology. The success of the global oceanographic survey carried out by the "HMS Challenger" greatly increased people's interest in the ocean.

At that time, survey vessels from different countries generally conducted individual and single-discipline marine surveys. Typical survey activities and their main achievements include the five-year global expedition from 1831 to 1836 by "HMS Beagle", which covered the Atlantic, Indian, and Pacific Oceans. Charles Darwin, the British scientist, participated in the survey. Based on the survey data, he explained the formation of coral reefs and proposed the theory of submarine movement, providing basis for his book *On the Origin of Species* published in 1859. The survey results were compiled into the four-volume report *The Voyage of the Beagle* by Captain Robert FitzRoy and Darwin. From 1839 to 1843, the British explorer James Clark Ross conducted expeditions in the waters of Antarctica, measuring a sea depth of 4,438 m and setting a record for deep-sea sounding at that time. He concluded that

Fig. 11 Oceanographic survey vessel "HMS Challenger"

the bottom-layer water of the entire ocean has similar characteristics and discovered the South Magnetic Pole. From 1842 to 1847, the US Navy Captain Matthew Fontaine Maury systematically studied the relationship between winds and ocean currents in the oceans, using the data to create ocean charts. In 1855, he published *The Physical Geography of the Sea*, the first classic work in oceanography. He also published the first bathymetric chart of the North Atlantic Basin in 1854, providing scientific basis for the laying of transatlantic submarine cables. The British marine biologist Edward Forbes first proposed the concept of zonal distribution of marine organisms, suggesting that the deeper the water, the fewer organisms, with a lifeless zone below 550 m. However, from 1836 to 1870, organisms were discovered repeatedly at depths of 550 m or deeper. From 1869 to 1870, the British vessel "HMS Porcupine" took samples 16 times at depths ranging from 1,800 m to 4,464 m, each time collecting various organisms, including sea urchins believed to have been extinct since the Cretaceous period. This shocking discovery completely overturned people's understanding of deep-sea organisms' life activities. After the 1850s, the laying of submarine cables promoted the development of oceanic sounding. In 1856, the cable-laying dedicated survey vessel "HMS Alcide" discovered the mid-Atlantic ridge, suggesting the possibility of laying submarine cables along this ridge.

After World War I, oceanographic research shifted from exploratory navigational surveys to specialized surveys of specific sea areas. From 1925 to 1927, the German vessel "Meteor" conducted hydrographic surveys on 14 sections in the South Atlantic, followed by supplementary observations on seven sections in the North Atlantic from 1937 to 1938. As a result, data from 310 hydrographic stations were collected. This survey focused mainly on marine physics, covering observations of hydrology, meteorology, biology, and geology with high precision. A major achievement of this survey was the elucidation of the fundamental characteristics of deep circulation and water masses in the Atlantic. In addition, it was the first survey to use echo sounders to explore the seabed topography, revealing that the seabed is as rugged as land and challenging the previous notion of a "flat seabed".

During this period, oceanographic survey vessels not only focused on the surface of the ocean but also on the space below the surface, along with changes in marine physics, chemistry, biology, and geology, such as ocean currents, temperature, and others. With the revival of oceanographic research and international oceanographic survey activities, as well as the invention of echo sounders and other instruments, oceanographic survey vessels were not only able to conduct multipurpose oceanographic surveys with a primary focus on biological surveys, but also tasks related to physical and chemical properties of seawater as well as geological and geomorphic features.

From 1947 to 1948, the Swedish oceanographic survey vessel "Albatross" (see Fig. 12) conducted a tropical ocean survey, which was praised by oceanographers as a model of modern multipurpose oceanographic surveys. This survey lasted 15 months and covered a total distance of 70,000 nautical miles. It involved 403 measurement points in the Atlantic, Pacific, Indian, Mediterranean, and Red Seas, with a focus on deep-sea observations in the doldrums of the three major oceans.

Fig. 12 Swedish oceanographic survey vessel "Albatross"

The oceanographic survey vessel "Albatross" filled the gap left by the "HMS Challenger" survey vessel, which was unable to observe in calm zones due to its reliance on wind power for navigation as a sailing ship. The "Albatross" conducted surveys on the equatorial ocean currents within 20 degrees north and south latitudes, explored the optical properties of the deep sea, obtained rock core samples up to 23 m long using a vacuum piston corer, and discovered records of Quaternary climate variations in deep-sea sediment layers. It also studied submarine structures using artificial seismic methods, measured the thickness of ocean sediments using bottom profiler, and determined the age and deposition rate of sediments using radioactive isotopes. Additionally, it also observed the turbidity currents, studied the chemical properties of subsurface seawater, and investigated the thermal effects of the oceanic crust, paving the way for deep-sea geophysical research. This expedition focused primarily on tropical deep-sea survey and geological sampling of the deep-sea bottom. After more than a decade of compilation, calculation, and analysis, the *Report of the Swedish Deep Sea Expedition* was published, comprising 10 volumes and 36 sub-volumes.

According to statistics, from the eighteenth century until the 1950s, approximately 300 single-vessel oceanographic surveys were conducted worldwide. These surveys offered people a basic understanding of the temperature and salinity distribution, large-scale water masses, and seabed topography of the world's oceans and major seas.

In the late 1930s, due to historical circumstances, 61 countries and regions worldwide were involved in WWII, resulting in a close integration of marine research with military affairs. Due to the significance of underwater acoustic detection research in locating submarines and the importance of wave research for special forces in amphibious combat missions, oceanographic survey vessels experienced prominent development in acoustic technology, wave observation and prediction technology, alongside combat capabilities for many survey vessels. After WWII, oceanographic survey experienced rapid progress. However, even with these efforts, oceanographic survey vessels were often retrofitted from old vessels. It was not until the late 1950s that the era of designing and building new oceanographic survey vessels began. Since then, the world has seen great improvements in survey objectives, methods, and equipment, as well as breakthroughs in vessel design and building technologies. The development of the world's oceanographic survey vessels can be summarized into two main periods:

The first period generally refers to the late 1950s to the late 1980s. With the application of electronic computers and the emergence of various advanced oceanographic equipment, modern and efficient oceanographic survey vessels gradually emerged and became popular. These vessels had significantly improved equipment, performance, general arrangement, and allocation of laboratories and specialized equipment compared to vessels retrofitted from old ships. As multipurpose survey vessels no longer met the in-depth survey needs of various branch disciplines of oceanography, specialized and special survey vessels were introduced gradually. Representative vessels during this period included the Soviet Union's 6,000-ton multipurpose oceanographic survey vessel "Akademik Lomonosov" (built in 1959), the US's 3,400-ton channel oceanographic survey vessel "Surveyor" (completed in 1960) and the oceanographic survey vessel "Atlantis II" (built in 1962), and the UK's 3,100-ton oceanographic survey vessel "Discovery". The channel oceanographic survey vessel "Surveyor", built by the US in 1960, was the first to be equipped with a multi-beam system, a deep-water and shallow-water echo system, a stable sonar navigation system, a hydrological data recovery system, a seismic reflection profile processing module, along with a dry/wet laboratory, a gravity laboratory, and a photography laboratory. The "Atlantis II", built by the US in 1962, was the first oceanographic survey vessel to be equipped with an electronic computer, marking the birth of modern and efficient vessel of this kind. The most significant feature of these vessels during this period was their employment of automotive, electronic, and computer systems.

The second period is from the late 1980s to the present. With the application of the electric propulsion system and dynamic positioning system for ships, the renewal of various specialized research equipment, the specialized and modular design of scientific research operations and laboratories, the automation of control support systems, and computer networking, among other advancements in technologies, ship design, especially for multipurpose oceanographic survey vessels, ushered in new design concepts. Coupled with the fact that some previously constructed ships were due for renewal, a new generation of advanced oceanographic survey vessels thus emerged.

In 1984, the US refitted and built the 18,000-ton ocean drilling vessel "JOIDES Resolution", and in 1997, it built the 3,600-ton deep-diving submersible mother ship "Atlantis III". Germany built the 6,300-ton ice zone multipurpose survey vessel "Maria S. Merian" in 2006, the UK built the 5,800-ton multipurpose research vessel "James Cook" in 2007, Japan built the 57,000-ton ocean drilling vessel "Chikyu" in 2007, and Russia built the polar research icebreaker "Akademik Tryoshnikov" in 2012. In 2013, the UK built the 6,000-ton “New Discovery”, Australia built the 6,000-ton multipurpose survey vessel “Investigator” in 2014, and Germany built the 8,000-ton new “Sonne” in 2015.

With the progress of the times, the improvement in technological levels, and the upgrade of survey instruments and equipment, as well as the development of oceanographic survey needs, newly built oceanographic survey vessels in foreign countries have made significant advancements and innovations in ship automation, computer networking, construction modularization, ship type diversification, survey discipline specialization, and ship vibration and noise reduction.

Chapter 2
Features of Research Vessels

1 Functions of Research Vessels

To study the ocean, it is essential to conduct investigations first. Oceanographic surveys are the foundation for the advancement of marine science. The ocean surface is covered by the atmosphere, and thus investigating the atmospheric factors such as temperature, pressure, radiation, and precipitation is necessary. Moreover, beneath the seawater lies the seabed, while land extends along the coast, making investigations into the geological structure being crucial. The ocean is home to a diverse range of organisms, necessitating investigations on their distribution. Furthermore, the dynamic nature of the ocean and its interaction with the surrounding environment can lead to phenomena like hurricanes and tidal waves, which necessitates extensive and complex investigations. These investigations encompass large-scale and section surveys over hydrology, chemistry, biology, meteorology, geology, acoustics, and geophysics observations, and even continuous and supplementary observations. While methods such as remote sensing and telemetry through satellites and aircraft, deployment of underwater instruments, fixed buoys, drift station automatic observations, unmanned aerial vehicle and unmanned surface vehicle observations, and synchronous comprehensive observations using multiple methods are employed, research vessels remain the primary and most effective means for multipurpose surveys.

Each research vessel is equipped with scientific research equipment and instruments based on specific tasks and the serviced area, allowing for various oceanographic research tasks to be carried out. These tasks can be categorized into water survey, atmospheric survey, seabed survey, acoustic survey, deep-sea survey, as well as sediment analysis, chemical analysis, biological analysis, and remote sensing information observation and verification. The research outcomes obtained fall into two main categories: electronic data, including various data, sounds, images, and videos, and physical samples, including water bodies, marine organisms, sediments,

L. Han et al., *History of the Research and Development of Chinese Marine Research Vessels*, https://doi.org/10.1007/978-981-96-8414-4_2

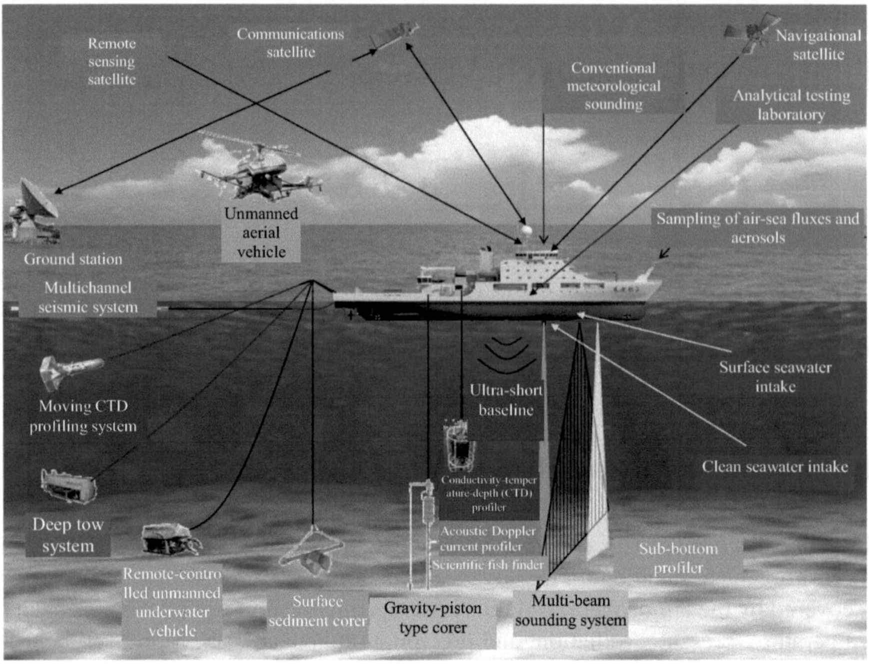

Fig. 1 Typical operations of research vessels

and rocks. These outcomes are utilized in the study of various disciplines within oceanography and serve as the primary data for analysis and research.

Typical operations of research vessels are shown in Fig. 1.

1.1 Water Survey

Water is the main component of the ocean. Water survey includes the measurement of depth, temperature, salinity, transparency, water color, seawater, as well as observations of luminescence, sea currents, and waves. It primarily involves the use of fixed-point (point) or moving (area) detection instruments to form a multi-element comprehensive measurement of basic oceanographic hydrological elements in surveyed areas. Such surveys enable a comprehensive understanding of the surface and deep-sea structures, and facilitate research on ocean circulation dynamics, air-sea interactions, and ocean internal waves.

Water surveys mainly involve the following shipborne detection systems.

1. CTD Detection System

The CTD (conductivity, temperature, depth) detection system enables continuous observations of vertical distribution of temperature, salinity, and depth of seawater.

Fig. 2 Water sampler and its carried CTD detection equipment before entering the water

The system can be deployed in either fixed-point or moving mode to provide basic data for the study of ocean circulation dynamics (see Fig. 2). The fixed-point system is equipped with a large water sampler carrying a CTD sensor, which can be deployed into the sea using the launch and recovery system on board to measure the CTD parameters at a single location. It can also collect water samples from different water layers for the study of marine biology or chemistry. The moving system, on the other hand, involves measuring physical and chemical elements along multiple profiles. Its advantage is the ability to conduct synchronized observations of underwater profiles over a large area while the research vessel is in voyage.

The moving CTD detection system can be further divided into towed and autonomous systems.

The towed system involves the use of conventional steel cables or communication cables to tow the CTD on the carrier, commonly referred to as moving vessel profiler (MVP). The MVP can adjust its position in the vertical direction in the ocean through instructions from an onboard computer or commands sent via a communication cable. This enables the instrument to conduct vertical profile observations while being towed up and down.

The autonomous system, on the other hand, observes the profiles of the target sea ares without the need of connection to the vessel but being controlled autonomously by an onboard computer. This form of systems are generally non-powered and rely on buoyancy and wing adjustment for underwater motion, such as underwater glider. At present, the systems are still designed into non-powered types, for this design enables greater movement performance.

China's scientific expedition team for the observation under million depth (Mariana Trench) set a new world record by using an underwater glider that reached a maximum depth of 10,619 m.

2. ADCP Detection System

The ADCP (acoustic doppler current profiler) detection system is among the most effective methods for observing multi-layer profiles of ocean currents. The system is known for its high accuracy, resolution, and ease of operation. ADCP can be used for moving or anchor-based observations of the vertical distribution of horizontal flow velocities in offshore and far sea areas. It contributes to the study of basic rules and variation mechanisms of ocean currents.

The system can be classified into two types: moving and lowering systems. Currently, moving ADCPs are more commonly used on research vessels and are typically fixedly installed at the bottom of the vessel. In order to observe currents of different depths and intensities, a research vessel usually needs to install three to four ADCPs with different acoustic performance characteristics (such as observation frequency and layer thickness). The sensors of lowering ADCPs are installed on a deployable instrument, which is deployed together with the sensors as a bundled unit. The lowering ADCP enables the measurement of flow velocity and direction of full depth profiles, which allows the pattern of current profile values with ocean depth for the entire water column.

3. Surface Multi-element Automatic Measurement System

The surface multi-element automatic measurement system can automatically measure and record surface ocean chemical (pH, dissolved oxygen), biological (chlorophyll), and hydrological (temperature, salinity) parameters under navigation conditions. The basic system consists of an open-water circulation cabin, which houses the necessary measurement sensors and chemical analysis equipment while performs data collection, recording, management, and transmission.

4. Scientific Fish Finder System

The scientific fish finder system is also called the bio-stock assessment echo integrator system or fish finder. For instance, the ultrasonic fish finder utilizes the emission, reflection, and reception of ultrasound by means of frequency modulation to detect the species, population size and quantity, spatial distribution, migration dynamics, and swimming speed of marine organisms in various water layers. This system is essential for conducting ecological surveys of marine ecosystems and studying the changes in global marine ecosystems and the sustainable utilization of biological resources. It is generally installed on the bottom or extendable fins of the vessel.

1.2 Atmospheric Survey

Atmospheric survey involves monitoring various variables in the atmosphere above the sea surface. Currently, shipborne survey equipment is capable of conducting comprehensive stereoscopic observations of the air-sea interface, the marine atmospheric boundary layer, and the upper atmosphere. By obtaining long-term, large-scale observational data and on-site monitoring data during extreme weather conditions such as typhoons, it scientifically unveils the crucial regulatory role of the air–sea coupling system and the ocean in climate change. These surveys meet the practical needs of marine meteorological and environmental forecasting, offshore weather forecasting, climate prediction, and atmospheric science research.

The main targets of atmospheric surveys include temperature, pressure, humidity, wind speed, wind direction, and visibility.

The shipborne detection systems used in atmospheric surveys consist of the following:

1. Conventional Meteorological Moving-type Continuous Detection and Data Collection System on Sea Surface

The conventional meteorological moving-type continuous detection and data collection system on sea surface, also called shipborne meteorological station, is essential for conducting marine research. It is capable of providing real-time and continuous records of temperature, pressure, humidity, wind speed, wind direction, and visibility data. This data is indispensable for analyzing the meteorological conditions in the surveyed area and studying the air-sea interactions. Additionally, it directly contributes to meteorological support during vessel moving and can be applied to establish and verify air-sea coupling models, validate meteorological and oceanographic satellite remote sensing, and carry out scientific studies on climate change. This system is typically installed at the front of the highest platform or on the mast of the research vessel to avoid obstructions that may affect the measurement of wind speed, wind direction, and other parameters.

2. GPS[1] Sounding System

The GPS sounding system is a device used to study and analyze the vertical distribution characteristics of atmospheric conditions. It can provide real-time measurements and record vertical distribution data of temperature, pressure, humidity, wind speed, and wind direction from the sea surface to an altitude of 30,000 m.

To enable the sensors to reach higher altitudes, balloons are needed. The balloons are usually filled with hydrogen gas, known as hydrogen balloons. Hydrogen gas is quite active, so for safety reasons, helium gas is sometimes used instead. Sensors carried on the balloon transmit the collected information back to the research vessel. Conventional meteorological observations usually involve the release of sounding balloons with sensors two to four times per day, and the frequency can be increased

[1] Global positioning system.

Fig. 3 Sounding balloon operations on China's polar icebreaker "Xuelong"

based on specific requirements. Figure 3 shows the sounding balloon operations on China's polar icebreaker "Xuelong".

3. Air-Sea Flux Observation System

The air-sea flux observation system provides data on the exchange of carbon dioxide and water vapor between the ocean and the atmosphere, forming the basis for studying the interactions between them. There is a very complex mass and momentum exchange between the ocean and the atmosphere. In the coupled atmosphere-ocean system, the atmosphere primarily exerts a dynamic influence on the ocean, known as dynamic flux, while the ocean primarily exerts a thermal influence on the atmosphere, known as heat flux. Together, dynamic and heat fluxes are referred to as air-sea flux, which serves as an essential mechanism for climate formation and change. During observations in voyage, efforts need to be made to minimize the influence of the vessel's hull on the atmosphere. The air-sea flux observation system is generally arranged on the foremast, away from the main hull of the vessel.

4. Online Gas and Aerosol Component Monitoring System

The online gas and aerosol component monitoring system is capable of providing online measurement of the content of soluble components in aerosols and gases in the ambient atmosphere during ship moving. The analyzed data of the system are of great significance for studying the offshore transport of land pollutants and the source of pollutants by wind, and the reflow transport of land pollutants by sea-land breeze circulation. This system is usually installed on the foremast.

5. Atmospheric Profiler

The atmospheric profiler, also known as wind profiler radar, performs a similar role to the ADCP (Acoustic Doppler Current Profiler) used in water surveys. It is able to measure the horizontal wind speed, wind direction, vertical wind speed, and atmospheric structural constants at different altitudes in the range of 50–3,000 m above the sea surface continuously and in real time. The data obtained from the system are crucial for studying air-sea exchange, wind-driven circulation, air-sea energy exchange, typhoon, marine meteorology, and atmospheric duct.

6. Ceilometer

Clouds play a vital role in weather system development, precipitation formation, and atmospheric radiation transfer, among other physical processes. Timely and accurate acquisition of cloud information is crucial for synoptic climatology research, meteorological forecasting, artificial weather modification, as well as various fields such as the national economy and military. The ceilometer can determine cloud base height and vertical visibility during rainfall events through laser ranging technology.

1.3 Seabed Survey

The seabed, which is in contact with the deep sea water, is generally composed of solid substances. Seabed surveys focus on studying the seabed and includes seabed topographical, geomorphological, and stratigraphic profile surveys, marine geophysical surveys, as well as sediment and benthic organism collection surveys.

1. Seabed Topographical, Geomorphological, and Stratigraphic Profile Survey

Seabed topographical, geomorphological, and stratigraphic profile surveys primarily utilize equipment such as single-beam and multi-beam echo sounders, sub-bottom profiler, and side-scan sonar. By utilizing acoustic signals at different frequencies, the surveys are conducted to probe the seabed surface and underlying geological layers. Such surveys also serve as the foundation for conducting marine geological research, providing essential environmental information for the development of submarine mineral resources and marine studies. With the continuous improvement in survey accuracy, these surveys have become indispensable components for the study of submarine tectonics, marine sedimentology, and deep-sea mineral deposits.

Seabed topographical, geomorphological, and stratigraphic profile survey can be further classified into two categories: topographic and geomorphic survey and marine geological sampling survey.

(1) Topographic and Geomorphic Survey

Topographic and geomorphic surveys involve the use of acoustic systems to conduct large-scale observations, and the resulting data is processed by specialized software to create comprehensive maps of submarine structures. The primary instruments used for such surveys include multi-beam depth sounder, single-beam depth sounder, and sub-bottom profiler, among others.

(2) Marine Geological Sampling Survey

Marine geological sampling surveys focus on understanding the composition of the seabed and its biological components. Analysis and testing of collected samples provide insight into the chemical elements present, as well as the ages of the formations. In some cases, to preserve the natural habitat of benthic organisms, sampling techniques that maintain temperature and pressure conditions are employed.

Sampling operations involve activities such as bottom trawling, cable-free grab sampling, large-scale gravity piston sampling, box coring, seabed sampling and shallow drilling by autonomous underwater vehicle (AUV) or remotely operated vehicle (ROV).

2. Marine Geophysical Survey

In marine geophysical surveys, gravity and geomagnetic fields are used to analyze and study deep stratigraphic textures, geological structures, and material structures below the seabed. They mainly include marine gravity surveys, marine magnetic surveys, and marine seismic operation surveys.

Marine gravity surveys employ marine gravimeters to investigate and study change rules of gravity in the ocean, providing fundamental data for applications such as missile flight, aerospace, and studies on Earth's material structures. The equipment for these surveys is typically installed in the lower part of the survey vessel.

Marine magnetic surveys are conducted based on the principle that rock layers beneath the seabed possess different magnetism and generate varying magnetic fields. In the survey, towed magnetometers are used to investigate and study the variations in magnetic fields, as well as the distribution of seabed mineral deposits and features of Earth's material structures.

Marine seismic operation surveys involve deploying seismic source system to provide sound signals and hydrophone to receive the reflected waves. The data is then analyzed using computer software to study the seabed mineral deposits and Earth's material structures (see Fig. 4).

3. Sediment and Benthic Organism Collection Survey

Sediment and benthic organism collection surveys mainly concentrate on sampling of seabed surface sediments, core sediments, and benthic organisms. The samples collected are primarily used for the study and analysis of seabed sediment structures, composition, sedimentation ages, and benthic organism communities.

4. Shipborne Detection Instruments for Seabed Survey

(1) Multi-beam System

The multi-beam system is used for seabed topography survey. This type of survey is of great significance for marine geology research, geophysics research, marine navigation safety, and oil and gas exploration. The system utilizes multiple array

Fig. 4 Twelve-Cable geophysical vessel "Haiyang Shiyou 720" in seismic operation

elements, wide angles, and narrow beam directional transmission and reception to generate wide-swath and high-density depth data after processing and calibration, hereby meeting the requirements of precise topographic mapping. Compared to traditional single-beam echo sounders, this system offers advantages such as 100% coverage of water depths, a wide measurement range, fast acquisition speed, high depth accuracy and resolution, and digital recording and real-time automatic plotting.

The multi-beam system functions like an underwater camera installed on the bottom of the vessel, providing coverage of up to 2,000 km on either side of the survey track and accurately recreating the seabed geomorphology.

The multi-beam system can be classified into deep-water and shallow-water systems, with a rough division at around 300–400 m. These systems are typically installed in fixed positions at the bottom of the vessel (see Fig. 5).

(2) 10,000-m Echo Sounder

The 10,000-m echo sounder is primarily used to measure water depth data during maritime surveys of various scientific research projects. This device utilizes multi-frequency technology, suitable for operations in sea areas cross different depths and with different bottom sediments. In addition to obtaining water depth data, it can also provide information on topography, geomorphology, seabed substrate types, and the thickness of surface sediment layers.

(3) Bottom Profiler

The bottom profiler is an instrument designed for continuous profiling of seabed layers. This device is widely used in marine geological science exploration, marine engineering, channel measurement, and other fields. It can detect seabed sediments,

Fig. 5 Multi-beam system in operation

shallow structures, and detailed stratigraphic distribution, providing fundamental data for submarine exploration. Based on the thickness of penetrated layers, it can be divided into sub-bottom profiler and mid-depth bottom profiler. The sub-bottom profiler generally detects depths of tens of meters, while the mid-depth bottom profiler can detect depths ranging from hundreds of meters to thousands of meters.

(4) Marine Seismometer

The marine seismometer is an instrument used to conduct marine geophysical surveys. It can continuously record natural seismic activities on the seabed or conduct artificial seismic detections.

(5) Marine Seismic Exploration System

The marine seismic exploration system is crucial for conducting geophysical surveys and studying marine stratigraphy, tectonics, and sedimentation. It is also vital for exploring underwater oil and gas resources and natural gas hydrates. After an air gun or electric spark is activated by the seismic streamer on the vessel, the reflected signals from the seabed and subsurface layers are received by the cable and transmitted to the seismometer for recording and memory. Processed data can then clearly reveal the structure of rock layers, the thickness of sedimentary layers, and the composition of the layers. This provides essential information for understanding the Earth's layer structure, plate tectonic evolution, hydrocarbon basin formation, engineering geological exploration, and analysis of geological hazards.

Fig. 6 Illustration of bottom trawling

(6) Sediment and Benthic Organism Collection System

The sediment and benthic organism collection system is used to collect samples of seabed sediments and benthic organisms. Equipment includes box sediment sampler, clamshell sediment sampler, multi-tube sediment sampler, gravity piston corer, and benthic organism collection system. Different sampling devices are used depending on the seabed conditions and sampling requirements. The collection of sediments and benthic organisms provides technical support for marine sediment research, submarine environment monitoring, and marine resource survey and research. It also provides a wealth of valuable samples, data, and information for national security and marine rights, geological and resource environments, and monitoring and prevention of marine disasters. Figure 6 is an illustration of bottom trawling.

(7) Marine Gravimeter

The marine gravimeter is used for continuous or fixed-point gravity measuring at sea or underwater, which is essential for conducting marine geophysical surveys. The gravity field is generated by the gravitational forces of the Earth and is an intrinsic physical characteristic. Due to the uneven distribution of various matters on Earth, gravity varies in different marine geological structures. Understanding and measuring the Earth's gravity field are important for comprehending the Earth structure, exploring the metallic mineral resources, and studying the aircraft guidance. Typically, the marine gravimeter is installed in the middle bottom compartment of the ship, as close to the ship's center of flotation as possible, away from the engine room and other sources of vibration.

(8) Marine Magnetometer

The marine magnetometer is one of the ocean geophysical survey devices that is used for measuring the Earth's magnetic field at sea. It operates on the principle that the sub-seafloor rock layers have varying magnetic properties, resulting in different intensity magnetic fields. Measuring the Earth's magnetic field is significant for understanding the Earth structure, exploring mineral resources, and studying aircraft guidance.

1.4 Acoustic Survey

Sound waves are the only carriers for long-distance information transmission in the ocean. Ocean acoustics, as an interdisciplinary field between oceanography and acoustics, has been continuously integrated into marine research and plays a positive role in advancing ocean studies. Over the past half century, ocean acoustics has become a foundation for the development of high-tech marine technologies and an important component, providing essential data parameters for the development and use of various advanced shipborne acoustic detection systems.

Compared to the continental shelf and offshore shallow sea areas, the study of underwater acoustic environmental characteristics and their relationship with physical ocean fields in deep and far sea areas is more important, as it is closely related to the national strategy for becoming a strong maritime nation and securing maritime rights and interests, and meanwhile possesses extremely high military utility value. Currently, major maritime powers are actively conducting joint data assimilation experiments in acoustics and oceanography. Significantly, advanced acoustic detection equipment is a prerequisite for collecting high-quality acoustic data.

Acoustic surveys primarily involve the following shipborne detection systems.

1. Deep-Sea Acoustic Emission System

The deep-sea acoustic emission system is one of the main installations for studying the characteristics of underwater acoustic environment. It provides information sources for studying the relationship of time, frequency, and spatial characteristics of the deep-sea acoustic environment (such as acoustic propagation and reverberation) with physical ocean fields. Furthermore, the system also supports the development of active acoustic detection technology in deep-sea areas.

2. Horizontal Towing Array System

As a basic device for sensing underwater acoustic signals, the horizontal towing array system is a type of high-tech integrated equipment for underwater acoustic environment measurement and recording. It features large aperture, high sensitivity, high precision, flexible configuration, and high reliability, providing direct technical support for conducting acoustic and oceanographic surveys in deep and far sea environments.

3. Deep-Sea Vertical Receiving Array System

The deep-sea vertical receiving array system is a type of high-tech integrated equipment for acoustic measuring and hydrologic synchronous recording in the deep and far-sea environments. Notably, the receiving performance of this system (center frequency, bandwidth, sensitivity, operating depth and other parameters) directly influences the quality of acoustic data obtained by the entire acoustic detection system.

4. Data Collection System

The data collection system is an essential component of the acoustic detection system. It is used to collect and store acoustic data sensed by acoustic detection equipment, providing a foundation for enriching acoustic environmental databases. Large-scale oceanographic surveys by acoustic technology and the development of active acoustic detection technology in deep and far sea areas require long-term, wide-range observation and data collection of sea ambient noise. Therefore, data collection and memory systems with high sampling rate, high memory capacity, high stability, and high dynamic range are indispensable for acoustic detection equipment.

1.5 Deep-Sea Survey

According to traditional understanding, deep-sea extreme environments refer to conditions in the deep sea that are not suitable for the survival of organisms, such as environments with no sunlight while with high temperatures and high pressures or with low temperatures and high pressures. However, with the advancement of oceanographic survey, scientists have discovered thriving communities of extremophiles in these environments, which do not rely on photosynthesis. This discovery has revolutionized the entire community of biology, fundamentally changing public understanding of traditional biology. Nowadays, studying the phenomena of life under these extreme conditions has become a forefront and hot topic in contemporary biological research.

For deep-sea extreme environments, such as hydrothermal fluid, conducting deep diving observation and sampling presents a significant challenge for ocean exploration technology. Human understanding of the deep-sea ocean largely depends on the progressiveness of survey equipment.

The deep-sea survey equipment mainly includes the following shipborne detection systems.

1. Deep-Sea Towing System

The deep-sea towing system consists of laboratory monitoring system, deck, cables, towed body, and various sensors installed on the towed body. During operation, the towed body is submerged in the water, deployed near the seabed. Equipped with hydrodynamic environmental element sensors and imaging systems, it allows for

Fig. 7 Deep-sea optical camera towed body

close-range measuring of the physical and chemical parameters of deep-sea bottom topography, sub-bottom, and near-seabed water, as well as observation of seabed organisms. Figure 7 shows a deep-sea optical camera towed body.

2. Unmanned Cable-Controlled Underwater Robot

Unmanned cable-controlled underwater robots are highly integrated modern technological equipment, which are of great concern to the world's marine research community today. They are able to carry out long-term underwater operations, particularly in high-intensity underwater tasks that are not feasible for divers or in deep and hazardous environments beyond the limits of human divers. Due to the harsh and dangerous underwater conditions and limited diving ability of humans, underwater robots have become critical tools for oceanic exploration.

Unmanned cable-controlled underwater robots provide technical means for operations such as oceanographic survey, ocean development, salvage, detection, sampling, as well as construction, installation and maintenance of underwater observation networks. Moreover, they also provide technical support for exploration and research related to deep-sea life phenomena, the Earth's interior, dynamics, and new marine resources, and other fields.

Unmanned cable-controlled underwater robots can generally include two levels: observation level and operation level. Operation-level robots possess more complex structures and use various mechanical arms to perform intricate underwater tasks. Figure 8 shows the operation-level unmanned cable-controlled underwater robot "Discovery" carried on the research vessel "Kexue".

Fig. 8 Operation-level unmanned cable-controlled underwater robot "Discovery" on the research vessel "Kexue"

3. Automatic Underwater Carrier

Automatic underwater carriers are one of the most advanced equipment in marine surveying. Originally developed for military purposes, they have been increasingly applied in civilian marine surveying and marine research in recent years. This type of carrier does not require umbilical cables and is capable of operating in water depths ranging from several hundred meters to several thousand meters, with operational durations exceeding 10 h. In the civilian sector, they can be used for pipeline laying, submarine survey, drilling support, underwater construction, data collection, and maintenance of underwater equipment, and other purposes. Figure 9 shows the deployment of an automatic underwater carrier on the research vessel "Tan Kah-Kee".

4. Truth-Preserving System for Sampling Hydrothermal Fluid, Near-Bottom Organism Larvae, and Sub-bottom Sediments

The truth-preserving system for sampling hydrothermal fluid, near-bottom organism larvae, and sub-bottom sediments enables truth-preserving sampling of volatile components in deep-sea hydrothermal fluid and bottom surface sediments, as well as truth-preserving sampling of barophilic near-bottom organism larvae. The maximum operating depth of the hydrothermal fluid sampler is 3,000 m, while the maximum operating depth for near-bottom organism larvae visual control towing sampler and sub-bottom sediment truth-preserving sampler can reach 6,000 m.

Fig. 9 Deployment of an automatic underwater carrier on the research vessel "Tan Kah-Kee"

5. TV Grab

TV grab, short for deep-sea cable visual sampling system, is a flexible and convenient system highly integrated with modern technology for underwater observation and sampling. Its grab enables the collection of a large quantity of seabed samples, and its built-in underwater camera captures images and videos, allowing for the observation of various submarine phenomena.

6. Acoustic Positioning System

The acoustic positioning system is an essential component on research vessels, primarily used for navigation positioning and data transmission for underwater operational facilities such as remote operated vehicles (ROVs), subsurface buoys, and tow cables. There are three main types of acoustic positioning systems: long baseline (LBL), short baseline (SBL), and super short baseline/ultra-short baseline (SSBL/USBL), or a combination of these three. The ultra-short baseline system is commonly used on ships. During operation, some parts of the system extends below the ship bottom to a certain distance, and retracts into the ship when not in use.

High-precision underwater acoustic positioning systems have played a crucial role in information transmission for ocean exploration and research, marine engineering, construction of underwater structures, oil exploration and extraction, underwater operation of divers, underwater archaeology, maritime defense construction, and underwater general positioning, and other fields.

1.6 Experimental Analysis

1. Sediment Experimental Analysis System

The sediment experimental analysis system includes equipment for suspended particle collection and testing, which is used for studying sediment dynamics and flux. It also includes equipment for sediment collection, processing, and analysis.

2. Chemical Experimental Analysis System

The observation and research of marine chemical elements mainly rely on sample collection, experimental analysis, and conducting simulated experiments. Some elements can be observed online or via sensors. The observation and research process requires high standards for sample collection, pre-processing, on-board analysis, and storage conditions. Some chemical elements must be analyzed on the ship within a limited time. Therefore, the chemical experimental analysis system is an essential consideration in the design of a research vessel.

The chemical experimental analysis system is primarily used for measuring seawater parameters, such as salinity and dissolved oxygen. It includes laboratory space and analytical instruments.

3. Biological Experimental Analysis System

The biological experimental analysis system is primarily used for collecting large, medium, and small plankton, as well as water samples or sediments. It is used for measuring primary productivity, chlorophyll concentration, and the photosynthetic activity of algae. It can also capture real-time images of plankton in water. Additionally, the system enables rapid and accurate classification, identification, and counting of nannoplankton samples. It is also used to observe the microscopic morphology of animals, plants, microorganisms, and more.

Some samples must be observed and tested on the ship within a limited time, while others need to be properly preserved and tested in onshore laboratories after disembarkation. Therefore, the biological experimental analysis system should consider methods and means for sample collection, processing, storage, and on-board analysis and testing.

The biological experimental analysis system includes laboratory, sample collection equipment, and analytical testing instruments.

1.7 Remote Sensing Information Observation Validation System

As a three-dimensional, multi-source observation system, the remote sensing information observation and validation system expands the conventional measuring parameters and scope of ocean surface and atmosphere. Meanwhile, it also serves as an effective means to validate satellite remote sensing data, providing technical support for the study of marine science, atmospheric science, and ocean exploration technology.

The remote sensing information observation and validation system achieves integrated observation of the sea, air, and space by utilizing ship-based measurements, unmanned aerial vehicle observation systems, and satellite data reception and processing systems. It can capture multi-source, multi-dimensional, and wide-range marine physical dynamics and atmospheric environmental parameters.

The main equipment in this system includes the marine optical observation system, which can systematically measure the optical properties of water and the atmosphere, as well as the spectral radiation in the water and underwater; the shipborne unmanned aerial vehicle remote sensing observation system, which can carry multiple sensors to conduct high spatio-temporal resolution regional observations of the deep-sea environment; the infrared radiometer, which can measure sea surface temperature; the Doppler lidar atmospheric profile detection instrument, which can continuously detect the wind field and aerosols in the marine atmospheric boundary layer as well as cloud profile structures in the troposphere; the satellite data reception and processing system, which enables real-time reception and processing of satellite data, providing weather decision support information to the research vessel, while automatically receiving, processing, and displaying marine and atmospheric parameters around the operating area.

China's first meteorological satellite, which was launched in September 1988, carried a five-channel scanning radiometer with two ocean color channels. Subsequently, the "Fengyun 2" ocean satellite was launched, further advancing the nation's marine observation and research endeavors.

2 Functional Requirements for Research Vessels

Marine research is an important strategic task for a country to transition from a major maritime country to a strong maritime nation. In order to gain a comprehensive understanding of the ocean, the missions of research vessels vary each time they set sail, and the research focus of each marine research institution is also different. The fundamental principle in type design of research vessels is to meet the requirements of users for specific functions. Regardless of the type of research vessel, the most basic requirements are: safety—the vessel should have a sturdy structure and strong wind resistance. It should be able to operate safely in adverse sea conditions.

Maneuverability—the vessel should be agile, with moderate speed and good control over its position. It should have good stability, seakeeping, and reliable deck retraction capacity, thus ensuring precise measurements and data repeatability. Seaworthiness—the vessel should provide excellent comfort for personnel, as well as good seaworthiness and wind resistance. It should be able to sail safely in various weather and sea conditions in different sea areas to fulfill its scientific investigation tasks. Environmental adaptability—low vibration and noise levels should be maintained to ensure the normal operation of acoustic instruments and equipment. The vessel should have sufficient and complete power supply capacity, good electromagnetic compatibility, and be environmentally friendly. Compatibility—the vessel should be able to adapt to different tasks and have the capability to temporarily carry multi-purpose platforms to meet the research needs of various disciplines. Economy—the vessel should have low fuel consumption, optimized endurance and self-sustaining period to meet the investment scale and design displacement or gross tonnage, with certain expandability.

To construct a new research vessel and fully utilize its platform advantages, it is essential to accurately grasp the functional requirements of the vessel. Generally, the following aspects are considered:

2.1 Performance Requirements

The research vessel should have appropriate stability, sufficient endurance and self-sustaining period, good seakeeping capability, and ability to reduce vibration and noise. Research expeditions typically involve long periods of sea travel. For example, a round trip to Antarctica usually takes six months, with half of that time spent at sea. Therefore, the vessel must be able to ensure safe navigation in strong winds and rough seas. Thus, in the design, stability and wind resistance requirements for research vessels are generally higher than those for conventional ships.

2.2 Profile Requirements

The performance of a research vessel is closely tied to its hull shape and main dimensions, particularly the length-to-breadth ratio (L/B), breadth-to-depth ratio (B/D), and breadth-to-draft ratio (B/T). For research vessels, there are certain requirements for their deck width and seakeeping ability, although an excessively large L/B ratio is not required. In the early days, research vessels often sailed in tandem with military ships. Due to the high speeds of military ships, they usually had larger L/B ratios and slender hulls. Subsequently, research vessels became capable of operating independently. In a bid to better support the scientific research equipment on board, the L/B ratio was reduced, allowing for increased deck area, enhanced stability, and smoother marine survey operations.

Economical analysis and main dimension evaluation of research vessels involve repeated demonstration and optimization of various technical and economic indexes. Since research vessels serve as platforms for numerous scientific research equipment and measuring instruments, in addition to the consideration of the rapidity, the profile design should also involve the optimization of the bow shape, and the matchability with the acoustic equipment, along side seakeeping performance and seaworthiness. The hull form analysis and profile design are interrelated, as different hull forms and profiles determine various performance factors such as speed, wave resistance, and ultimately impact the effectiveness of underwater acoustic equipment, propulsion system selection, and main engine configuration. Therefore, in hull form analysis and profile design, it is vital to focus on key aspects and optimize the overall balance.

In general, new research vessels often exhibit characteristics including low *L/B* ratio, low block coefficient, flared bow, smooth bilge, upward expanded hull, short parallel midship body, flat bottom, and square stern, among others. In addition, if a research vessel is equipped with a bow thruster, careful consideration needs to be given to the design of the bow thruster's outlet profile, in order to prevent interference from bubbles with the acoustic equipment. Furthermore, the size and installation method of bottom-mounted acoustic equipment must be carefully considered during the initial profile design phase to avoid significant deviations in ship speed estimation due to non-embedded equipment installation. If the dimensions of the set multi-beam echo sounder exceed the profile limits for embedded installation, the multi-beam echo sounder installation solution must be selected as early as possible from the traditional dome type, suspended type and other forms. Suspended installation may present challenges in structural design, shipyard construction, and dock maintenance, as well as increased appendage resistance. In comparison, embedded installation is preferable, while it imposes limitations on the measurement angle and detection depth of the multi-beam echo sounder. If the traditional dome installation method is adopted, the vortex generated by the dome may affect the testing effectiveness of the multi-beam echo sounder and increase appendage resistance. Based on the mechanism by which the acoustic transducer is affected by the hull boundary layer and the bubbles generated by waves during sailing, the research and development team from the Marine Design and Research Institute of China (MARIC) has created multiple hull form options in terms of bow shape, bow profile inclination, and multi-beam configuration. While integrating "traditional dome installation" with "embedded installation", the team made efforts to repeatedly compare and optimize profile schemes, finally achieving the "new dome" which exhibits a streamlined total appearance with gradually widened sections after considering changes in the hull form based on calculations and optimizations for the whole profile using computational fluid dynamics (CFD). This "integrated design technology for hull economy and multi-beam anti-bubble interference" is a key technology in the profile design of research vessels. Notably, the design has successfully passed experimental validation and has been widely applied to similar ship types.

2.3 *Safety Requirements*

According to regulations, research vessels are neither classified as passenger ships nor maritime work ships. They are considered as specially arranged ships with a focus on their capability for marine operations and scientific research. These vessels are classified as special purpose ships and must comply with the safety regulations for such vessels, as stipulated in the *International Convention for Safety of Life at Sea* (SOLAS) and domestic inspection rules.

2.4 *Electric Propulsion*

Research vessels are designed to carry out various scientific research tasks in complex and dynamic conditions at sea. Some research projects require the vessel to remain stationary, while others require low-speed or full-speed navigation. When departing, research vessels need to reach the research area as quickly as possible, which requires high speed to achieve that. However, during towing operations, a speed of 2 to 4 knots is sufficient. Low-speed operation has a significant impact on the main engine and may reduce its efficiency. In the past, when electric propulsion was not available, research vessels achieved low-speed navigation through low-power auxiliary power units, active rudders, or variable pitch propellers.

With the advent of electric propulsion systems, the placement of the propulsion motor does not have any special requirements and can be positioned anywhere. As a result, integrating the propeller with the steering mechanism solves the problem of generating thrust in different directions.

Currently, there are three main propulsion forms used on research vessels: propeller propulsion, podded propulsion, and all-direction rudder propulsion. Placing the propulsion motor and propeller in the water, with the steering gear inside the compartment, this type of propeller is known as podded propeller. The propeller is installed in a small pod suspended in the water, which can rotate 360 degrees using steering gears. This allows the propeller to generate thrust in any direction (see Fig. 10).

The podded propulsion provides excellent maneuverability and allows for propulsion in all directions, significantly improving the vessel's maneuverability in quick response and emergency situations. Specifically, podded propulsion is achieved by using an underwater motor to drive a coaxial propeller directly, without the need for reduction gears. There is no mechanical transmission involved, resulting in lower mechanical vibration and noise levels. Due to its simple structure, convenient steering, and small space requirement, podded propeller is increasingly widely used in electric-propelled research vessels.

The all-direction rudder propeller involves the propulsion motor placed inside the vessel, the propeller submerged in water, and the steering gear located between the propulsion motor and the propeller. Depending on whether the propulsion motor is

Fig. 10 Podded propeller

placed vertically or horizontally inside the vessel, it is referred to as L-type propeller or Z-type propeller, respectively. An L-type propeller has only two shafts, which are positioned at a right angle to each other, while a Z-type propeller has three shafts, with the upper and lower shafts parallel to each other and the middle shaft perpendicular to them (see Fig. 11).

The all-direction rudder propeller utilizes a larger number of bevel gears to achieve the transmission between different angles, enabling 360-degree rotation. Therefore, the structure is more complex. The all-direction L-type propeller is often used as an auxiliary propeller and is typically installed at the bow of the research vessel. When not in use, it can be stowed inside the vessel, making it a retractable propeller. There are two common types of retractable modes: telescopic mode, where the propeller extends up and down, and folding mode, where the propeller extends forward and backward (see Fig. 12). The channel side propeller is a modified version of the L-type propeller without steering gears. It is fixedly installed in a pre-made opening in the underwater hull that runs from left to right, and is commonly referred to as a "transverse thrust" unit.

2.5 Dynamic Positioning

Ships generally rely on anchoring system for positioning at sea. The conventional method is to deploy an anchor on the larboard and starboard of the bow, respectively. Once the anchor is released, the anchor claws grip onto the seabed and connect the ship to the anchor using anchor chains or cables, thereby anchoring the ship in a specific position. However, this method has two major drawbacks. Firstly, it can only be used in shallow waters, where the depth does not exceed 80 m, usually around

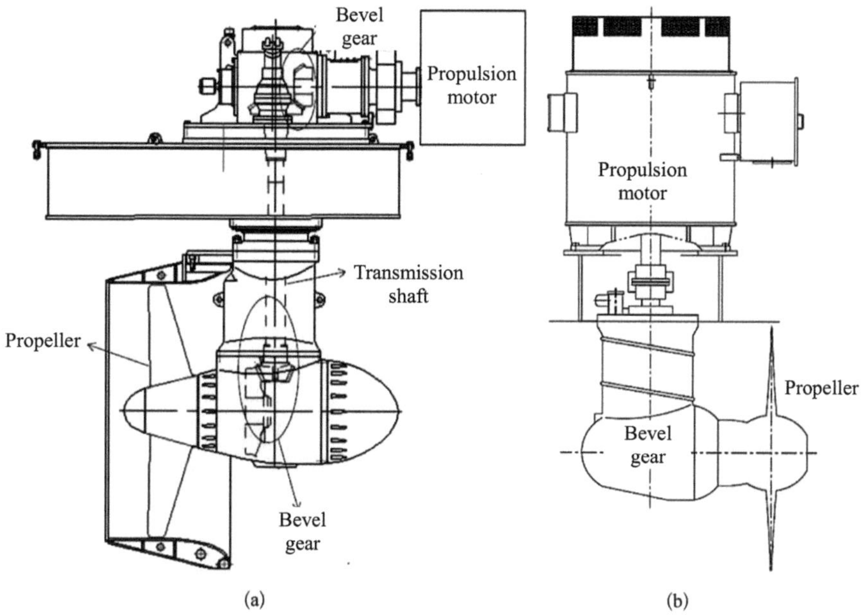

Fig. 11 Z-type propeller and L-type propeller. **a** Z-type **b** L-type

50 m. Secondly, it is not very precise and is prone to anchor dragging. Therefore, for ships that require high positioning accuracy, such as ocean platforms, additional anchors are required to be installed on both sides and the stern in order to improve the accuracy of positioning.

Prior to the advent of dynamic positioning, research vessels had to rely on another anchoring method, called deep-water anchoring, in order to position themselves in deep waters. The principle of deep-water anchoring is similar to regular anchoring, but it has additional challenges due to the increased water depth, such as calculation of external forces and selection of anchors and cables. In deeper waters, the anchor chains become impractical due to excessive weight, so alternative options such as steel wires are used as substitutes. China's first-generation multipurpose survey vessel for non-restricted waters, "Shijian", was equipped with a 500-m deep-water anchoring device. The oceangoing scientific survey vessel "Xiangyanghong 10" was equipped with a 1,000-m deep-water anchoring device. However, the deep-water anchoring device had notable limitations in terms of accuracy and was prone to anchor dragging. Moreover, the system was complex and lacked reliability. As a result, it was quickly replaced by dynamic positioning.

The dynamic positioning system is primarily composed of various sensors, computers, and propulsion components. Its operation involves real-time measurement of environmental parameters using position sensors, heading sensors, attitude sensors, current sensors, and other instruments. These parameters and information are then transmitted to the computer for processing. The system then compares them

(a) (b)

Fig. 12 Two types of retractable propellers. **a** Telescopic type for up-and-down-extension **b** Folding type for back-and-forth extension

with stored data regarding the intended docking position, identifies differences, and promptly sends instructions to the individual thrusters to adjust their thrust accordingly. This allows for differential corrections to be made until the predetermined docking position is reached.

Although the requirements for speed is not demanding, research vessels require accurate positioning for operations such as data collection, water sampling, and deep-diving submersible deployment. To minimize the impact of complex environmental factors like wind and waves, new-built research vessels are all equipped with dynamic positioning systems. This system is a distinctive feature of contemporary research vessels. It does not rely on anchoring system for positioning and can automatically detect any deviations between the ship's position and the predetermined position. Based on the influences of external forces such as wind, waves, and currents, the system calculates the thrust required to push the ship back to the predetermined position. This thrust is then distributed to various propellers, which generates differentiated thrust forces to keep the ship as close as possible to the predetermined position. Dynamic positioning systems are classified into different levels according to the levels of redundancy, with higher-level ones capable of maintaining positioning even if some equipment in the system is damaged.

2.6 Vibration and Noise Reduction

Vibration and noise have a significant impact on the quality of life for personnel on board and the normal functioning and lifespan of equipment. Therefore, reducing vibration and noise is a top priority for engineers involved in the design and building of ships. Research vessels are equipped with water detection system, atmospheric and spatial weather detection system, biological research and collection system, acoustic research and underwater communication system, and remote sensing data validation system. They also carry various detection devices, precision instruments, and numerous acoustic devices that rely on sounds for information transmission. These devices and instruments has high requirements for working environments, as vibrations and noise in laboratories and living spaces directly affect the normal operation of these equipment and the comfort of onboard personnel. Therefore, research vessels have higher requirements for ship vibrations and noise compared to transport and working ships. To achieve effective vibration and noise control is therefore a critical and systematic task for research vessels.

The main sources of noise on board are hydrodynamic noise, mechanical noise, and propeller noise.

Hydrodynamic noise is directly proportional to the fifth or sixth power of the vessel's speed. When the vessel's speed doubles, hydrodynamic noise increases by 15 to 18 decibels. However, hydrodynamic noise only affects the underwater radiation noise when the speed exceeds 12 knots. In practical, research vessels usually operate at speeds of 8 to 11 knots for underwater radiation noise assessment, thus hydrodynamic noise is not the main source and has minimal impact on underwater radiation noise. Additionally, since hydrodynamic noise is related to the hull profile and underwater appendage profile, it becomes necessary to make the hull smooth, reduce openings on the hull or reduce the number and size of appendages, as well as optimize the design to highlight the profile of transducer compartments and domes on the hull exterior. Moreover, there should be a smooth transition between the closed cover of the bow thruster and the hull.

Mechanical noise is generated by the transmission of vibrations from the ship's electromechanical equipment and power transmission systems to the ship structure, causing structural vibrations and underwater radiation noise. In quiet sailing conditions, mechanical noise is the primary source of noise of the entire ship. To address this, efforts should be made at the source. This includes strict control at accuracy and acceptance criteria for products from mechanical equipment suppliers, as well as the adoption of appropriate vibration and noise reduction measures during the installation of various devices based on their specific characteristics.

Propeller noise is generated by the rotation of the propellers during high-speed navigation and is a major source of noise on ships. Noise reduction can be achieved by lowering propeller rotation speeds, reducing blade loads, and improving the machining precision of propellers. One solution is to adopt silent propellers.

In order to achieve effective vibration and noise reduction, analysis and assessment of the ship's vibration and noise levels should be conducted during the early stages of the design and building process. Starting from the design phase, predictive analysis of vibration and noise should be carried out to put forward specific measures for vibration and noise control. This involves considering various aspects, including the hull structure, propulsion system, compartment layout, mechanical equipment selection, and air-conditioning ventilation system design. Such measures include using floating rafts or double-layer installations for diesel generator sets to reduce transmission of the vibrations to the ship's bottom, and installing mufflers in the exhaust pipe to reduce exhaust noise from the diesel generator sets. Elastic mounting can be adopted for fans and air compressor units. Fans can be equipped with streamlined air outlets and coated with sound absorption materials. For exhaust pipes, tension brace and flexible support can be used to provide support. Isolation compartments can be used for fans and air ducts. The generator base can be coated with damping materials, which can also be coated on the ship's side walls, fore and aft compartment walls, and ceiling to reduce vibration and noise levels on the ship.

3 General Arrangement Requirements for Research Vessels

3.1 General Arrangement

As specialized ships, research vessels have higher requirements and unique considerations in terms of performance and general arrangement compared to transport ships. The general arrangement of a research vessel should reflect a rational layout in the flow of personnel, samples, and data. It should be equipped with spacious working decks and laboratory spaces in order to meet the long-term needs of all personnel, including scientific researchers, for work and living at sea. To adapt to the exploratory and multi-discipline natures of marine operations, modular conversion and expandability should be considered. According to the principles of modern research vessel design, under the premise of meeting regulatory requirements, the maximum functionality for scientific research should be achieved while maintaining a balance between the ship system and the scientific research system, with due consideration for the comfort of personnel onboard. The layout of the deck and laboratories should consider comprehensive utilization and interchangeability, while achieving advanced levels in the unit indicators of "deck working area/displacement" and "laboratory area/displacement" for the entire ship.

The scientific research system of a ship serves as the platform for the ship's detection and experimental systems. During the preliminary design and demonstration stage, the design unit should collaborate with the user to discuss and analyze the future mission and tasks of the ship, the requirements for onboard equipment and installations, the operational conditions and procedures for offshore operations, the functionality and spatial division of laboratories, modular design, and provisions for future modifications. Ultimately, the design should fully realize the survey function and maximize survey efficiency, accuracy, and stability for the user.

The overall layout of the scientific research system should take into account the development trends of mainstream international research vessels. It generally adopts a left-arranged chimney and L-shaped configuration on the starboard side. Dry and wet compatible laboratories should be designed, in addition to special laboratories. Furthermore, a large low-freeboard working space should be reserved on the aft deck. This not only facilitates scientific exploration operations and future space expansion but also allows for the installation of containerized laboratories by reserving space for fasteners embedded on the deck. Moreover, the bow science mast and the lifting fins in the middle of the vessel both ensure essential support for scientific research operations on the vessel.

Typically, research vessels accommodate more personnel than other civilian ships except for passenger ships. Despite their relatively smaller tonnage compared to some cargo vessels, research vessels require a significant number of experimental compartments. Therefore, to improve the living and working conditions for numerous personnel within limited space becomes a challenge. Compared other ship types, living conditions on research vessels are relatively primitive. Moreover, research vessels generally spend longer periods at sea. Under the premise of balancing the needs of scientific research, two approaches are commonly adopted to improve living conditions. One approach is to maximize the per capita living area and increase the provision of necessary equipment in order to enhance the convenience and comfort of personnel. However, this may result in fewer and smaller public spaces. For example, only a multifunctional hall is provided, and when not in use, residents can engage in activities within their living quarters. The other approach involves meeting the requirements of the Maritime Labour Convention (MLC) for living compartment area and equipment allocation. In this case, numerous leisure and recreational areas, such as gyms, reading rooms, cafes, and cinemas, are included on board. During leisure time, personnel are encouraged to relax in these areas.

3.2 *Laboratory and Its Special Installations*

1. Laboratory

Laboratories are the primary locations for scientific experiments on research vessels and serve as working spaces for scientists and researchers. The current trend in scientific research and survey emphasizes on-site analysis, leading to significant changes in the types and functionalities of laboratories. The range of experiments that can be conducted in onboard laboratories is now comparable to those performed in land-based laboratories. However, the compact layout, muli-unit and multi-disciplinary collaboration, and the "freshness" of data in onboard laboratories are unmatched by their land-based counterparts. With the advancement of networking technology, the work efficiency and outcome accuracy of onboard laboratories have experienced significant leaps forward. In the near future, onboard laboratories will surpass land-based laboratories to become the primary birthplace of academic achievements in marine sciences.

(1) Laboratory Layout

Laboratories on research vessels serve as centralized spaces for various indoor experiments and are equipped with precision instruments to cater to different marine disciplines. In general, laboratories are positioned on a single deck as possible, usually located towards the aft section of the main deck. The laboratory should be located as close as possible to the respective operational areas, taking into account the comprehensive flow of samples, personnel, and data based on the support system, operational areas, and various operational methods during the scientific research process. This ensures convenience, efficiency, and safety in sample collection, distribution, and processing. Considering the comprehensive functionality, universality, and applicability of the laboratories, large general laboratories should be designed to achieve maximum utilization rates. Depending on the specific requirements of different research projects, the laboratories should be equipped with both general-purpose and specialized equipment, with designated exclusive work areas to maximize the spatial utilization of the laboratories.

(2) Laboratory Classification

Laboratories on research vessels can be classified into general laboratories, special laboratories, and container laboratories. General laboratories are typically large in size and designed to accommodate research needs of multiple disciplines. The interior layout of general laboratories features flexibility, allowing for adjustment based on specific scientific expedition voyages. Special laboratories, on the other hand, are specifically dedicated to conducting research in a particular discipline or on a research subject. Container laboratories are set up to cater to infrequently used survey projects and experimental instruments.

General laboratories are positioned as multi-functional workspaces. They can be used for temporary storage, separation, and processing of samples from the working decks, including geological samples, sediments, captured plankton, and fish.

Given that research personnel on board come from different disciplines such as atmosphere, water, biology, and geology, it is rare for all disciplines to simultaneously utilize the laboratory at any given time during a voyage. Allocating a dedicated area for each discipline would result in wastage of space. Experience has shown that sharing larger general laboratories among multiple disciplines is a more efficient approach. As a result, most research vessels, apart from having laboratories with special functions, possess larger general or comprehensive laboratories. In particular, research vessels with large tonnage are even equipped with several large-scale general laboratories. The interior of such laboratories is not equipped with specific instruments but rather includes interfaces where various equipment can be easily installed, enabling plug-and-play functionality. In some cases, large laboratories on research

Fig. 13 General laboratory on China's multipurpose research vessel "Kexue"

vessels are sectioned off using removable partitions, creating separate workspaces for different research groups and disciplines to avoid mutual interference.

General laboratories can also be further classified as general dry laboratories or general wet laboratories, or combined as general dry/wet laboratories. The main distinction lies in the fact that general wet laboratories involve handling of marine samples containing water, thus necessitating water-resistant flooring, furniture, and electrical fittings. General laboratories generally include walkways and doors with large clear opening to allow convenient movement of bulky samples. Figure 13 illustrates the general laboratory on China's multipurpose research vessel "Kexue" built in recent years.

Due to the drawbacks of general laboratories, such as personnel and environmental clutter, separate laboratories, known as special laboratories, are set up to undertake special experimental tasks and accommodate precision instruments. Main types of special laboratories include atmospheric laboratory, in-situ direct seawater analysis laboratory, temperature-controlled laboratory, geophysical laboratory, gravimeter room, electronic laboratory (see Fig. 14), and clean laboratory.

Atmospheric laboratories primarily focus on conducting atmospheric experiments, serving as the terminal for all atmospheric sensors, the meteorological data collection devices, as well as as the terminal for satellite remote sensing systems. Some research vessels further categorize atmospheric laboratories into air-sea flux laboratories, atmospheric environment laboratories, and atmospheric detection laboratories.

Moving laboratories for seawater analysis are designed to measure the characteristics of surface seawater while the vessel is in motion. They function as the terminal and analysis center for surface multi-element testing instruments.

Fig. 14 Electronic laboratory

A gravimeter is a specialized instrument used to measure gravity. A gravimeter room have specific requirements for temperature and humidity. Generally, the gravimeter room is located near the center of flotation on the ship.

Container laboratories are temporary laboratories that can be easily loaded onto ships. Given that the scientific research tasks undertaken on different voyages may vary, some projects might require the use of experimental instruments. However, refitting the existing laboratories on the ship repeatedly would be impractical. Instead, container laboratories are capable of accommodating the specific requirements for special tasks and equipment on board. Due to their cost-effectiveness and ease of use, most research vessels reserve space for these laboratories. Once securely fixed on board, these container laboratories can be readily utilized. Definitely, it is essential to consider that container laboratories necessitate specific interfaces with the vessels, such as connections for cold/hot freshwater, seawater, compressed air, electricity, and network access.

2. Special Installations in Laboratory

(1) Anchor Bolt

Anchor bolts are bolt bases that are embedded into the deck. They are arranged in a matrix pattern on the open working deck, workshop, or laboratory, which are versatile and can be used to install, secure, and tie temporary scientific research instruments and equipment loaded on board (see Fig. 15).

If the anchor bolts are only used for temporarily tying and securing ordinary lightweight instruments and equipment, they are not subjected to particularly large forces, so there is no need for specific force calculations. However, if they are

Fig. 15 Tying of scientific research instruments and equipment temporarily loaded on board using anchor bolts

used to secure heavy equipment for cross-deck operations, relevant calculation and verification efforts are required.

(2) C-shaped Groove

C-shaped grooves involve ribs, which are similar to light steel keels, set on the walls and ceilings of laboratories. They can be used for temporarily hanging and installing instruments and equipment, facilitating the switching of laboratory functionalities (see Fig. 16).

C-shaped grooves mainly support loads such as displays, mainframe boxes, which are not heavy. However, considering the inertia forces caused by the ship's sway in rough seas, it is advisable to connect the C-shaped grooves and the internal lining plates to the installation ribs and weld them to the structural framework when installing C-shaped grooves. This helps to enhance the strength and rigidity of the C-shaped grooves.

(3) Additional Cables Bracket

Given the expandability of research vessels and the addition of different research tasks on different voyages, various instruments and equipment are needed in different areas. These instruments and equipment require cables for power and signal. Due to the unpredictability of future equipment installations, these cables cannot be pre-embedded (see Fig. 17). The additional cables brackets are therefore set up to facilitate the temporary routing of cables and provide support.

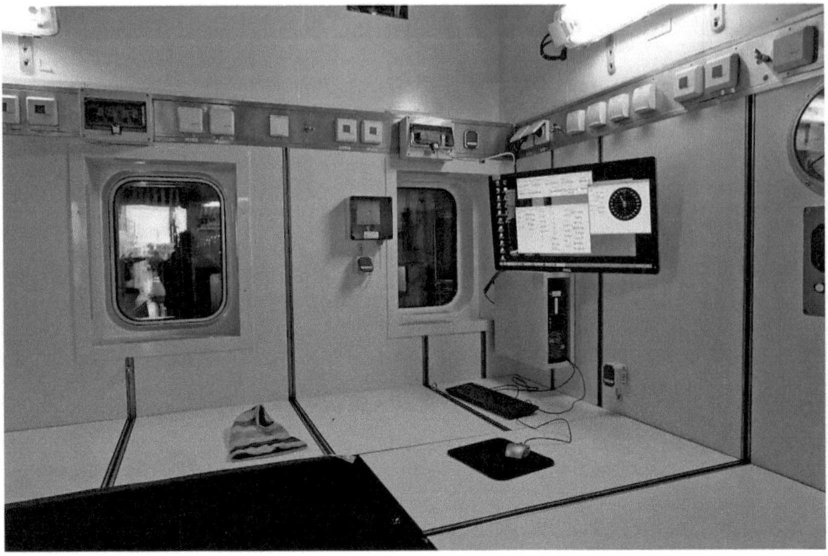

Fig. 16 Mainframe box and display installed using C-shaped groove

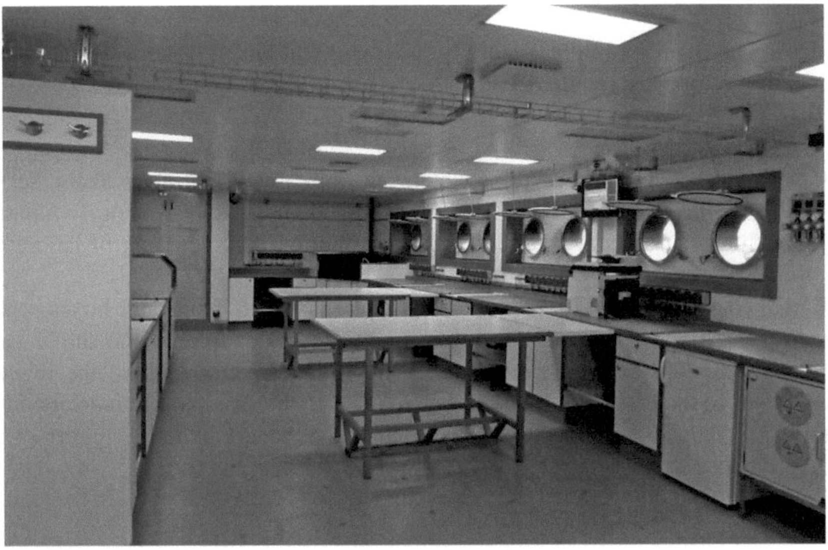

Fig. 17 Additional cables bracket on the ceiling of the laboratory

3.3 Arrangement of Working Decks

The working deck is the main outdoor area for scientific research operations on the research vessel. It is generally divided into three sections: stern working deck, the compass deck, and the bow deck working area.

1. Stern Working Deck

The stern working deck is the primary location for open-air research operations on the vessel, housing most of the launch and recovery systems. The deck is generally divided into left and right zones. Research operations typically take place on the starboard, while drainage, exhaust pipes, and other pipelines are arranged on the larboard. Generally, the larboard is used for regular operations, such as launching and recovering workboats, ROVs, and towed acoustic equipment. The stern working deck should be spacious, with minimal protrusions. Mooring equipment typically uses compact vertical capstans, and even detachable ones in some cases. Some vessels also adopt detachable bulwarks and mooring bollards.

Anchor bolts are typically installed throughout the whole stern working deck. Dedicated container pedestals are installed for commonly used container laboratories, while temporary onboard containers are secured with anchor bolts. The work area should have necessary interfaces for water (seawater/freshwater), electricity (power supply and signal transmission), and gas (compressed air/special gases).

From front to back, the typical arrangement on the stern working deck is as follows:

(1) Main crane: This is the crane with the largest lifting capacity and coverage area. It is used for launching and recovering work boats, and lifting research supplies and container laboratories, among other operations.
(2) Stern control room: It is often referred to as the "octagon house" for its octagonal shape and glass structure on all eight sides. The control console in the stern control room is equipped with control devices for all winches, A-frames, gravity piston corers, and other equipment on board. The room also has a display screen hung on the walls to monitor the operating conditions of winches in the winch compartment, lifting equipment on the deck, and the position of in-water equipment. The stern control room typically provides a 360-degree view. In some research vessels, the stern control room also includes a ship control system for direct ship handling. It has become a "standard configuration" of multipurpose research vessels. Figure 18 shows the stern control room aboard the research vessel "Tan Kah-Kee".
(3) Stern A-frame: The A-frame has the highest load-bearing capacity and is used for towing large objects at low speeds in deep sea conditions. The A-frame is typically stored in an upright position perpendicular to the working deck and is brought close to deck level only during operations (outboard) or maintenance (inboard).

Fig. 18 Stern control room aboard the research vessel "Tan Kah-Kee"

2. Compass Deck

The compass deck is mainly used for atmospheric operations and sky remote sensing. It is typically equipped with anchor bolts and interfaces for water, electricity, and gas connection. This deck usually includes a sheltered area for marine organism observations, atmospheric sampling equipment, movable winches, various antennas and radars, among others. It should ideally be as large as possible to facilitate the installation of future electronic equipment. If feasible, an observation room can be set up on the compass deck to create favorable conditions for research personnel in atmospheric and biological sciences and other disciplines. In addition, some support brackets can also be installed on both sides of the deck to satisfy the needs for observing the sea surface, ice surface, and so on. Figure 19 shows the antenna arrangement on the compass deck of the research vessel "Dongfanghong 3".

3. Bow Deck Working Area

The bow deck working area serves as a storage and layout area for activities such as launching and landing unmanned aerial vehicles (UAVs), placing biological culture tanks, and setting up mobile laboratories. The deck should be equipped with anchor bolts, container pedestals, and interfaces for water, electricity, and gas connection. For research vessels involved in atmospheric sampling operations, a sampling platform (see Fig. 20) should be installed at the top of the foremast (also called the scientific mast). As this location is exposed to the first encounter with sea airflow and is far from upper structures, it is ideal for positioning the sampling and measuring equipment for air-sea flux, atmospheric sampling, and wave height. Corresponding meteorological or atmospheric laboratories can be arranged inside the mast or on the lower deck.

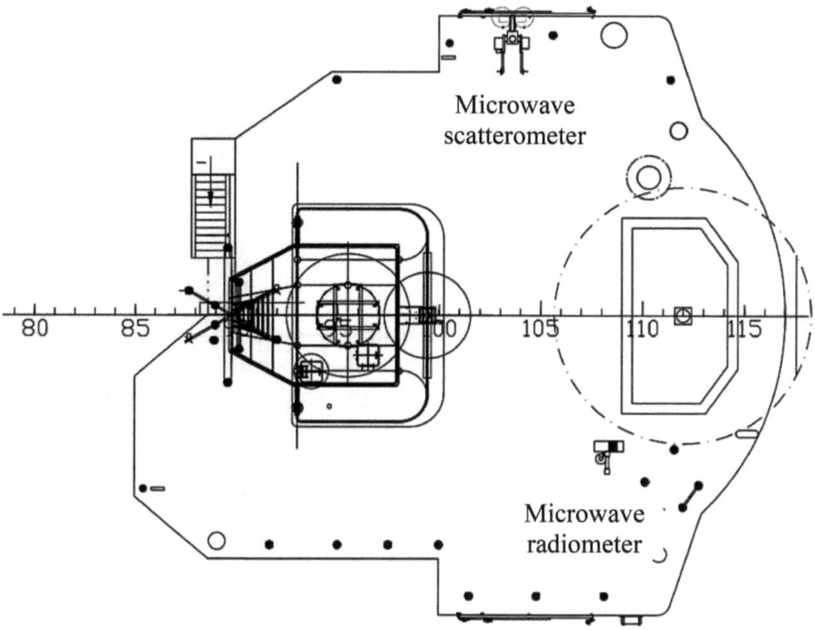

(a)

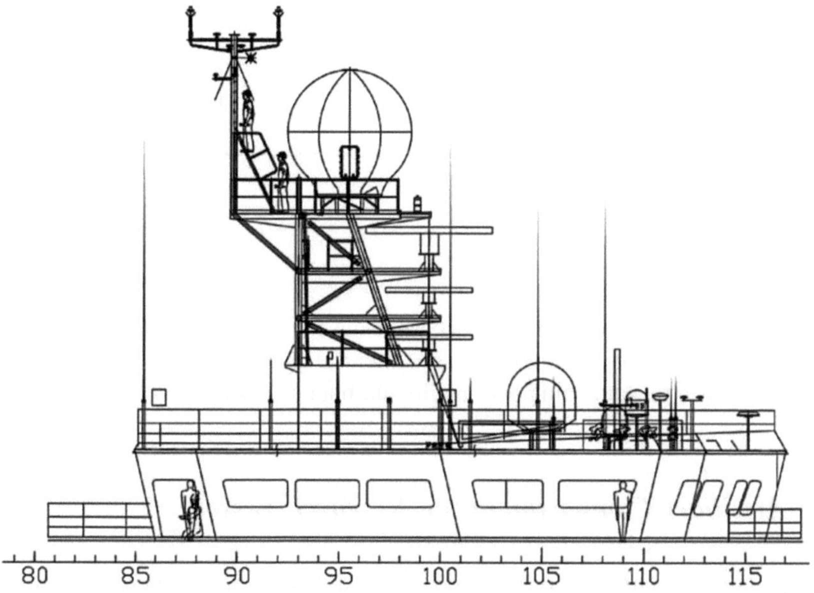

(b)

Fig. 19 Antenna arrangement on the compass deck of the research vessel "Dongfanghong 3". **a** Top View **b** Side View **c** Front View

Fig. 19 (continued)

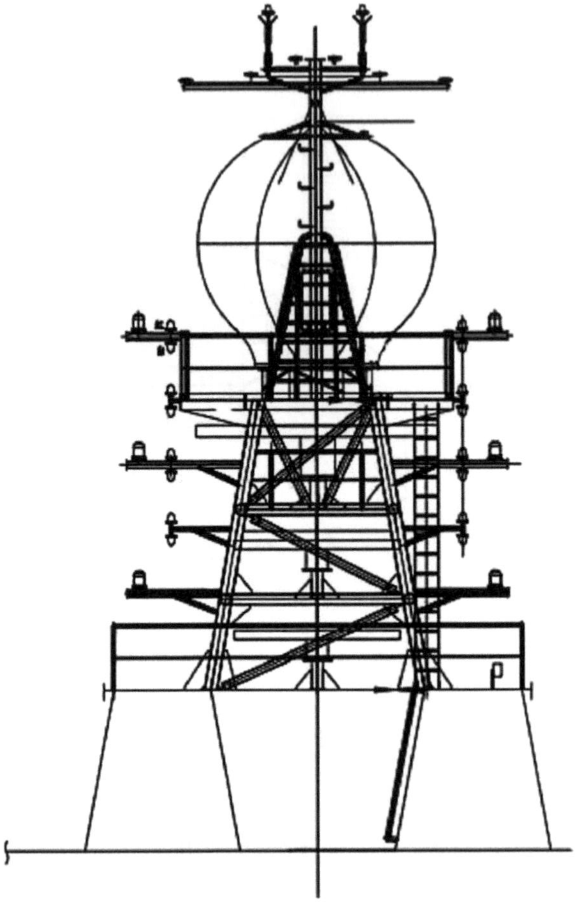

(c)

Since the bow deck working area does not have large A-frames and other large devices, and it does not possess as many antennas as the compass deck, it is also a suitable location for launching and landing unmanned aerial vehicles.

3.4 Auxiliary Compartments on Research Vessels

The auxiliary compartments on research vessels mainly include winch compartments, scientific research equipment compartments, acoustic equipment compartments, and workspaces.

The winch compartment is primarily used to house scientific research winches that do not need to be positioned on the open deck. In addition to the winches, accessories

Fig. 20 Scientific mast on the bow deck of the research vessel "Tan Kah-Kee"

for operation support systems, such as frequency converters, brake resistors, and hydraulic pump stations, may also be installed here. The advantage of housing large and bulky mechanical equipment in the compartment is that it does not occupy valuable space on the working deck, improving the deck's usable area and safety while enhancing the appearance of the ship.

In scientific research operations, the compartments that store equipment such as seismic air compressors and scientific research water pumps (excluding large winches) are collectively referred to as scientific research equipment compartments. They are commonly arranged near the engine room or auxiliary engine room.

The acoustic equipment compartment contains equipment that converts the acoustic signals from underwater transducers into electrical signals for long-distance transmission. Since shorter cable lengths result in better signal quality, the distance between the acoustic equipment compartment and the transducers should not be arranged too far.

The workspace is also known as the pre-processing workspace or shielded working deck. Some research vessels that focus on CTD water sampler operations may have a separate CTD sampling room. Vessels equipped with a moonpool system may have a dedicated workspace for this system.

The workspace, a space for equipment storage and sample pre-processing, plays a crucial role as a key node in the sample flow throughout the ship. Thus, space efficiency should be considered to ensure that it neither impedes the equipment and sample transfer routes for nor compromises on storage space. Some workspaces may also include maintenance workbenches and equipment such as bench clamps and drills, which are used for processing and assembling instruments, consumables, and other materials.

4 Scientific Research Operations Support System

The scientific research operations support system mainly consists of winch equipment as well as launch and recovery equipment for scientific research, making it one of the core systems of research vessels. During scientific researches and investigations, these equipment need to work together to safely and smoothly transport and deploy the detection equipment between the ship and the ocean, so as to provide support for various oceanographic studies, particularly in the fields of hydrology, biology, and geology.

4.1 Scientific Research Winch

The scientific research winch refers to the collective term for various types of winches used to launch and recover cables for scientific research operations. Most scientific research equipment deployed in the sea are released using various types of cables. Some cables serve solely as connectors, without the ability to transmit signals, while others are responsible for transmitting electricity and signals to the scientific research detection equipment. These cables are stored and deployed using corresponding winches. Winches can be categorized into the following types.

Depending on the type of operation, there are geological winches, deep towing winches, TV sampling winches, hydrobiological winches, CTD winches, ROV winches, seismic gun winches, and seismic cable winches (see Fig. 21).

Fig. 21 Seismic cable winch

Depending on the type of cable, scientific research winches can be classified as steel cable winches, fiber cable winches, coaxial cable winches, and optical cable winches.

Depending on the form of installation, scientific research winches can be classified as fixed installation winches and mobile installation winches. Fixed installation winches include geological winches, deep towing winches, coaxial cable winches, CTD winches, and hydrobiological winches. Mobile installation winches include ROV winches, seismic cable winches, seismic gun winches, moving vessel profiler winches, and magnetometer winches. Fixed installation winches refer to winches that are securely installed on the ship through welding or bolted connections. Based on their installation position, these winches can be installed in a specialized chamber or installed in specific positions on the deck. Figure 22 shows a winch installed on an open deck. Whenever possible, winches should be installed in a winch chamber for the sake of equipment protection and maintenance. Mobile installation winches are typically secured with bolts or container pedestals, making them easy to install and dismantle. Therefore, for temporary survey projects, winches required on board are all installed in this mobile manner.

Depending on the function of operation, scientific research winches can be divided into cable storage winches and traction winches (see Fig. 23).

Depending on the form of traction, scientific research winches include combination winches (cable storage + traction) and direct pull winches.

Depending on the form of driving, scientific research winches can be classified as hydraulic winches and electric winches.

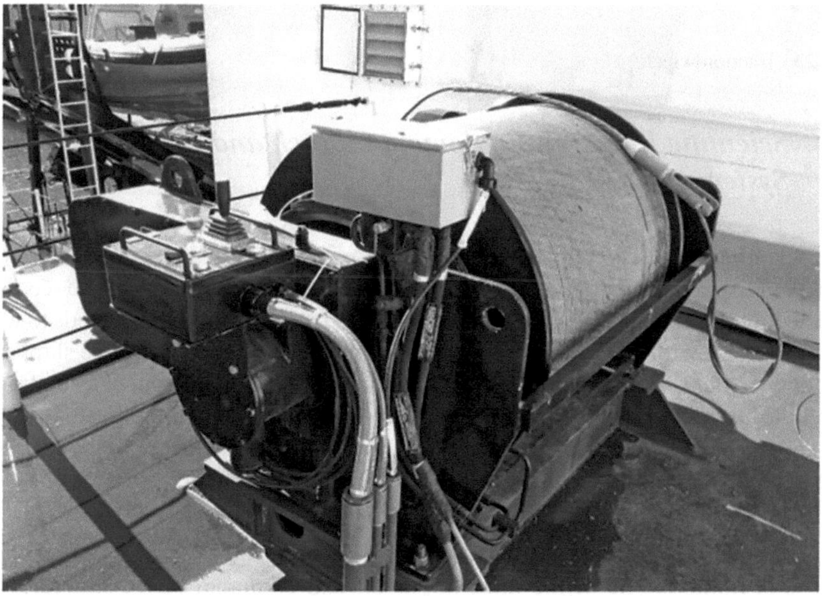

Fig. 22 Winch installed on an open deck

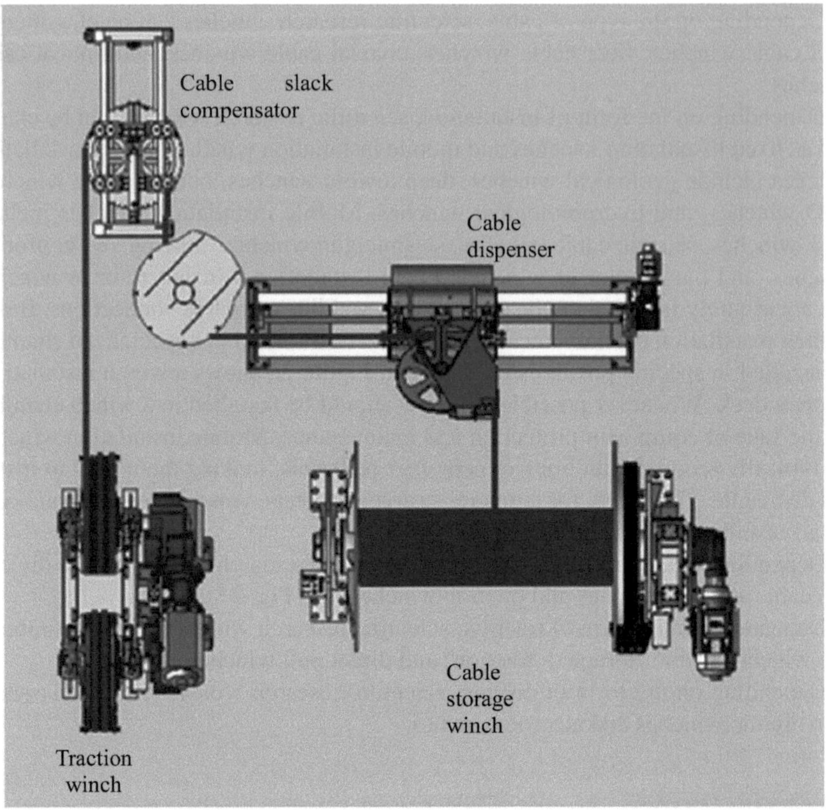

Fig. 23 Traction winch

4.2 Scientific Research Equipment Launch and Recovery System (LARS)

The scientific research equipment launch and recovery system (LARS) refers to a variety of lifting and operating equipment used for the hoisting, launch, and recovery of detection equipment. In general, the launch and recovery equipment needs to work in conjunction with the winch system to ensure effective operation of the shipborne detection system. The scientific research equipment LARS mainly includes A-frame, scientific research crane, CTD crane, core sampling device, moonpool system, lifting fin, and movable deployment equipment. Figure 24 shows the layout of the scientific research equipment LARS on the stern deck of a research vessel.

1. A-Frame (see Fig. 25)

The A-frame is one of the most important and commonly used structures for launching and recovering scientific research deployment equipment, primarily for

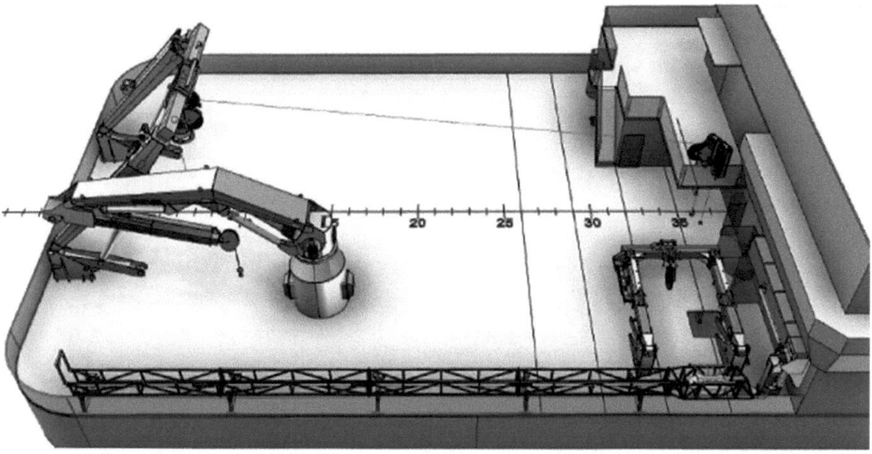

Fig. 24 Layout of the scientific research equipment LARS on the stern deck of a research vessel

Fig. 25 A-frame on the stern of the vessel "Kexue"

support scientific research operations at the stern and sides of the ship. Therefore, it is generally installed at the stern or side area for the launch and recovery of underwater detection equipment. A-frames come in various shapes, such as L-shaped frames (see Fig. 26) when equipped with a single support leg or π-shaped frames when the top beam extends on both sides. They are available in different types, including

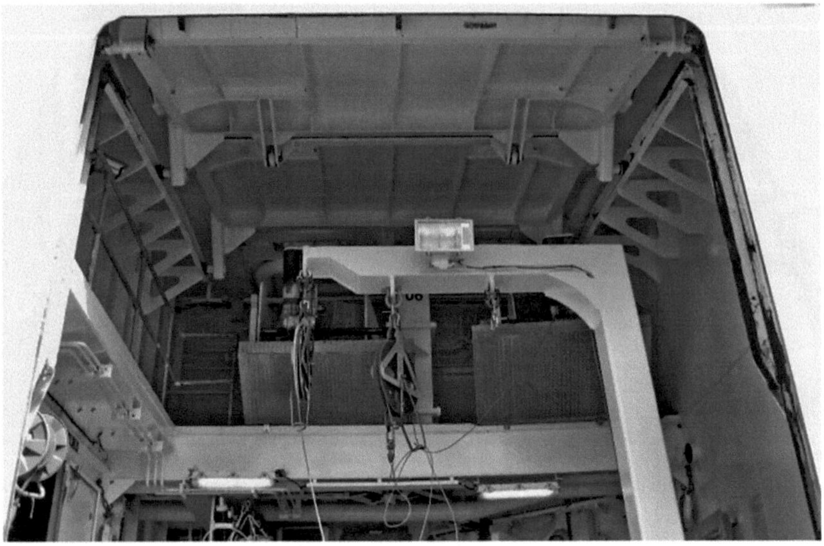

Fig. 26 Reversed L-shaped frame on a research vessel

retractable, non-retractable, foldable, non-foldable types, with or without stabilizing frames. A typical A-frame consists of a load-bearing frame structure, derricking cylinders (including a telescopic cylinder and a folding cylinder), a pulley block, an auxiliary winch, a control system, a hydraulic pump station, and a stabilizing frame.

For retractable A-frames, the telescopic cylinders can extend the load-bearing frame structure during operation, allowing for increased lifting height and expanded boom outreach. For foldable A-frames, the folding cylinders enable the load-bearing frame to fold when not in use, significantly saving the deck available space. For A-frame structures with a small range of swing angles, the stabilizing frames serve as an indispensable component to ensure the safe and stable launch and recovery of the detection equipment outboard, as they can effectively reduce the swing amplitude caused by the ship's motion during lifting.

The operation of an A-frame involves three main steps: lifting, swinging outward, and deploying. Lifting means to hang the main lifting cable, which includes cable lifting and hooking. Swinging outward means to swing the A-frame outboard. Deploying means to release the cable and unhook to deploy the equipment into the sea. Among them, operations such as cable lifting, hooking, cable releasing, and unhooking are particularly critical. Therefore, proper coordination between each step is crucial to ensure the equipment remains under control and prevent any swinging or collision that could result in damage to the equipment or pose a risk to personnel on the deck.

The top of the A-frame is typically equipped with a pulley block and an auxiliary winch. The pulley block is used to guide the cables, ensuring a tight connection between the A-frame and the scientific research winch, thereby facilitating efficient coordination between the two during operation. The auxiliary winch can provide assistance during equipment launch and recovery. Larger A-frames have ladders and guardrails installed on their sides and tops to facilitate maintenance and rope replacement. Some A-frames are designed to be collapsible, allowing them to be laid flat on the deck for maintenance and rope replacement. However, this design requires more deck space and a larger power source.

Underneath the A-frame structure at the stern, a long roller device is usually installed. By utilizing the rolling motion, this roller device allows for minimized friction, making it easier to deploy and retract equipment into and from the sea. When the equipment is being towed, the steel cable rests on the stern roller, which can rotate freely depending on the applied forces, thereby avoiding frictional damage to the cable. To prevent the equipment from getting stuck on the ship's bottom or sides or being severed by the side structures, the turning point at the stern is typically designed in an arc shape for better protection.

2. Scientific Research Crane

Scientific research cranes are one of the most common types of equipment for launching and recovering scientific research equipment. Here, scientific research cranes refer to common general-purpose cranes, excluding cranes specifically designed for CTD equipment. Such cranes can be used for the transportation on deck of various detection equipment, auxiliary equipment, and containers. They feature flexible operation, wide range of motion and good adaption to multiple working conditions, which are essential to research vessels. They can also assist in the launch and recovery of detection equipment at the shipboard. With different sizes, they are distributed in different areas, such as the bow, midship and stern, and can be used independently or in coordination to meet different operation requirements.

To withstand the various movements caused by wind and waves at sea, these cranes usually require dynamic compensation design, such as constant tension, passive compensation, servo compensation, and active compensation.

Constant tension is the simplest form of compensation, primarily involving comparing the actual tension of the cable with the set tension and adjusting the cable reel's movement accordingly to achieve compensation. Passive compensation primarily relies on compressed air to drive hydraulic cylinders, utilizing the extension and retraction of hydraulic cylinders to achieve rapid cable releasing and retraction, thereby providing compensation. This method offers faster compensation response compared to the constant tension method.

Servo compensation involves cranes with a main hook and a compensation hook, with the compensation hook connected to the detection equipment. When the compensation function is activated, the compensation hook adjusts the cable releasing in line with the relative movement between the detection equipment and the ship. At this time, the main hook adjusts the cable releasing accordingly to maintain synchronization with the compensation hook, thus achieving compensation.

Active compensation is the most technologically advanced and effective form of compensation. It utilizes ship motion attitude sensors to acquire real-time data on

Fig. 27 Folding-arm crane for scientific research

ship motion and controls the cable reel's movement based on the ship's motion state to achieve compensation.

Scientific research cranes generally come in different types, such as folding-arm cranes (see Fig. 27), telescopic cranes, and telescopic folding-arm cranes (see Fig. 28). The telescopic folding-arm cranes are particularly advantageous in saving deck space. Telescopic cranes can be divided into cylindrical telescopic cranes and deck horizontally-mounted telescopic cranes.

CTD cranes are specifically designed for the launch and recovery of CTD water samplers at the shipboard. Sometimes, they can also be used for organism vertical trawling and sampling operations during hydrological survey. The CTD cranes can be mounted below the deck or on the outdoor deck. With an under deck-mounted CTD crane, the CTD water sampler can be lowered in and retracted form water surface at a time, with high efficiency. The crane can be installed on the top of shielded operation area, with the ability to extend outboard for increased reach. Outdoor deck-mounted CTD cranes (see Fig. 29) usually have side passages. As they cannot lift the CTD water sampler into the compartment, the sampler needs to be rotated at the side of the ship and then pushed into the compartment using a trolley. Alternatively, the A-frame can be used for the launch and recovery of CTD water samplers. Figure 30 shows the Chinese polar icebreaker "Xuelong" using the A-frame at the stern to deploy a CTD water sampler.

Fig. 28 Telescopic folding-arm crane for scientific research

3. Core Sampling Device

Core sampling devices are used for the collection of column samples of seabed sediments. A device consists primarily of a gravity corer, a flip/rotation mechanism, and a corer support frame, among other components.

The gravity corer is the core component of this device and is essential for successful and high-quality sampling. Its basic principle relies on its own weight to penetrate the seabed. The depth of penetration depends on the hardness of the seabed and the structure and weights of the corer. Typically, the corer can penetrate to depths of 10 m, 20 m, and even up to 50 m.

There are various forms of flip/rotation mechanisms. Depending on different structures, they include vertical rotating arms, bulkhead slide-rail rotating arms, multi-point flip mechanisms, and other forms. However, they share same basic function, which is to transition the corer from a horizontal storage state to a vertical working state outside the board.

The corer support frame consists of storage trusses and brackets and is used for storing and disassembling the corer.

Core sampling devices for seabed sediment sampling are primarily used in the exploration of submarine mineral resources, marine engineering geology survey, and marine geology research. Depending on different principles, core sampling devices can be categorized into two types: one with a flip mechanism, mainly adopting a side deployment of the sampling pipe (see Fig. 31); and the other without a flip

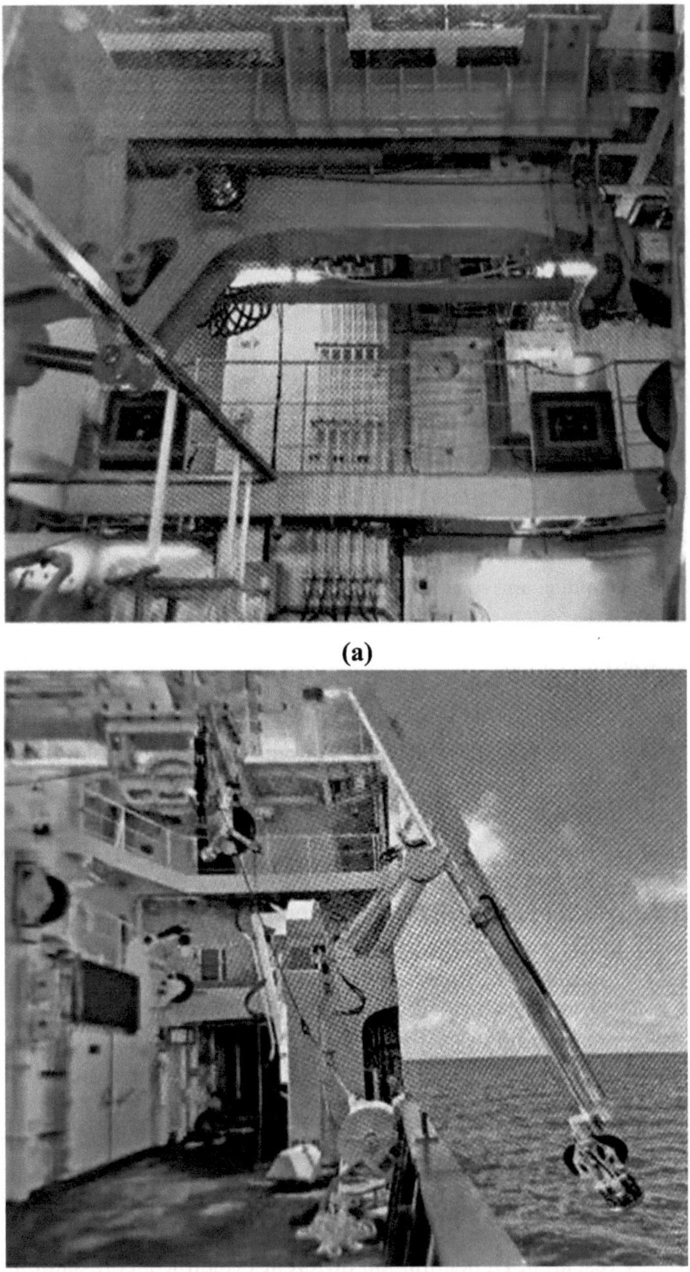

(a)

(b)

Fig. 29 CTD crane. **a** Under deck-mounted form **b** Outdoor deck-mounted form

Fig. 30 The Chinese polar icebreaker "Xuelong" using the A-frame at the stern to deploy a CTD water sampler

mechanism, instead utilizing tracks and the stern A-frame for deploying the sampling pipe.

4. Moonpool System

The moonpool system comes in a trunk structure which is vertically penetrated into the main structure of a ship or offshore structure in the middle. It allows for the installation, deployment, and recovery of related operation equipment, protecting the equipment from the impact of larger waves near the waterline on the exterior of the hull. This system is commonly found on drilling ships, life-saving ships, catamarans, floating production storage and offloading oil tankers, and research vessels. Large research vessels are generally equipped with moonpool systems, which are used to deploy or install large-volume underwater detection equipment or equipment that is not suitable for use in surrounding sea areas, such as ice zones or adverse sea conditions. The moonpool system consists of several parts, including the moonpool well, wave-absorbing structure, launch/recovery mechanism, moonpool cover, hydraulic pump station, and control system. The moonpool well is the opening, whose size is determined based on the dimensions of the equipment to be used. The moonpool wave-absorbing structure is usually a double-hull structure. The moonpool launch/recovery mechanism is used to lower the detection equipment from the moonpool workspace inside the ship into the water or raise it into the moonpool. It includes lifting frame, lifting winch, vertical guide rail, and other components. A simulated picture of the moonpool system (excluding the moonpool well) is shown in Fig. 32.

Fig. 31 Core sampling device with a flip mechanism

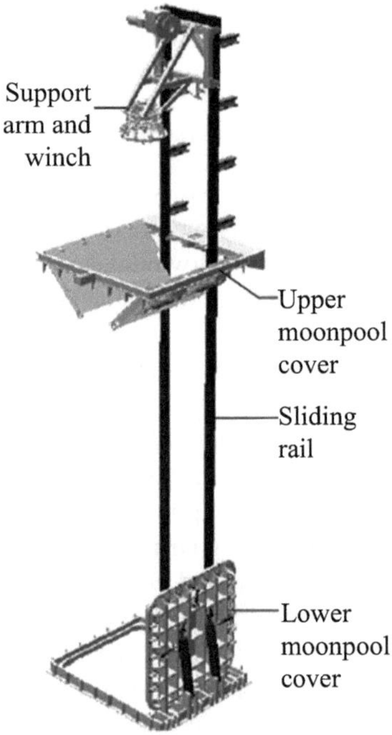

Fig. 32 Simulated picture of moonpool system (excluding moonpool well)

When conducting surveys in the polar regions, due to the limitations of weather conditions, some outboard scientific research operations (such as the deployment of hydrological and biological underwater instruments) need to be carried out using the moonpool. Therefore, polar icebreaking research vessels need to be equipped with a moonpool system.

The positioning of the moonpool should consider the compatibility with other equipment on the ship, such as the CTD winch and hydrobiological winch. To improve the safety of the moonpool structure and equipment operations, the following aspects should also be considered:

(1) Double-Hull Structure

The moonpool well motions in two ways: vertical ascending/descending along the well and horizontal sloshing perpendicularly to the well. If the frequency of incoming waves is close to the natural frequency of the water motion, resonance can occur, which would damage the equipment in the moonpool well and affect the operational safety. To reduce the impact of water motion in the moonpool well and mitigate the noise during navigation, a common practice is to install wave-absorbing structures on the moonpool well walls, which is achieved by applying double-hull structures (see Fig. 33), with perforations inside for wave absorption.

(2) Moonpool Cover

Based on different locations, moonpool covers are classified as bottom covers and top covers. The moonpool is opened through several decks from ship bottom, all the way up to the moonpool workspace. The bottom cover of the moonpool is subjected to the impact of water flow during navigation, which requires a high level

Fig. 33 Double-hull structure in the moonpool

of watertightness of the ship. When using a closure device at the bottom, the lifting forces and slamming caused by bottom water flow and ice flow should be considered, as strong upwelling forces often lead to damage and breakage of the hatch cover. The installation of a bottom cover is not always necessary and it varies from ship to ship. The top cover of the moon pool is typically a folding embedded weather-tight hatch cover. The hatch cover is flush with the deck and can be used for storage when it is in the lowered position.

(3) Auxiliary Safety Facilities

To ensure the safe launch and recovery of equipment, prevent collisions with the inner walls of the moonpool during operations, avoid friction between steel cables and the edges of the moonpool bottom opening, ensure the safety of personnel, and provide appropriate de-icing facilities in case of ice formation within the moonpool, it is advisable to incorporate auxiliary safety facilities in the moonpool system.

5. Lifting Fin

The lifting fin, a liftable plate-like device, is used to carry detection equipment. Serving as the carrier of shipborne detection equipment for installation, this device is typically arranged inside a vertical well located in the midship that is accessible to the sea. Some ships may have only one lifting fin, while others may have them in pairs.

Modern research vessels usually carry and install sophisticated detection instruments. Many of these instruments have strict requirements for vibration and noise. The lifting fin can effectively reduce the impact of bubbles in the turbulent boundary layer at the bottom of the ship on the measuring accuracy of the carried detection equipment, significantly enhancing the quality and precision of the collected data.

The fin plate system mainly consists of the fin plate itself, lifting mechanism, locking mechanism, hydraulic power unit, and control system. The shipborne detection equipment installed on the fin plate commonly includes fish finders, ADCPs, cameras, and clean seawater sampling tubes. The working position can be lowered to 2–3 m below the ship's baseline. During cruising, the bottom plate of the fin plate is aligned with the ship's baseline, ensuring that it does not affect the normal sailing speed of the vessel.

The fin plate is the core component of the system. To minimize the resistance caused by the fin plate during deployment, its profiles is designed in a drop shape, resembling a ship rudder. The interior of the fin plate is equipped with vertical and horizontal ribs, which provide sufficient strength to withstand the operational conditions. The fin plate can be designed as fully enclosed or with lightening holes in the upper part and drainage holes in the lower part. Due to the stringent positioning requirements of the detection equipment mounted on the fin plate, it is crucial to consider how to effectively "secure" it to the hull, ensuring that no loosening occurs. Otherwise, in adverse sea conditions, any movement or shaking of the fin plate could result in deviations in the collected data. A notable advantage of the fin plate is that it enables easy maintenance or replacement of equipment during navigation. This significantly saves time and resources, thereby improving the efficiency of research vessel operations.

There are two methods for deploying the fin plate: gravity lowering and lowering via hydraulic cylinder. Gravity lowering involves the participation of the lifting winch and sliding track. The fin plate moves up and down along a sliding track within the well. The upward movement is facilitated by a lifting winch, while the downward movement relies on the gravity of the fin plate itself. This method features a relatively simple structure and operation. However, there lies a drawback that when the waterline in the lifting well is relatively high, fin plate lowering can be affected by buoyancy forces. Lowering via hydraulic cylinder involves the participation of the lifting winch, telescopic cylinder, and sliding track. This method addresses the issue of impact from buoyancy forces in gravity lowering during high waterline in the lifting well. However, the structure is more complex, which increases the difficulty of equipment maintenance.

Chapter 3
Overcoming Challenges and Building the Foundation

1 Overview

China, a major maritime country, borders on the vast Pacific Ocean, with four linked adjacent seas, namely, the Bohai Sea, the Yellow Sea, the East China Sea, and the South China Sea. A winding curve, demarcating the boundary between land and sea, forms China's coastline. It begins at the Yalu River estuary in Liaoning province, and stretches southwards to the mouth of the Beicang River in Guangxi province. Together with the coastlines of various offshore islands, China's coastline spans approximately 32,000 km in total. Despite elusive marine meteorology and sea conditions, there lies incredibly enormous resources in the ocean, making understanding the ocean a long-standing and unwavering pursuit of the Chinese people living alongside it. Countless historical examples have fully demonstrated that understanding and utilizing the ocean plays a vital role in the survival and prosperity of a nation and its people. Ancestors of the Chinese nation had once achieved remarkable accomplishments in the field of the ocean at a level surpassing the rest of the world. However, from the eighteenth century until 1949, due to the weakening of national strength and ineffective governance, China fell far behind the West in terms of technology, equipment, and other aspects.

The historical mission of understanding, utilizing, and developing the ocean has now fallen upon the marine scientists, ship technologists, and construction workers of New China. History has shown that the scientific and technological level of a country or a nation greatly influences its future and destiny. The Party and the government have always regarded science and technology as the primary productive forces. Since the establishment of New China, efforts have been made for the development of marine science and technology. In August 1950, the first domestic marine research institution, the Qingdao Marine Biological Research Laboratory of the Chinese Academy of Sciences (CAS), was established. Subsequently, the Ministry of Agriculture and Forestry established several local fisheries research institutes,

L. Han et al., *History of the Research and Development of Chinese Marine Research Vessels*, https://doi.org/10.1007/978-981-96-8414-4_3

and the Ministry of Geology established the Marine Geological Survey. Nonetheless, as the ancient saying goes, "To accomplish a task, one must first sharpen the tools. "At the beginning of the establishment of New China, due to factors such as the country's limited economic, technological, and industrial levels, there were no specialized survey vessels to support marine research work. As a result, the research work of marine biological research laboratories could only be conducted on seashores and intertidal zones, collecting specimens like today's people leisurely collecting seashells on the beach. At most, research tasks had to rely on fishing vessels, cargo vessels, or even sampans to be completed. However, as the international situation evolved, there was an urgent need for China to possess dedicated, modern research vessels. The Party Central Committee and the State Council always kept a firm grasp on the correct direction of China's ship development. They included shipbuilding as an industry that would drive the development of the national economy, and provided supports for the research and development of research vessels. As a result, a shipbuilding technology under the leadership of the Party was set up, which was dedicated to serving the national economy and defense, boosting development of research vessels. China's research vessels have gone through three distinct phases of development. The first phase took place from the 1950s to the 1970s, followed by the second phase from the 1980s to around 2010, and the third phase starting from 2011 onwards. During this time, both the quality and quantity of China's research vessels experienced a rapid growth, resulting in the construction of several state-of-the-art research vessels that matched global standards. In 1956, the Chinese government formulated the *Long Range Development Program of Sciences and Technology*, which designated marine science as the seventh priority and included "China's multipurpose oceanographic survey and development plans". In June 1957, the Qingdao Marine Biological Research Laboratory of the CAS embarked on a scientific expedition focused on marine biology using the modified oceanographic survey vessel "Jinxing", marking the beginning of China's oceanographic survey and exploration efforts. In September 1958, the State Scientific and Technological Commission of the CPC coordinated a scientific survey involving over 600 scientific and technological personnel form more than 60 institutions, including the Navy, Chinese Academy of Sciences, Ministry of Fisheries, and Shandong University, among others. With the research vessel "Jinxing" as the main vessel, supplemented by four temporarily modified naval frigates, this expedition marked the largest-scale national marine scientific survey. The survey covered most of China's seas, with multiple survey sections, extensive cruise observation points and continuous observation points established in the Bohai Sea, Yellow Sea, East China Sea, South China Sea, and along the coasts of Zhejiang and Fujian. The observation encompassed various aspects such as marine hydrometeorology, marine chemistry, marine biology, and marine geology.

Through this extensive marine survey, Chinese scientists obtained a wealth of marine data for China's offshore waters, gaining a preliminary understanding of the fundamental characteristics of marine hydrology, chemistry, biology, geology, among other aspects. In 1963, multiple volumes of *General Oceanographic Survey Data of Bo Hai, Huang Hai, Dong Hai and Nan Hai* and an *Atlas of Multipurpose*

Marine Survey of China were published, laying a solid foundation for the further research of marine science and the development and utilization of the oceans.

With the increasing economic activities in our country's marine sector, inadequate understanding of marine hydrology, meteorology, geology, and topography has resulted in major accidents and significant losses of lives and properties. Notable incidents included over 30,000 wooden sail fishing vessels capsizing due to inaccurate coastal weather forecasts and the stranding and sinking of the ten-thousand-ton cargo vessel "Yuejin" on its journey to Japan due to inaccurate sea chart markings. These incidents highlighted the urgent need for scientific research of the oceans, prompting discussions on the construction of modernized oceanographic survey vessels as part of the country's industrial development plans.

In November 1950, the Shipbuilding Industry Office of the Ministry of Heavy Industry of the People's Republic of China was established, encompassing an design group in the technology department. This design group was the first ship design unit in New China, and later developed into the Second Product Design Office of the Ship Products Design Institute of the Ministry of the First Machine-Building Industry. By 1954, this design unit already had a workforce of over 500 and enhanced technical capabilities, capable of designing civilian ships ranging from 4,000 to 5,000 tons, as well as small high-speed gunboats. Its research and development capacity could basically meet the requirements of the ship users at that time, completing the design tasks of research vessels such as "Qixiang 1", "Shijian", and "Dongfanghong".

In 1959, with the approval of the State Planning Commission, approximately 10 million yuan was allocated to design and construct the 2,500-ton marine training survey vessel, "Dongfanghong", for Ocean University of Qingdao (now Ocean University of China). The construction of the vessel commenced in December 1962 and was completed in January 1965. The vessel was used for the training and offshore practice of students in marine disciplines at higher education institutions, aiming to cultivate professional talents for marine research.

Also in 1959, the South China Sea Institute of Oceanology of the CAS commissioned the Second Product Design Office of the Ship Products Design Institute to design a multipurpose oceanographic survey vessel "Shijian". According to the technical requirements, this vessel should be able to conduct professional surveys in marine physics, chemistry, biology, and geology, among other disciplines in the South China Sea, with the potential to extend its operations to the neighboring Western Pacific when required. This project was listed as a national key product project and was directly managed by the State Planning Commission.

The "Dongfanghong" and "Shijian" were both designed by the Second Product Design Office of the Ship Products Design Institute. After learning, exploration, and practice, a strong foundation was laid for the design of various research vessels. The Office later became a part of the MARIC. Together, they forged ahead as stalwarts in the design of various types of research vessels and figured as New China's backbone forces in this field.

2 Classical Research Vessels

From 1949 to 1978, China's oceanographic survey vessels underwent significant changes, evolving from nonexistence to existence, from refitting to self-design, from simplicity to complexity, and from a single purpose to multiple purposes, witnessing the development of classical oceanographic survey vessels in China.

2.1 The First Oceanographic Survey Vessel "Jinxing"

"Jinxing" holds the distinction of being China's first oceanographic survey vessel refitted from an old ship (see Fig. 1). In 1950, following the establishment of the Qingdao Biological Research Laboratory of the Chinese Academy of Sciences (CAS), researchers were in urgent need of data on coastal marine meteorology and resources to support China's ocean development and utilization. However, to achieve success in any endeavor, it is necessary to have the right tools at hand. Without a vessel, venturing out to sea remained elusive. Given the nation's economic difficulties at the time, building a new vessel was costly. Given this, the State made the decision to refit an old vessel into a research vessel exclusively for the Qingdao Biological Research Laboratory. Originally, this vessel was named "Shengchan 3", an ocean-going rescue tug purchased from the US by the old China's Marine Department in the early twentieth century. With a full displacement of 1,700 tons, this refitted vessel was equipped with six laboratories (specializing in physics, chemistry, biology, geology, etc.) and one meteorological observation room, allowing for various ocean research activities to be conducted onboard. The vessel was outfitted with state-of-the-art instruments and equipment of the time, including self-recording water temperature gauges and radio direction finders, to facilitate automatic recording of temperature, current, and depth of the ocean.

On June 28, 1957, "Jinxing" departed from Qingdao Port, embarking on its maiden expedition voyage to the Bohai Sea. This significant event unveiled China's first multipurpose oceanographic research in history. Accompanying the vessel was a marine research team from the Qingdao Marine Biological Research Laboratory of the CAS. Mao Hanli, a marine physicist, served as the team leader, with four Soviet marine biologists involved in the expedition. "Jinxing" collected a wealth of valuable data during this voyage for China's development and utilization of marine resources. This expedition marked the country's preliminary systematic survey of offshore waters, ushering in a new era of marine research. In 1958, the State Scientific and Technological Commission (SSTC) organized more than 600 scientists and experts from over 60 institutions, including the Navy, Chinese Academy of Sciences, and Ministry of Transport, to conduct an extensive marine survey. "Jinxing" took center stage during this large-scale survey, which covered almost all of China's territorial waters, excluding the Taiwan Strait and the South China Sea. The survey covered

Fig. 1 Oceanographic survey vessel "Jinxing"

marine physics, marine chemistry, marine biology, marine geology and geomorphology, among other disciplines. The purpose of the survey was to conduct comprehensive and integrated surveys of China's coastal systems, compile marine atlases and gazetteers, prepare survey reports, develop marine resource exploitation plans, establish marine hydrological and meteorological forecasts, fishery forecasts, and provide fundamental marine environmental data for national defense and maritime transportation.

This survey covered a large portion of China's seas, with multiple survey sections and extensive cruise observation points as well as continuous observation points set up in the Bohai Sea, Yellow Sea, East China Sea, South China Sea, as well as along the coastal areas of Zhejiang and Fujian. The observation included water depth, temperature, salinity, water color, transparency, sea luminescence, sea wave, air temperature, humidity, air pressure, wind, cloud, and visibility in terms of marine hydrology and meteorology; dissolved oxygen, phosphates, and pH in terms of marine chemistry; layering and vertical sampling of planktons, benthic organisms sampling and geological sampling in terms of marine biology; and seabed surficial sampling, core sampling, suspended particle sampling, and continuous sounding in terms of marine geology.

The survey yielded over 90,000 original records and various data reports, more than 70,000 charts (including various marine element plane and vertical distribution maps, cross-section diagrams, temperature-salinity curve charts, and temperature-depth recording charts), and over 10,000 biological specimens and samples (including sediment samples from the bottom surface layer, bottom vertical sediment samples, and suspended particle samples). The collected data and charts were compiled and published as several volumes of the *General Oceanographic*

Survey Data of Bo Hai, Huang Hai, Dong Hai and Nan Hai and the *Atlas of Multipurpose Oceanographic Survey of China.*

Through this extensive oceanographic survey, Chinese scientists obtained a relatively comprehensive set of data on our country's offshore waters for the first time. This allowed for a preliminary understanding of the fundamental characteristics of marine hydrology, chemistry, biology, geology, among other aspects in our offshore areas. It laid a solid foundation for further marine research and the development and utilization of the oceans.

In 1980, after more than 20 years of seafaring service, "Jinxing" bid farewell. As a pioneer in China's oceanographic survey vessels, its decommissioning marked the pivotal moment when China began its journey towards the ocean, embarking on real oceanographic survey and research.

2.2 The First Self-Designed and Built Meteorological Survey Vessel "Qixiang 1"

In 1957, an unanticipated devastating storm with violent winds and roaring waves caused the tragic capsizing of over 30,000 wooden fishing vessels, resulting in severe loss of lives and property. Upon hearing the news, Premier Zhou Enlai was deeply shocked and promptly sought an explanation from the Central Meteorological Administration regarding the absence of an advance forecast. The leaders of the Administration reported that due to the lack of meteorological monitoring stations at sea, they had to rely solely on land-based information from the monitoring stations on land, resulting in inaccuracies in marine forecasting. Timely and accurate weather forecasting for the ocean required meteorological data from monitoring stations at sea. Recognizing that, Premier Zhou immediately ordered the construction of a dedicated meteorological survey vessel. With the full cooperation of various government departments, the Shipbuilding Industry Bureau entrusted the Second Product Design Office of the Ship Products Design Institute with the task of design, while the Qiuxin Shipyard was responsible for its construction. The vessel needed to be built and delivered as quickly as possible.

This vessel marked China's first self-designed and built vessel dedicated to meteorological surveys. At that time, there was limited data available and the design team had little familiarity with meteorology. However, in order to ensure the safety of fishermen's lives and property, as well as the accuracy of catastrophic weather forecasts, efforts were made from the central to local levels to overcome various challenges. When the design team initially received the task of designing a marine meteorological survey vessel, they were unsure of how to proceed, given their lack of previous experience in this field. The design unit promptly formed a design team, driven by a sense of loyalty to the country and its people, as well as a strong urge to explore the ocean. The team collected relevant information on marine meteorological survey vessels, conducted investigations on the requirements at the Shanghai Meteorological Service, and visited manufacturers to gather information on the equipment. By referring to foreign journals, the design team sought foreign information on designing similar vessels. Meanwhile, they adapted the land-based meteorological survey methodologies to sea conditions. The Meteorological Administration shared the practice of releasing a two-meter-diameter hydrogen balloon equipped with a transmission device simultaneously from various meteorological stations across the country at 6:00 am to collect meteorological data such as high-altitude wind speed, wind direction, and temperature. The data obtained would then be transmitted to the National Meteorological Center in Beijing for scientific analysis and weather forecasts. Considering the risk of hydrogen balloons being blown away by sea winds if placed on open deck, the design team decided to house them indoors. After research, they designed a dome-shaped cabin cover, which was closed normally and opened during balloon release. The construction of this dome-shaped cover posed challenges, but they were successfully resolved through the efforts of shipyard technicians. Another challenge was hydrogen supply for the hydrogen balloon, as carrying hydrogen cylinders onboard could pose safety risks. To address this, the design team purchased dedicated equipment for on-board hydrogen production at any time. With all these technical difficulties overcome, a meteorological survey vessel that met the specified requirements was successfully designed within a short span of time.

Construction of the vessel commenced in August 1959 at the Qiuxin Shipyard. The construction was completed successfully through close collaboration between the design team and the shipyard. The vessel was designed with a total length of 55.30 m, a molded breadth of 9.0 m, a molded depth of 6.10 m, a designed draft of 3.50 m, a trial sailing speed of 13.37 knots, an endurance of 5,000 nautical miles, and a capacity of 62 persons, with 6ДР-type diesel engine as its main engine. The vessel was equipped with four hydrographic winches, one depth sounder, as well as meteorological, hydrological, biological, and chemical laboratories, along with a hydrogen production room.

After its completion, the vessel was delivered to the Central Meteorological Administration in September 1960 and named "Qixiang 1". After a trial voyage, the vessel was verified to met all performance requirements specified in the technical assignment.

This vessel marked China's first endeavor in independently and successfully designing and constructing a modern vessel dedicated to meteorological surveys, which provided valuable experience for the design and construction of more similar ships.

The vessel played a crucial role in raising awareness among national and social departments regarding the challenges posed by the marine environment during the process of ocean development. Therefore, it is essential to rely on research vessels for detailed and precise surveys in order to better understand the dynamics of the ocean, and thus to facilitate ocean development.

In 1963, the responsibility for marine hydrometeorological surveys was transferred to the Chinese Academy of Sciences. Accordingly, the vessel "Qixiang 1" was renamed as "Shuixing" (see Fig. 2).

Fig. 2 Meteorological survey vessel "Shuixing"

2.3 The First Marine Training Survey Vessel "Dongfanghong"

In order to carry out systematic oceanographic surveys in the 1960s, it was necessary to expedite the training of a group of professionals for maritime scientific research while constructing research vessels. In anticipation of this need, the Ministry of Education proposed to the State Scientific and Technological Commission (SSTC) the construction of a 2,500-ton marine training survey vessel for the Ocean University of Qingdao. The aim was to provide students with practical opportunities to conduct maritime scientific research alongside their theoretical studies and prepare them for future scientific research tasks after graduation. In August 1959, the SSTC included the construction of a new marine training survey vessel in its plan, and commissioned the design to the Second Product Design Office of the Ship Products Design Institute under Shipbuilding Industry Office.

Through analysis, the design team recognized that given the limited economic capabilities of the country at the time, this vessel that required a huge investment should be given the greatest efficiency. This entailed maximizing the scientific research items that could be conducted onboard to offer students broader opportunities for practice while also ensuring the vessel's function to undertake actual research tasks. History has proven that this design concept has played a significant role, as the vessel has taken on and completed numerous scientific research tasks in the context of limited research vessels in the country.

This vessel was primarily used for scientific research tasks in China's adjacent sea areas, enabling a wide range of marine research activities.

Construction of the vessel commenced in the Hudong Shipyard in December 1962. In January 1965, the vessel was delivered and named "Dongfanghong" (see Fig. 3).

This vessel was designed with a length of 86.8 m, a molded breadth of 13.2 m, a molded depth of 7.4 m, a draft of 4.4 m, a service speed of 14.6 knots, and a capacity of 139 persons (including 45 crew members and 94 passengers).

"Dongfanghong" successfully completed 115 maritime teaching and training tasks, involving more than 6,500 teachers and students from the major of oceanography of Ocean University of Qingdao (now Ocean University of China). Additionally, from 1982 to 1988, it also undertook teaching and training for teachers and students from four other institutions of higher learning, including East China Institute of Water Conservancy (now Hohai University), Xiamen University, Tongji University, and Nankai University. The vessel played a crucial role in cultivating talents for marine research in China.

Fig. 3 Marine training survey vessel "Dongfanghong"

"Dongfanghong" also undertook various scientific research tasks. These include participating in surveys of the East China Sea continental shelf, collaborating with the US on the investigation of "sediment dynamics in the Bohai Sea's central-southern region and the Yellow River estuary", and conducting joint surveys with France and Japan. Moreover, this vessel also received foreign leaders such as Chairman Ho Chi Minh of Vietnam and Prime Minister Penn Nouth of Cambodia.

Between 1981 and 1995, the vessel safely sailed over 180,000 nautical miles and successfully completed observations at 142 sections, 28 survey lines, 2,282 major observation stations and 408 continuous observations stations. It also conducted surveys in the Yangtze River estuary, Jiaozhou Bay, the Yellow and Bohai Seas, the coastal zone of Shandong, the East China Sea, the Liaodong Bay, and the South China Sea. The vessel made four visits to Japan for friendly exchanges and marine research tasks and was decommissioned in 1996.

The design and construction "Dongfanghong" proved that the research and development concept of combining teaching and scientific research tasks was in line with China's national conditions. This concept was further exemplified in the development of the second and third generations of marine training survey vessels, namely "Dongfanghong 2" (1995) and "Dongfanghong 3" (2019), by the MARIC for the Ocean University of Qingdao.

2.4 *The First Multipurpose Oceanographic Survey Vessel "Shijian"*

In the late 1950s, China faced a challenging situation in terms of national economic development and defense construction. This underscored the growing importance of marine research, and conducting more extensive oceanographic surveys became an urgency. Given the significance of the South China Sea in various aspects, the construction of a multipurpose oceanographic survey vessel for the South China Sea was approved as a national key project by the State Council in 1959.

In 1960, the Second Product Design Office of the Ship Products Design Institute was commissioned by the South China Sea Institute of Oceanology of the Chinese Academy of Sciences to design a multipurpose oceanographic survey vessel. The technical specifications for this vessel are stringent, with a total of 13 specialized laboratories and intricate equipment involved.

After receiving the design task, the Second Product Design Office conducted thorough arguments and completed the preliminary design in February 1960. Throughout the process, the Office had multiple consultations with the user to refine the technical requirements. With approval from the State Council, the technical design was finalized after three meetings. The construction design was completed in October 1964, and the shipbuilding began at the Hudong Shipyard in 1965. In September 1967, the vessel, named "Shijian", (see Fig. 4) was delivered to the Chinese Academy of Sciences (CAS).

Fig. 4 Multipurpose oceanographic survey vessel "Shijian"

This vessel was designed with a total length of 94.73 m, a molded breadth of 14.00 m, a molded depth of 7.80 m, a draft of 4.75 m, and a displacement of 3,000 tons, a speed of 16.0 knots, an endurance of 7,500 nautical miles, a self-sustaining period of 45 days, and a capacity of 105 persons (including scientific researchers). The vessel was classified by the Zhongguo Chuanjian (ZC).

The construction of the vessel took eight years, providing a glimpse into the limited economic strength of China at that time when the country's manufacturing industry was undergoing tough development. It was seen as a reflection of the arduous journey of building such a modern and multipurpose oceanographic survey vessel. "Shijian" possessed outstanding seagoing features and well-equipped scientific research facilities, which met the basic standards of a modern multipurpose survey vessel. In 1977, it won the Shanghai City Award for Significant Scientific and Technological Achievements, and in 1978, it received the National Science Conference Award.

Firstly, as the first modern multipurpose oceanographic survey vessel not only for the CAS but also for China as a whole, the vessel undoubtedly represented a gem in the marine science community, with high expectations placed upon it. After completing the preliminary design according to the original technical assignment, the CAS held multiple meetings to amend the technical assignment, requiring the addition of laboratories and specialized equipment, as well as an increase of the vessel's main dimensions and displacement for enhancing its resistance against wind and waves and expanding the navigation areas. In response to this, the MARIC took it upon themselves to develop a research vessel that would satisfy the marine science community. They conducted in-depth analysis and deliberation, drawing on advanced international experience while closely integrating with the level of manufacturing in China at that time. During the design process, they visited renowned scientists in the field of marine science, seeking their opinions and incorporating their suggestions into the design.

Secondly, given the ship's importance and significant investment as a major national project, its implementation was directly supervised by the State Planning Commission (SPC).

On April 18, 1962, the SPC issued a document stating: "When considering the supporting equipment, the CAS and design departments should prioritize the use of domestically produced equipment. In cases where new equipment are needed, their design and trial production should be arranged early. "Subsequently, relevant departments of the State Council, the CAS, the Shanghai Local Committee Office, and related trial production units worked diligently to implement the 331 items related to equipment selection and trial production for the vessel. With the exception of a few items that were ordered from abroad, all other products were purchased or produced as part of tasks assigned by the SPC to various departments.

In order to avoid the situation of "idle vessel" caused by the inability to keep up with the supporting equipment after the completion of the vessel, on August 25, 1962, the SPC issued a document instructing the project to "prioritize equipping over shipbuilding", and meanwhile assigned plans and tasks for the development of 53 piece of supporting equipment to various collaborating units and made arrangements for the funding required for the equipment. The CAS required the MARIC and the Institute of Oceanology of the CAS to provide specific technical specifications, models, drawings of the supporting equipment, in order to avoid any modifications or changes in design and models after the contract was signed, which could affect the progress of trial production and result in economic losses.

This vessel was China's first self-designed multipurpose oceanographic survey vessel. With strong support from the South China Sea Institute of Oceanology of the CAS and various equipment research units, a high-performance multipurpose oceanographic research vessel was successfully designed after numerous technical challenges were resolved. Its main features include:

(1) Excellent performance in speed, stability, seaworthiness and other aspects: To improve the ship's speed and seaworthiness and ensure stability against winds of up to Force 12, extensive calculations, analyses, and optimizations were carried out in selecting the hull parameters and form. The vessel's capacity for ballast water was increased, and anti-rolling tanks were added to achieve excellent performance in speed, stability, and seaworthiness.
(2) Arranged from an overall perspective to ensure safe and effective survey operations: The requirements for the hull, ship system, power, and electrical system of the multipurpose oceanographic survey vessel were different from those for transport vessels. Therefore, it was necessary to consider comprehensively from the tasks and missions of the ship during the design and demonstration stage, so as to ensure the safety and efficiency of survey operations.

(3) Scientific designed draft: The size of the designed draft is closely related to various performance aspects of the ship. Multipurpose oceanographic survey vessels, in particular, typically require a high endurance and capacity. These vessels are also equipped with a substantial amount of fuel, fresh water, and provisions. As a result, when fully loaded for departure, they have a larger draft. To ensure optimal performance during survey operations, the designed draft for the vessel is determined by taking the corresponding draft value at 75% of its carrying capacity. This approach was pioneering in the design of multipurpose oceanographic survey vessels and was subsequently adopted in the development of similar vessels.

(4) A self-designed active rudder device: Electric motor-driven small propellers were used as an auxiliary propulsion system for low-speed navigation, which proved to be effective during trial navigation. In hydrographic surveys, it was required to measure parameters such as water temperature and water quality at different depths in the same location. This required the ship to remain stationary. At that time, there were not yet dynamic positioning units, and conventional anchors could only be used in shallow waters. Therefore, the use of deep water anchors became necessary. However, the deep-water anchoring device was still a research topic both domestically and internationally. To address this, collaboration among several institutions was established, leading to the successful research, development, and installation of a deep-water anchoring device capable of operating at a water depth of 500 m. This marked China's first successful attempt in conducting anchoring operations in waters this deep. To study the forces and calculation methods involved in deep-water anchoring, dedicated anchor tests were conducted at depths of 100 m, 300 m, and 500 m using the deep-water anchoring device on this vessel. The data collected provided a basis for future research and design of deep-water anchoring devices.

(5) Equipped with self-developed sewage treatment unit and advanced equipment: Experiments conducted on the survey vessel had strict requirements on the water quality, necessitating the prevention of water contamination at the sampling points. Therefore, sewage discharge from the vessel should be strictly controlled. At that time, domestic sewage treatment systems did not yet meet the necessary technical standards. To address this, two approaches were taken in the design. First, sewage was discharged from one side of the vessel and the other side was kept clean. Second, focus was put on developing a sewage treatment system. Space was reserved on the vessel for the installation of this system once it was developed successfully. Later, the system was developed successfully and passed all the necessary land-based tests. With this independently developed sewage treatment system, the accuracy of the experimental data were guaranteed. Additionally, several other advanced equipment at that time, such as the shallow-water hydrographic winch, deep-water medium winch, and sonar system, were also equipped.

After its delivery, "Shijian" participated in numerous domestic and international oceanographic surveys. In 1978, together with survey vessel "Xiangyanghong 09", it participated in the First GARP Global Experiment (FGGE) organized by the United Nations. This expedition involved surveys on the Pacific region, two voyages along the 160–165 degrees east longitude of the Pacific Equator and the 5 degrees north latitude area. The expedition lasted for seven months and endured challenges of intense heat and strong winds up to Force 11. Despite these difficulties, the mission was successfully completed, garnering positive feedback both domestically and internationally. Subsequently, the vessel also participated in the China-Japan Cooperation for laying of submarine cables.

As the first-generation domestically designed and built multipurpose oceanographic survey vessel, "Shijian" carried the glorious legacy and dreams of the older generation of marine workers to "fully explore China's seas and venture into three oceans". Over the course of more than four decades, the vessel undertook various oceanographic survey missions, including International Collaborative Survey on Global Air-Sea Interactions, China-Japan Kuroshio Cooperation Research, as well as a series of surveys in China's offshore and oceanic regions.

The construction of the "Shijian" multipurpose oceanographic survey vessel marked a significant milestone in the history of research vessel development. Firstly, it had evolved from "survey" to "scientific survey". Secondly, it achieved interdisciplinary and multipurpose scientific research, a qualitative leap in this field.

In January 2013, the "Shijian" underwent major repairs and renovations by Wuchang Shipbuilding Industry Group Co., Ltd. (former Wuchang Shipyard). After more than a year of refitting, it was reintroduced as the "Xin Shijian" (see Fig. 5) in May 2014 and joined the fleet of survey vessels.

Fig. 5 Multipurpose oceanographic survey vessel "Xin Shijian"

2.5 *The First Multipurpose Coastal Survey Vessel "Xiangyanghong 07"*

In 1969, the former State Oceanic Administration (SOA) made a request to the MARIC to modify the existing marine hydrometeorological vessel ("Xiangyanghong 02") to design and construct a multipurpose oceanographic survey vessel for use in coastal areas.

At that time, the SOA commissioned the construction of two multipurpose oceanographic survey vessels at Jiangnan Shipyard, with the MARIC responsible for the construction design. In the first quarter of 1970, the MARIC completed the modification of the drawings and delivered them to the shipyard. Due to changes in the plan, the Ministry of the Sixth Machine-Building Industry decided to transfer the construction of the vessel to Wuhu Shipyard and the ordered and manufactured equipment and components from Jiangnan Shipyard to Wuhu Shipyard.

Construction at Wuhu Shipyard began in May 1970, and on September 17, 1973, the vessel was signed for delivery. The vessel, named "Xiangyanghong 07", (see Fig. 6) belonged to the North China Sea Branch of the SOA.

This vessel was designed with a total length of 74.00 m, a displacement of 1,170 tons, a full load draft of 3.4 m, a maximum speed of 17.3 knots, a molded depth of 4.7 m, a capacity of 104 persons, an endurance of 3,000 nautical miles and a self-sustaining period of 40 days.

The main power plant consisted of two Д39-type main engines, with a rated speed of 500 revolutions per minute and a rated continuous power of 1,850 kilowatts.

The vessel mainly solved two technical problems:

Fig. 6 Multipurpose coastal survey vessel "Xiangyanghong 07"

Firstly, the vessel was based on the design of a Japanese patrol vessel. During profile design process, insufficient consideration was given to the interaction between the propeller wake and the hull form. During its trial voyage, high wave peaks were found to form at the stern, especially during navigating in shallow waters. This caused waves to hit the stern deck. To address this, a 0.7-meter wave suppression plate was added to the stern after multiple modifications and tests, significantly reducing the wave height and moving them further back.

Secondly, considering that both the original marine hydrometeorological vessel and other vessels equipped with Д39-type main engines experienced severe vibrations during trial voyages, the main focus of the vessel's trial voyage was to test the vibrations of the hull and the main engine. On October 15, 1973, several vibration tests were carried out to verify free and local vibrations on the hull, vibration on the main engine body, as well as torsional and lateral vibrations on the main shaft.

Based on the preliminary analysis of the test results, the main sources of vibration were identified: Firstly, the main engine itself exhibited a second-order pitch unbalanced moment. According to calculations, the second-order pitch unbalanced moment reached 9 ton-force·meters at the main engine's rated speed of 500 RPM. To eliminate the vibrations caused by the main engine to the hull and the main engine body, efforts should be made to balance the internals of the Д39-type main engine. Secondly, the propeller also posed a challenge. Particularly, when the main engine was disengaged from the clutch, the frequencies recorded at various points were twice the main engine speed. The vibrations became more pronounced at a main engine speed of 500 RPM, with local vibrations in the spare parts compartment reaching a level causing body numbness, and noise in the engine compartment exceeding 120 decibels at speed of 450 RPM. Following this test, the design team and shipyard technicians improved the plan and developed measures for optimization, significantly improving the hull's vibration problem compared to its mother vessel.

As this vessel was refitted from the original marine hydrometeorological vessel, it had inherent deficiencies. The design team suggested that for future refitting, the second-order pitch's unbalanced moment within the Д39-type main engine should be eliminated, and the propeller blades should be changed from four to five, with a test to verify its impact on sailing speed.

In 1981, a refit was necessitated for "Xiangyanghong 07" as required by the SOA. This refit aimed to ensure the vessel's participation in the China-US Joint Survey and Deep-Bottom Profiler Experimental Mission. On January 12, 1981, a meeting was held in Qingdao. The meeting focused on the refitting plans for repairing the vessel's low-speed electric propulsion system and installing the deep-bottom profiler. The following plans were established:

(1) Repairing the low-speed electric propulsion system: Prior to February 20, 1981, the North China Sea Branch was responsible for inspecting and maintaining the low-speed electric propulsion system to prepare for the testing.

Before the end of February 1981, the SOA was responsible for organizing the MARIC to conduct the tests, and representatives from Shanghai Pioneer Electrical Machinery Factory and Shanghai Electric Set Factory were invited to participate. Based on the test results, this institute needed to provide modified drawings to the shipyard before the end of March 1981. If there were equipment replacements, the shipyard would be responsible for the installation. After completion, the institute was responsible for the testing.

(2) Installing the deep-bottom profiler: The items requiring refitting include: replacing the 3# diesel generator set to increase the power of the power station; adding a sound source compartment, along with various bases and cables; installing one small boom on the larboard and starboard at the stern deck, and one triple-roller cable guide and three cable winch bases at the stern; and refitting the receiver, recording and monitoring devices in the geological laboratory.

Design of all technical drawings required for the refitting was commissioned to the MARIC. After approval, these drawings needed to be delivered to the North China Sea Shipyard before the end of March 1981, and the construction drawings should be drawn by the shipyard. The SOA should provide the equipment required for replacing the diesel generator set to the shipyard before the end of March 1981. The acceptance of the refitted diesel generator set should be conducted in line with the ship inspection specification.

(3) Installing the instruments and equipment required for the China-US Joint Survey: This included adding an AC 110-V, 3-kilowatt power source to the geological laboratory, and installing the sub-bottom profiler, side-looking sonar, and single sound seismometer. The MARIC should provide specific requirements before the end of February 1981 to the SOA, and dispatch technicians to the shipyard to assist with the construction.

With the collaborative efforts of all parties, the vessel was successfully delivered on July 20, 1981, ensuring the timely execution of the China-US Joint Survey. This collaboration fully demonstrated the superiority of the cooperation and coordination between China's participating departments in upholding national reputation and working towards a common goal.

2.6 The First Geophysical Exploration Vessel "Fendou 1"

Petroleum is an important strategic resource for a country. In the late 1950s, with the development of China's economic and defense construction, the demand for petroleum continued to increase. In 1963, with an aim to enhance the exploration of offshore oil and gas reserves, the Ministry of Geology wrote to the MARIC, stating: "In order to address the need for a specialized marine vessel for geophysical exploration and facilitate offshore oil exploration, we hereby entrust you with the design of a 500-ton geophysical exploration vessel. "The MARIC immediately formed a design team to discuss the basic operating mode of the vessel with the Fifth Geophysical Prospecting Brigade of the user, the Ministry of Geology, which involved the use of explosives to create sufficiently strong seismic waves in the water. These seismic waves would reach the seabed layers and reflect back, and the reflected waves would be received by cables connected to floating crystal detectors at a specific depth on the sea, which would then transmit the waves to the seismometer for analysis by marine geologists to study the underwater geological structure. The Bureau of Geophysical Prospecting of Ministry of Geology organized site visits for the design team to better understand the operational requirements. The Fifth Geophysical Prospecting Brigade considered this vessel as the most fundamental means for conducting marine geophysical surveys and attached great importance to it. During the design phase, they provided specific requirements for the equipment and layout of each operational area on board. In addition, they also established a dedicated office in Shanghai to facilitate coordination and cooperation with the design and construction units at any time.

After thorough discussions, the design team put forward an alteration during the design review meeting. In order to meet the operational requirements specified in the technical assignment and comply with the requirements of the *Rules for the Stability of Sea-Going Ships in China* for Class I sea-going ships, as well as the design requirements for the number of personnel and endurance, the currently designated 500-ton vessel determined by the Ministry of Geology was too small, and the displacement should be increased to 640 tons instead. The central government attached great importance to this consideration as it was the first specialized vessel for offshore oil exploration. Representatives from the State Planning Commission, State Scientific and Technological Commission, Ministry of the First Machine-Building Industry, Ministry of the Sixth Machine-Building Industry, Ministry of Fisheries, and Chinese Academy of Sciences, attended the meeting. After being agreed upon at the meeting, the design proposal was approved by the Ministry of Geology, where the 500-ton geophysical exploration vessel was changed to a "medium-sized geophysical exploration vessel".

The MARIC completed the construction design in early 1965, and the Hudong Shipyard delivered the vessel in 1966. The vessel was named "Fendou 1" (see Fig. 7).

This vessel was designed with a length between perpendiculars of 48 m, a molded breadth of 8.8 m, a molded depth of 4.1 m, an average draft of 3.0 m, a displacement of 640 tons, an endurance of 3,000 nautical miles, a self-sustaining period of 20 days, and a capacity of 48 persons (including 28 researchers).

Given the characteristics of marine geophysical exploration, the following measures were considered:

(1) The vessel was equipped with a passive anti-rolling tank to improve seaworthiness and enhance geophysical explorers' adaptability to offshore operations.
(2) Two bombing work boats were specifically designed for the vessel. When operating, the vessel would send a work boat carrying one ton of explosives to the sea and detonate them according to instructions. To ensure operational efficiency, the work boats were capable of withstanding winds of Force 6–7. They were equipped with fixed explosion units and the explosives compartment was equipped with seawater-proof facility. The structural strength of the work boats could withstand the shock wave within 100 m of the explosion source.
(3) An explosives storage room capable of storing 10 tons of explosives was arranged away from the engine room and fuel tank on the vessel. The room

Fig. 7 Geophysical exploration vessel "Fendou 1"

featured anti-vibration, insulation, and heat insulation measures, and temperature gauges and alarms were installed, whose signals could directly reach the pilothouse and command platform. A detonator storeroom, capable of storing 2,000 detonators, was set up in the captain and political commissar's room for safety purpose.

(4) The vessel adopted adjustable pitch propellers manufactured by the Shanghai Fishing Wheel Factory. The propellers allowed a low-speed navigation at no more than 2 knots to meet the requirements of geophysical exploration.

(5) The deck area at the stern was increased to accommodate cable winches and small cranes for facilitating the work of geophysical explorers.

(6) The exploration vessel was designed according to passenger ship standards as required by the user. Due to difficulties in meeting fire protection requirements, the vessel finally passed the inspection after taking corresponding measures. This ensured that the geophysical explorers would have not only better resting conditions but also sufficient energy to dedicate themselves to the challenging exploration operations.

The successful development of this vessel was the result of close collaboration between various departments. Its construction and use provided crucial data for the subsequent development of similar vessels. Soon after, the SOA proposed appropriate modifications to the design of this vessel and commenced the construction of two more.

Two insights can be drawn from the development process of this vessel. Firstly, during the 1960s, a vessel with a displacement of only 640 tons was defined as "medium-sized", indicating China's limited financial resources and shortage of vessels at that time. Secondly, as a geophysical exploration vessel, the use of bombing work boats to deploy explosives for obtaining seismic reflection waves was inherently dangerous, but the geophysical explorers were fearless. They detonated explosives every four minutes, making significant contributions to obtaining geological data. Therefore, the vessel was aptly named "Fendou" (meaning struggling).

Chapter 4
Forging Ahead and Heading Towards the Deep Blue

1 Overview

In response to the requirements of national economic development and oceanic surveys, explicit demands and a series of significant strategic deployments have been made at every stage of the development of scientific research vessels. Oceanic exploration and research has been assigned a more vital mission to serve the national economy, national defense, and the protection of the country's maritime rights and interests.

In the mid-1960s, China successfully detonated atomic and hydrogen bombs, followed by the launching of artificial satellites into space carried by the rocket "Chang Zheng 1 (Long March 1)". However, it was only when the both were combined and the capability for launching nuclear missiles over long distances was ensured, that the nation would truly possess strategic weapons, which could deter potential aggressors. In the 1960s, China conducted short and medium-range missile experiments in the unmanned areas of the northwest desert. After that, intercontinental missiles were developed, which required long-distance launch tests. Although China has a vast territory, measurements on land cannot meet the spatial span required for intercontinental missile tests at any point. The only option is the ocean, specifically the high seas, where the test range can be set. Therefore, the CPC Central Committee decided to establish an oceangoing fleet to conduct surveys and investigations in the designated ocean areas. Only after obtaining firsthand information on ocean conditions, weather, hydrology, and physical fields can the specific location of the offshore target range be determined. The oceangoing fleet also needs to undertake tracking and recovery tasks during missile launch tests. The State Council decided to task the MARIC with the selection and refitting design of an in-service cargo vessel into an oceangoing survey vessel. The MARIC attached great importance to this task. It assigned a R&D team headed by Academicians Xu Xueyan and Zhang Bingyan to participate in the vessel selection and undertake the refitting design. Owing to the team's over 10 years of experience in the design of various types of

L. Han et al., *History of the Research and Development of Chinese Marine Research Vessels*, https://doi.org/10.1007/978-981-96-8414-4_4

coastal survey vessels, the refitting design was completed smoothly. The Guangzhou Shipyard completed the refitting work in 1972, naming it oceangoing survey vessel "Xiangyanghong 05". This vessel subsequently conducted four surveys in the South Pacific for carrier rocket splashdown area, successfully completing the preliminary survey tasks.

The China's first independently designed and built oceangoing survey vessel, "Xiangyanghong 10", came out along with the space tracking ship "Yuanwang". It marked a milestone in the history of the development of oceanographic survey vessels in New China.

To support intercontinental missile tests, the establishment of an offshore tracking ship fleet became necessary. Alongside the emergency refitting of existing oceangoing survey vessels, the central government also decided to independently design and construct new ships. Despite the challenges posed by the "Cultural Revolution" at the time, the central government ordered to organize a specialized working team. The team was mainly composed of leaders and scientists from the MARIC, along with personnel from relevant units, with the provision of exemption from the "movement". The team members regarded national interests as the highest principle and were fully aware of the significant mission they shouldered, devoting themselves wholeheartedly to the design work. During the design process, they maintained timely communication and coordination with equipment manufacturers and shipyards to ensure the orderly and systematic progress of the vessel's development. In October 1979, "Xiangyanghong 10" was successfully delivered at Jiangnan Shipyard.

With a displacement of 13,000 tons, "Xiangyanghong 10" not only had excellent ocean navigation capabilities but also overcame prominent technical challenges in oceangoing survey vessels, including transoceanic communication, meteorological center system, underwater acoustic power source, and helicopter deck system. Finally, the vessel was fitted with comprehensive capabilities for oceanographic survey, research, and resource acquisition. Its successful development signified that China's capability for developing oceanographic survey vessels had extended beyond coastal and near-sea areas to the deep blue ocean. It demonstrates that China has achieved world-leading levels in overall design technology and construction capabilities. In 1985, "Xiangyanghong 10" received the National Science and Technology Progress Special Award, and in 2006, it was rated as one of China's Top Ten Famous Vessels.

On May 18, 1980, the vessel participated in an intercontinental missile sea range test mission. At 10 am that day, a dazzling arc of light rose from the ground, flying eastwards towards the South Pacific resembling an arrow. It connected on one end to the launch center in Jiuquan and on the other end to the offshore target range at 171 degrees 33 min East Longitude and 7 degrees 42 min 23 s South Latitude. The rocket rapidly flew towards the predetermined splashdown area. In the high-altitude airspace several thousand meters above the sea surface, an instrument compartment containing important rocket flight parameters automatically ejected from the rocket warhead. The parachutes deployed and the instrument compartment descended gently onto the sea surface. The helicopter on "Xiangyanghong 10" quickly took off, hovering vertically over the instrument compartment at a height of 30 m above the ocean.

Divers descended from the helicopter using a rope ladder and connected slings to the instrument compartment, which was then lifted and transported back to the survey vessel by the helicopter. With this operation, China achieved a complete success in launching carrier rockets to the Pacific Ocean from its own territory, making it the fourth country after the US, the Soviet Union, and France to possess offshore tracking technology in the field of aerospace science and technology. In this experiment, the oceangoing survey vessel played a pivotal role.

A major country without a freely navigable maritime domain is impossible to become a world power. During the 1978 National Science Conference, the SOA put forward the strategic goal of "fully exploring China's seas, venturing into three oceans, and reaching Antarctica" to the marine science community. Positive response was received from practitioners in the shipbuilding industry. They expressed their hope to closely communicate and collaborate with the marine science community, in order to develop oceanographic survey vessels that would bolster this strategy.

In order to "fully explore China's seas", the MARIC designed the "Xiangyanghong" series of multipurpose survey vessels, as well as individual survey vessels for hydrometeorological survey and electric power technology detection, and other purposes. These vessels were constructed for the North China Sea Branch, East China Sea Branch, and South China Sea Branch under the Ministry of Natural Resources, among other marine research entities. The "Xiangyanghong" series includes "Xiangyanghong 04", "Xiangyanghong 05", "Xiangyanghong 06", "Xiangyanghong 07", "Xiangyanghong 09", "Xiangyanghong 10", "Xiangyanghong 14", "Xiangyanghong 16", and "Xiangyanghong 21". These vessels were successfully built by Jiangnan Shipyard, Hudong Shipyard, and Wuhu Shipyard. Moreover, during the same period, the MARIC, together with other overall design research institutes, developed various types of survey vessels specifically for the geological survey divisions of the marine science community, such as the "Shiyan 2" and "Dayang 1". With these survey vessels, marine research and survey organizations were able to gather a wealth of information on hydrology, meteorology, geology, geomorphology, marine chemistry, biology, and physical characteristics of the Yellow Sea, Bohai Sea, East China Sea, and South China Sea.

To "venture into three oceans", the marine science community engaged in the first Global Atmospheric Experiment hosted by the World Meteorological Organization (WMO) in December 1978, as required by the State. In response to the specific research requirements of the experiment, the SOA commissioned the MARIC to design the 4,000-ton multipurpose oceangoing survey vessel "Xiangyanghong 09", which was completed by Hudong Shipyard in November 1978. Within just one month of its commissioning, it represented China's participation in the first global atmospheric experiment organized by the WMO. This marked the first global cooperative experiment China participated in after regaining its legitimate status in the United Nations. The vessel conducted two research voyages in the Pacific Ocean, operating in the region between 160–165 degrees east longitude and 5 degrees north latitude. Over the course of conducting surveys at 168 stations, this vessel provided abundant research data on the sea, hydrology, physics, geology, biology, and other disciplines to the WMO. This included the collection of a geological core sample measuring up to 4.3 m in length. The vessel successfully completed the tasks of World Global Atmospheric Research Program assumed by China, receiving commendation from the head of the atmospheric experiment of the WMO. Out of the more than 50 survey vessels participating in the experiment, "Xiangyanghong 09" stood out and gained international recognition.

According to the provisions of the international *Law of the Sea*, regardless of the nation, if a country discovers accurate data and samples of seabed mineral resources in international waters, it enjoys the right to exploit them in that area. In 1993, "Xiangyanghong 09" was sent to the Pacific Ocean on an seabed geological and mineral resource survey. During the long voyage and strenuous exploration operations, scientists obtained solid evidence of exploitable mineral resources in a specific seabed area, granting China the right to exploit a 75,000 square kilometer mining zone in the international waters of the Pacific Ocean. Scientific estimates suggest that with an annual extraction of 3 million tons of ore, the resources could be mined for at least 20 years, bringing significant benefits to future generations. Following numerous special missions of "Xiangyanghong 10" venturing into the Pacific, the ocean mining survey carried out by "Xiangyanghong 09" marks the beginning of China's penetration into three oceans.

Antarctica is considered the "untouched virgin land" on Earth. Major Western powers have had their eyes on it for a long time, and countries such as the US, France, and the Soviet Union have successively established scientific research stations there. They occupied the first three of the four crucial points in Antarctica: the pole, the ice point, the magnetic point, and the highest point. Professor Knox, Chair of the Scientific Committee on Antarctic Research, once said, "We are approaching an era of economic interests, followed by political interests that dominate all activities in Antarctica." This demonstrates the intense competition for maritime rights in Antarctica. In June 1983, China became a signatory to the *Antarctic Treaty*. On June 25, 1984, the central government approved the establishment of a research vessel fleet organized by the National South Pole Investigation Commission, the SOA, and other major marine research institutions, to embark on a journey to Antarctica. This fleet was composed of "Xiangyanghong 10" and oceangoing salvage lifting vessels, involving the participation of a total of 591 personnel with nearly a hundred scientists and engineering technicians. On December 26 of the same year, the fleet crossed the area with prevailing northwest winds (where gales of Force 8 or more occur for more than 300 days each year, with wind speeds often reaching 40–50 m per second) and arrived at King George Island's Civil Defence Bay in Antarctica. On New Year's Day in 1985, China's first Antarctic research station, the Great Wall Station, was inaugurated amidst the immaculate ice and snow. This filled a gap in China's scientific undertakings, marking a new phase of China's Antarctic scientific research. The fleet overcame numerous challenges, making significant contributions to the oceanographic research in Antarctica.

To facilitate Antarctic research missions, in 1990, the SOA decided to construct an Antarctic research icebreaker. The MARIC was entrusted with the task of advanced research. The institute accorded great importance to this task and appointed Zhang

Bingyan, an academician of the Chinese Academy of Engineering who had participated in the first voyage to Antarctica aboard the vessel "Xiangyanghong 10" as its chief designer, to lead the advanced research. He put forward two proposals: refitting "Xiangyanghong 10" or building a new vessel. Due to limited economic strength of the country at the time, the plan for a new vessel did not materialize. Coincidentally, Ukraine had a new vessel under construction, but due to the collapse of the Soviet Union, the former was unable to make the payment and was prepared to abandon it. After careful analysis, Zhang Bingyan believed that it had a suitable carrying capacity and could be transformed into an Antarctic multipurpose research vessel through moderate refitting. With the approval from the SOA, Zhang Bingyan, accompanied by a delegation, embarked on a challenging journey to Ukraine for the negotiation, ultimately purchasing the vessel and bringing it back to Shanghai in April 1993. The first refit was carried out in 1994 with certain facilities optimized, and the vessel gained its name "Xuelong". After ten years of service, it underwent subsequent upgrades in 2004 and 2012. The first refit was completed by the Shanghai Merchant Ship Design and Research Institute (SDARI) and the Shanghai Shipyard, and the second refit was accomplished by the MARIC and the Jiangnan Shipyard. Since its inaugural Antarctic research mission in October 1994, "Xuelong" had made a total of 34 trips to Antarctica by 2018. It plays a pivotal role in China's expedition to the Antarctica, the establishment of the Zhongshan Station, and the conquest of the last remaining highest point among the four contested points in Antarctica. Besides, "Xuelong" has also made nine voyages to the Arctic, contributing to the establishment of the Yellow River Station in the Arctic. Additionally, the experience and lessons learned from its extensive operations have provided invaluable guidance for the development of "Xuelong 2", a new polar research icebreaker, by the MARIC.

At this point, China's oceanographic survey vessels have provided strong support for the rapid development and deep-sea exploration of the country's marine research and survey.

With the experience from earlier refitted ships and survey vessels, China has gradually mastered the key technologies for designing and constructing survey vessels. Coastal surveys could no longer meet the needs of the country's marine science and national defense construction. With the emergence of national major marine projects and defense engineering needs, such as long-range carrier rocket launch experiments, China began to develop various types of oceangoing survey vessels, ushering in an era of independent design and mass production of large oceangoing survey vessels. To address the need of national defense, the survey vessels developed during this phase were mostly dual-use for both military and civilian purposes. These vessels had a slender and elongated hull, low superstructure, strong wind resistance, and a large capacity.

During this phase, the construction of research vessels included multipurpose, specialized, and special survey vessels. The survey areas expanded from coastal waters to far waters, and ship tonnage increased from hundreds of tons to thousands or even tens of thousands of tons. This provided China with sufficient technical equipment for marine research and survey. During this period, some survey vessels reached Antarctica for establishing research stations and conducting scientific research, while others ventured into the deep sea to collect manganese nodules. Some vessels conducted extensive high-altitude meteorological surveys in the South Pacific, while others participated in the China-Japan Kuroshio Cooperation Research in the northwest Pacific Ocean. There were also some that participated in the China-US Joint Survey in the South pacific. Since then, China's survey vessels were no longer restricted to domestic research in coastal areas but gradually participated in international cooperation.

After more than ten years of offshore operations, marine scientists gradually realized that oceanographic survey vessels needed to accurately position themselves at sea, and meanwhile observe multiple points within the same area and process information and data in real time. Thus, only through integration and cooperation could ideal results be achieved. Therefore, it was necessary to develop multipurpose survey vessels that combine various disciplines, as well as efficient specialized survey vessels that complement each other. The 1980s witnessed rapid development of China's oceanographic survey vessels, both in quantity and quality. In this period, survey vessels formed several series, such as the "Dongfanghong" series and the "Xiangyanghong" series. Representative vessel types include the "Xiangyanghong 05", "Xiangyanghong 09", and "Xiangyanghong 10" oceangoing survey vessels, which were refitted or built for long-range carrier rocket full-range tests, as well as the "Dongfanghong 2" multipurpose oceanographic survey vessel. By the end of the twentieth century, China possessed survey vessels of over 40 types. The refitting and commissioning of vessels "Dayang 1" and "Xuelong" propelled China's marine research into deep and far sea areas, as well as the polar regions.

The survey vessels designed and built during this period exhibited the following features:

(1) To improve speed, the vessels had slender and elongated hulls with large length-to-breadth ratios.
(2) To enhance wind resistance, the superstructures were relatively low.
(3) The stability of the survey vessels was required to withstand Force 12 winds, and thus a standardized stability calculation method for withstanding wind force of up to 12 was developed, which became an industry standard.
(4) Most survey vessels during this period were equipped with low-power auxiliary power units, active rudders, or variable pitch propellers, in order to accommodate complex and changing working conditions.
(5) To enable fixed-point measurements, some survey vessels were equipped with deep-water anchoring devices.
(6) Various measures were taken to reduce vibration and noise.

2 Classical Research Vessels

2.1 *Shallow Sea Petroleum Seismic Exploration Catamaran*

The shallow sea petroleum seismic exploration catamaran is mainly utilized for operations in shallow sea areas with water depths of five meters or less. In the case of the Bohai Sea, for example, the operational line in this marine region exceeds 8,000 km. However, the existing survey methods of the surveying department could only cover less than 200 km per year, emphasizing the urgent need for a modernized shallow sea petroleum seismic exploration vessel to expedite the exploration progress. Therefore, in order to accelerate the exploration of submarine oil resources and seismic activities in coastal areas of China, the 641 Geological Survey Command of the North China Oil Campaign Command of the Ministry of Fuel Chemical Industry submitted a design task to the MARIC for a shallow sea petroleum seismic exploration catamaran in September 1974.

The institute formed a special design team in October 1974, which subsequently carried out field surveys and actual ship inspection in Tianjin, at the Bohai Sea 641 Geological Survey Command, and in Guangzhou. Given the requirement for a sizable deck area on the exploration vessel, a catamaran design was selected with high-pressure air guns as the primary operation method, supplemented by explosive sources. In April 1975, a design review meeting was held in Tanggu, followed by the technical review meeting in Shanghai in November.

Construction design of the vessel was conducted at the Dagu Shipyard between June and July 1976. The vessel was finally delivered after a trial voyage on August 27, 1979.

This vessel was a catamaran with a steel structure featuring a single deck. The two hulls were interconnected through a cross structure. Powered by diesel engines, it was equipped with twin propellers, twin rudders, a forward leaning bow column, and a cruiser-style stern.

This vessel was designed with a total length of 38.2 m, a molded breadth of 12.60 m, a molded depth of 3.0 m, a draft of 1.50 m, a displacement of 240 tons, a single hull width of 4.0 m, a center spacing between hulls of 8.6 m, a trial sailing speed of 11.5 knots, an endurance of 1,000 nautical miles, a self-sustaining period of 10 days, and a capacity of 36 persons.

Stability: Complying with the requirements of the *Seagoing Vessel Stability Regulations* for Class III sailing areas issued in 1974.

Main engines: Two 12V1135ZC diesel engines, each with a rated power of 365 kilowatts and a rated speed of 1,500 RPM.

The stern of the vessel's main deck was equipped with a streamer winch with a maximum pulling force of 1,500 kg, a cable length of 1,500 m, and a drum diameter of 1,100 mm. It also had a gun winch with a pulling force of 150 kg, a drum diameter of 309 mm, and a gun cable length of 300 m. A winch control room was located on the main deck. The lower deck housed an explosives storage room with a capacity of approximately 8 tons. Two 4-meter-long work boats were positioned on each side of the upper deck, with each equipped with a manual rotating small boom with a lifting capacity of 600 kg. There was a suspension bridge at the stern and on each side, which was operated manually. The bridgehead was equipped with seats and gun cable hooks. The suspension bridge measured 3 m in length with a triangular structure.

Design features of the vessel:

(1) Due to the vessel's operation in shallow water areas for petroleum seismic exploration, it was restricted by water depth, requiring a maximum draft not exceeding 1.5 m. Moreover, the user expected a large deck area, thus a catamaran design was chosen. However, the weight of a catamaran hull was around 30% heavier than that of a monohull of the same displacement. Therefore, to maintain the same draft, it was necessary to increase the main dimensions and the block coefficient to enlarge the displacement. Considering the reduction of resistance, it was more appropriate to increase the block coefficient.
(2) To control weight as much as possible, multiple layout schemes were developed for this vessel. However, the height between decks was relatively low, with a height of 2.15 m from the upper deck to the navigation deck and 2.00 m from the navigation deck to the master compass platform, which had some impact on operation. Therefore, the second vessel increased the heights by 150 mm.
(3) Ship model tests were conducted to assess resistance and seaworthiness using variable-scale models. Through analysis and comparison, the optimal main dimensions were selected.
(4) In the design of propellers, due to the absence of resistance tests on ship models, the resistance of the monohull was calculated using the Taylor formula. Then, the interference resistance between the two hulls was added. Considering several factors during the trial voyage, a 20% margin was included. Subsequent ship model tests demonstrated close alignment between the tested resistance data and the calculated resistance data.

The design of this vessel met the requirements of the technical assignment. It served as the first exploration vessel capable of conducting operations in shallow sea areas as the only vessel of its kind at the time, receiving high satisfaction from the user.

At the time, the 641 Geological Survey Command was engaged in joint exploration activities with Britain and France in the Yellow Sea and the Bohai Sea. While China had priority over the lease of exploration equipment, there was no equipment available for lease at that time. As a result, they leased German vessels, with rental fees ranging from USD 250 to 1,000 per kilometer, with the cost assumed by the foreign parties. However, since German vessels could only operate in water depths greater than five meters, there were still no vessels available for areas with water depths of less than five meters. Therefore, subsequent exploration activities in shallow sea areas were carried out using domestically developed exploration vessels. This not only accelerated the progress of operations but also generated considerable foreign exchange income.

2.2 1000-Ton Scientific Survey Vessel "Shiyan 2"

Due to the urgent need for geological data necessary for offshore oil development, the Ministry of Geology, the SOA, and the Chinese Academy of Sciences all put forward an urgent need of a 1,000-ton marine geological exploration vessel. In addition to the equipment required by the Chinese Academy of Sciences for scientific surveys in other associate projects, the requirements of each department were basically the same. Upon approval from the Ministry of the Sixth Machine-Building Industry, designed by the MARIC, three vessels were simultaneously constructed at the Guangzhou Shipyard in 1977. "Fendou 4" and "Fendou 5" were assigned to the SOA, while the scientific survey vessel "Shiyan 2" (see Fig. 1) was delivered to the South China Sea Institute of Oceanology of the Chinese Academy of Sciences in June 1980.

Unlike "Fendou 1", "Fendou 2" not only increased in tonnage but also improved in operational efficiency and expanded in operating range. Moreover, the seismic wave generation method was changed from explosive detonations to high-pressure air blasts. The previously dangerous practice of detonation controlled by operators on bombing work boats was abandoned, showcasing the vessel's modernity.

The vessel measured 68.45 m in length, 10.0 m in molded breadth, 5.4 m in molded depth, with a draft of 3.6 m, a displacement of 1,100 tons, an endurance of 3,000 nautical miles, a self-sustaining period of 30 days, and a capacity of 61 persons.

Design features of the vessel include:

(1) It adopted high-pressure air blast operation method, which was relatively new internationally at that time. It was equipped with four air compressors onboard, with an exhaust pressure of 200 kg/cm^2. Correspondingly, 10 high-pressure air bottles with a capacity of 410 L each were equipped to meet the maximum air consumption requirement (60 L/min) of the air guns.

Fig. 1 Scientific survey vessel "Shiyan 2"

(2) Its equipment could ensure a clear and safe workflow. During operations, high-pressure air passes through filters, air–water separators, and other accessories, with the working pressure regulated to 150 kg/cm^2 through a stop valve. The air then enters the inlet of the air gun solenoid valve, and the detonator signal is sent from the seismometer through the air compressor distribution board to control the solenoid valve action. Each time the solenoid valve is activated, an air blast occurs. Besides, two air guns were mounted on both sides outboard the vessel with a distance of 4.5 m, which were deployed by air gun lifting frames. The intervals between air blasts were as short as 10 s, significantly enhancing work efficiency compared to detonating explosives once every four minutes.
(3) The vessel was equipped with a magnetometer, enabling operation through deploying magnetic probes by the magnetometer winch. A gravimeter and other instruments were also equipped for exploring the marine gravity field. A magnetometer room and a gravimeter room were also arranged on the vessel, providing means for investigating the geological structure of the seabed.
(4) The vessel was equipped with a dual-channel automatic satellite navigation system to accurately determine the vessel's position during operations, thereby enhancing operational efficiency.
(5) In order to expand the vessel's use functions, hydrological, meteorological, and other survey equipment were installed onboard, enabling the vessel to conduct various scientific investigations.
(6) Due to the accommodation of four air compressors, strong local vibrations might occur when operating. To solve this problem, an appropriate design for the hull structure and a complement of vibration dampers and other measures were adopted. Finally, the effectiveness in reducing vibrations was verified by the former South China Institute of Technology, which ensures the air compressor units can operate at full load under normal conditions.

Following delivery, the vessel successfully completed tasks including air blast experiments, investigations of winter warm currents in the South China Sea, surveys of the Xisha reef area, and gravity surveys during a voyage spanning nearly 15,000 nautical miles, showcasing a good performance. During navigation in the South China Sea and adjacent ocean margins, the vessel withstood the test of severe gales of Force 8 and 9 and huge waves. Then, it was recognized by the user for its good performance.

In October 1982, the "National Acceptance Appraisal" for the vessel was presided over by the Guangzhou Branch of the Chinese Academy of Sciences. According to the acceptance leading group:

The design and construction of "Shiyan 2" met the requirements specified in the technical assignment. The technical condition of the vessel complied with national ship inspection standards. The shipbuilding drawings and main equipment documents were complete. The installation and commissioning of scientific research equipment, including the introduced "Satellite Navigation" system, generally met technical standards. Through geophysical measurements and winter warm current investigations, the seagoing capability of the vessel was proven reliable.

2.3 *Multipurpose Oceangoing Survey Vessel "Xiangyanghong 05" Refitted for the "718" Project*

In order to support China's intercontinental missile launch experiments, the National Defense Science and Technology Commission and the Navy organized discussions among relevant scientific and technological personnel. After repeated deliberations, it was decided to construct a total of seven vessels of five types, including the multipurpose survey vessel type, select a target range, conduct aerospace measuring, and form an oceangoing salvage lifting fleet.

To prepare for the survey work in advance, a decision was made to first adopt the refitting of an existing oceangoing transport vessel into a multipurpose oceangoing survey vessel.

In 1969, the SOA commissioned the MARIC to join a ship selection team composed of the Administration, the General Staff Meteorological Administration, and other units. The Institute was tasked with the refitting and design of the selected ships. The ship selection team conducted actual ship inspections at the fleet under the Ministry of Transport and China Ocean Shipping Company, Guangzhou (COSCOGZ). A total of three passenger ships and nine cargo ships were investigated, and a large amount of data was reviewed, followed by the inspection of the real ships and in-depth discussions. Zhang Bingyan of the ship selection team recommended the "Changning" cargo vessel, which was built in Poland in 1967, purchased by the Ministry of Transport, and was owned by COSCOGZ at the time. The main reason was that the vessel was basically new. It had a displacement of 11,000 tons, a length of 152 m, with moderate tonnage and main dimensions. The electromechanical equipment is basically complete, providing sufficient usable space and the ability to navigate globally. This proposal was approved by the ship selection team and reported to Premier Zhou, who approved the allocation of the vessel to the SOA for refitting.

Receiving the command, the MARIC immediately launched the refit design. The key focus of the refit was to meet oceangoing requirements by adding survey instruments for hydrology, meteorology, acoustics, geology, waves, biology, and sounding, as well as numerous supporting equipment and laboratories. It was also necessary to install additional communication systems such as radar stations, shortwave radios, and secure telephone systems. The fuel, water, and food carrying capacity should be enlarged to improve the vessel's endurance and self-sustaining period. It also needed to increase some solid ballast and adjust the weight distribution to improve the stability and seakeeping capability of the vessel and to meet the wind resistance standards for Class I sailing areas, further to ensure the capabilities for long-distance multipurpose surveys.

In May 1970, the Ministry of Transport officially delivered the cargo vessel "Changning" to the SOA. At the end of September 1970, the refit of the vessel commenced at Guangzhou Shipyard and was completed in 1972. After a trial voyage, the vessel was successfully delivered, named "Xiangyanghong 05" (see Fig. 2). The

Fig. 2 Multipurpose oceangoing survey vessel "Xiangyanghong 05"

vessel immediately began conducting offshore experiments and training for long voyages.

After the initial refit, "Xiangyanghong 05" undertook four oceangoing scientific surveys in the Pacific Ocean from 1972 to 1978. From March 8 to May 20, 1977, this multipurpose survey vessel carried out an extensive survey and research in certain areas of the Pacific Ocean, which lasted for over 20 days. From March 15 to June 1, 1978, it embarked on its longest voyage in the Pacific Ocean, obtaining China's first geological sample of manganese nodule from the ocean depths. From August 21 to October 21, 1978, the vessel went on its fourth voyage to validate the data collected for the missile target ranges selected during the previous three voyages. This time, it collected a substantial amount of hydrological, meteorological, gravitational, magnetic, marine environmental, and geological data from an altitude of 32,000 m and a depth of 2,600 m in the deep sea, filling the gap in China's oceangoing survey records.

After four long voyages of more than three years, it became evident that the limited working and living conditions on the vessel had affected smooth progress of oceangoing surveys. To enhance the vessel's scientific research capabilities, the SOA applied to the Special Committee of the Central Government for a further refit. In September 1978, the MARIC was authorized to take on the vessel's redesign, while Guangzhou Shipyard was entrusted with the refitting work.

The refitting followed the principles of unified planning, unified layout, and phased construction to ensure measuring tasks can be completed. The refit design included:

Adding landing point measurement facilities, improving oceangoing survey equipment, and substantially increasing experimental apparatus, including 24 laboratories, a 352-A radar, a 272-underwater acoustic antenna, a 6,000-m hydrological winch, and a 6,000-m geological winch.

Adding an auxiliary engine room, and refitting the No. 4 and No. 5 cargo holds as well as the superstructure.

Meeting the specifications of far sea sailing areas (Class I sailing areas), with the wind pressure standard in Class I sailing areas increased from 1,302 Pa to 2,354 Pa to align with the wind pressure standard of Force 12 gusts.

The design was completed in April 1979, and the refitting was carried out from March to November of that year. From December 9 to December 29, 1979, the vessel underwent trial voyages in the seas near Langya Bay, Hainan Island, before being officially delivered at Guishan Island.

In May 1980, the vessel served as the command vessel during the execution of the "580" experiment. This experiment was undertaken by a special joint fleet, comprising 18 ships, 4 helicopters, over 5,000 personnel from the navy, army, air force, and scientists spanning the fields of aerospace, oceanography, and communications. They sailed to the South Pacific for the mission of recovering the missile data capsules. Throughout the mission, the vessel's radar and underwater acoustic measuring system detected targets accurately. The weather forecasts proved precise, and the communication systems and various instruments operated smoothly. The vessel's crew made outstanding contributions throughout the experiment and received a collective second-class merit. In 1986, the vessel was awarded the third prize in the Science and Technology Progress Award by the National Defense Science and Technology Commission.

2.4 Multipurpose Oceangoing Survey Vessel "Xiangyanghong 09"

In 1972, the SOA commissioned the MARIC to design a 4,000-ton multipurpose oceangoing survey vessel for participation in the first Global Atmospheric Experiment organized by the World Meteorological Organization in 1978. According to the analysis by the design team, it was determined that the designated displacement of this vessel by the SOA was 4,000 tons. Compared to the "Shijian" vessel, which also carries out multipurpose oceanographic survey missions, it was expected to exhibit better operational capabilities and navigational performance. Based on the accumulation of previous research vessel design experience, the design was completed in June 1975. Hudong Shipyard started construction in September 1977. Considering the significance of the vessel's first mission to the nation's reputation, the user, the designer, and the constructor all worked in close coordination. To ensure smooth completion of the vessel, the shipyard made a lot of preparations and completed the construction in just 14 months. In November 1978, the vessel was delivered and fulfilled its mission in the Global Atmospheric Experiment as scheduled. The vessel was received by the North China Sea Branch of the SOA, named "Xiangyanghong 09" (see Fig. 3).

Fig. 3 Multipurpose oceangoing survey vessel "Xiangyanghong 09"

The vessel was designed with a total length of 112 m, a molded breadth of 15.20 m, a molded depth of 8.20 m, a draft of 5.50 m, a displacement of 4,400 tons, a speed of 18 knots, an endurance of 10,000 nautical miles, and a self-sustaining period of 60 days. It was equipped with two 9ESDZ43/82B low-speed diesel engines, each with a power (MCR) of 4,500 kilowatts at 200 revolutions per minute. The vessel was classed by the China Classification Society (CCS).

Design features of the vessel:

(1) Enhanced stability: As the ship was slated to operate in non-restricted waters and severe sea conditions, its safety and reliability were crucial. Given the requirements of the survey mission, 12 winches, one 2.5-ton crane, one 5-ton crane, and various laboratories were situated on the upper deck, all of which would potentially impact the vessel stability. Additionally, the decreasing of slop from full load to empty load during sea operations would significantly affect stability. To solve this problem, deep slop tanks and ballast tanks were adopted to regulate stability. In addition to satisfying the requirements for Class I sailing areas, the vessel also underwent stability check under a sea condition of Force 12 winds as per the design requirements of two-compartment floodability.
(2) Improved seaworthiness and low-speed navigation: The vessel incorporated a passive anti-rolling tank and an enlarged bilge keel to reduce yawing amplitudes. The bow sheer was also augmented to minimize deck waves. In line with the demands of survey operations, two 250-kilowatt diving motors were installed onboard, enabling low-speed navigation from 0.5 to 5 knots.
(3) Maneuverability and course stability: The vessel featured dual engines, twin rudders, and double active rudders capable of steering up to 90 degrees. These elements ensured both nimble maneuverability and dependable course stability.

(4) Expanded deck and laboratory area: To optimize conditions for scientific research, the vessel provided spacious fore and aft decks for operational purposes. It incorporated a combined area of 262 square meters, consisting of 16 laboratories, as well as a 66-square-meter sample cabin and a 64-square-meter spare cabin.
(5) Hull structure: The vessel employed transverse framing system for its hull. Accounting for its anticipated operations in adverse weather environments (such as fragmented thin ice regions), it adopted class 3 ice strengthening.
(6) Winch equipment: The vessel was equipped with 12 special deck winches of 9 different types, including domestic most advanced equipment such as the 6,000-m geological winch, the 11,000-m deep-water hydrological winch, and the 13,700-m deep-water benthic winch. With the 6,000-m geological winch, the vessel successfully retrieved a core sample of more than four meters long from a water depth of 5,000 m. This was the highest level of core sample collection achieved domestically at that time.

The vessel, while engaged in international duties, also conducted survey tasks in the "718" project sea area. For the remarkable performance in fulfilling its duties, it received the collective third-class merit and commended by the National Defense Science and Technology Commission, as well as the Navy Command.

To meet the oceanographic research needs, following the success of "Xiangyanghong 09", the SOA invested in two additional vessels of the same type, namely "Xiangyanghong 14" and "Xiangyanghong 16".

In November 2007, the South China Sea Branch of the SOA commissioned the China Ship Development and Design Center to undertake the refit design of the vessel, with Shanghai Lifeng Shipyard responsible for the construction. This aimed to refit the "Xiangyanghong 09" into a 7,000-m manned submersible mother ship.

(1) Refit Principles and Contents

The refit prioritized the safety of the surface support system. Equipment layout in the submersible's surface support system must consider applicability, ease of operation and maintenance. Compatibility of this support system with existing equipment systems on the mother ship, as well as deck area and cabin space usage must also be considered. This was to ensure the safety and reliability of both the mother ship and the support system.

To ensure a safe distance between the submersible and the stern during launch and recovery of submersible, an A-frame was installed and modifications were made to the stern lines. Give the fact that the stern A-frame base would be subjected to a bending moment of 1,000 ton-force·meters during the lifting process, and the stern A-frame itself weighed 50 tons while the submersible weighed 25 tons, the structure of the stern transom plate on the upper deck within nine meters above the waterline towards the bow was redesigned and replaced.

(2) Layout of Submersible Tracks and Other Parts

Two tracks were positioned on the centerline of the stern of the upper deck, with a three-meter distance between them. To ensure the safe movement of the submersible on the tracks while considering the significant sheer of the stern without modifying the upper deck structure, the the tracks were set at an angle of less than 15 degrees with the vessel's baseline and with a bow trim angle. This maintained the appropriate stern trim of the vessel throughout the launch and recovery process.

The vessel's original galley was transformed into a submersible support compartment, divided into four rooms: a hazardous goods storage, a mechanical hydraulic operation room, a charging room, and an electronic/electrical operation room. The hazardous goods storage was sealed with insulation materials and equipped with standard three-tier shelves to store items like one-ton mercury and carbon dioxide absorbents. The mechanical hydraulic operation room was arranged with one hydraulic pump, one refueling vehicle, one frame-type air compressor, one oxygen filling pump, and four barrels of hydraulic oil. It was also furnished with a workbench and a water basin. The charging room had a workbench with a drawer-style toolbox between two doors. This room was capable of storing one silver-zinc battery charger, one lead-acid battery charger, one main battery, and one auxiliary battery box. The electronic/electrical operation room housed a surrounding workbench and supporting toolbox.

The generator sets and distribution panels were replaced. Three 600-kilowatt generator sets were replaced, while one 400 kilowatts set was retained as the harbor generator. The engine room monitoring console and bridge control console were also replaced. Additionally, the air conditioning system was renovated.

(3) Functionality of Launch/Recovery System

Lifting function: Under specified ocean conditions, it could hoist the 7,000-m manned submersible from the mother ship and deploy it into the sea, as well as lift the submersible from the sea surface and securely place it on the deck storage rack.

Towing function: Once the manned submersible completed its underwater tasks and surfaces, it could be towed by a towing cable to get close to the stern of the mother ship using a towing winch. The position of the submersible could be controlled during the lifting process, enabling prolonged towing of submersible in emergency situations.

Stabilizing function: In specified ocean conditions, when both the mother ship and the submersible experienced their own sway and heave movements, hydraulic servo devices and hydraulic damping stabilizing devices would be employed to keep the submersible's motion under control during the launch and recovery process, ensuring safe, smooth, and reliable operations.

Compatibility function: Apart from launching and recovering the 7,000-m manned submersible, the launch/recovery system could collaborate with specialized winches to deploy and retrieve heavy remotely operated vehicles (ROVs) equipped with relay TM5, acoustic and optical tow bodies, and the acoustic system of manned submersibles. It was also capable of conducting deep-sea towed survey operations at depths exceeding 6,000 m.

Following the refitting and delivery of the "Xiangyanghong 09" vessel (see Fig. 4), it had successfully completed multiple missions with "Jiaolong" manned deep-diving submersible for tests at various depths. On June 3, 2012, the vessel, carrying

Fig. 4 Manned submersible mother ship "Xiangyanghong 09"

"Jiaolong", embarked on a 7,000-m deep-sea diving mission from Nansu Terminal, Jiangyin, Jiangsu, heading to the Mariana Trench region. "Jiaolong" successfully overcame the challenge of a maximum operating depth of 7,062 m, setting a record for China's deep diving endeavors. Over 130 dives, "Jiaolong" hit the deepest diving record, during its exploring submarine geomorphy, discovering new species, and returning valuable biological samples. This greatly advanced China's deep-sea research.

2.5 *Xiangyanghong 10, One of China's Top Ten Famous Vessels*

Alongside refits of civilian transport vessels into multipurpose oceangoing survey vessels, China's ministries and commissions also embarked on the development of new types of oceangoing survey vessels of better performance that would be more suited for extensive scientific surveys in the open seas, meeting the requirements of the "718" project. Given the huge workload, substantial investments, and tight timeframe, developing a new state-of-the-art research vessel independently was undoubtedly an arduous task. It mandated a step-by-step approach, from feasibility analysis and exploring various proposals to studying foreign similar ships and domestic equipment manufacturing capabilities, before the design could be completed and delivered to the shipyard for construction. To this end, the ministries and commissions instructed the MARIC, with its leaders and scientists along with personnel from relevant departments, to form a specialized working team for the design. Even during "Cultural Revolution", the team remained focused on positive education, exerting utmost effort to advance the design work.

The working team was established in February 1971. Under the guidance from relevant ministries and commissions, the tireless efforts of the entire team, as well as the support from dozens of research and manufacturing organizations nationwide, the technical design was finally completed in the first half of 1975. Subsequently, the MARIC organized an assembly of nearly a hundred individuals who formed a tripartite collaborative team, comprising the user, the shipyard, and the designer, to undertake the construction design at Jiangnan Shipyard. In July 1975, the construction commenced at Jiangnan Shipyard. To ensure completion on schedule, a parallel design and construction approach was adopted. The vessel was finally delivered in October 1979 and named "Xiangyanghong 10" (see Fig. 5).

The vessel was designed with a total length of 156.2 m, a molded breadth of 20.6 m, a draft of 7.75 m, a displacement of 13,000 tons, a maximum speed of 20 knots, and an endurance of 18,000 nautical miles.

The vessel's primary mission was to conduct surveys and select offshore experimental target ranges; provide medium and short-term weather forecasts and hazardous weather warnings for its designated sea area, and provide hydrological and meteorological support for experimental fleets and missile flight experiments; participate in investigations of the Earth's gravity and magnetic fields, and provide essential data for ballistic corrections; guarantee all-weather long-distance shortwave communication and long-term digital transmission for experiments, as well as communication frequency forecasting; conduct surveys of marine acoustics to contribute marine hydrological data for the arrangement of underwater sound arrays during the recovery of missile data capsules; and undertake helicopter telemetry tasks during missile impact events.

The vessel was equipped with multipurpose oceangoing research capabilities. It was capable of conducting multidisciplinary marine surveys and research in hydrology, meteorology, acoustics, geology, geomorphology, geophysics, marine

Fig. 5 Oceangoing survey vessel "Xiangyanghong 10"

physics and chemistry, marine biology, and microbiology across all major oceans. It could amass a wealth of marine-related data while conducting extensive investigations into marine meteorology, ocean waves, tides, currents, sea ice, physical and chemical properties of seawater, sea depth, seabed topography, minerals, geology, submerged objects, seabed resources, and crustal structures. Its operations could provide valuable insights for the development of marine resources, furthering marine research, and facilitating technological advancements.

Compared to refitted vessels, this vessel featured an enhanced suite of research facilities, including:

(1) It was equipped with a helicopter pad and a hangar. The pad spanned an area of 533 square meters, while the hangar had a command tower and a navigation platform at the top, meeting the requirements for sustained deployment of a "Super Frelon" helicopter at sea.
(2) The fore deck was equipped with a 29-ton crane, facilitating the lifting and recovery of deep-diving submersibles.
(3) A complete meteorological system was configured, encompassing meteorological rocket launch system, sounding balloon system, wind and rain radar, satellite cloud image receiver, and weather fax equipment. These elements enable medium- and short-term weather forecasting as well as hazardous weather warnings.
(4) A high-power transmitter communication system and other communication devices were arranged, enabling simultaneous long-distance communication in two directions, as well as general communication between ships and ships, ships and shores, and ships and aircrafts. The long-distance communication antennas were arranged on the top of the foremast and aftermast, with numerous whip antennas erected around the compass deck and navigation deck.
(5) It was fitted with advanced navigation equipment, including a satellite navigation positioning system, enabling safe navigation worldwide and meeting the high-precision positioning requirements for oceanographic surveys. The satellite navigation antennas were arranged on the deck above the hangar.
(6) The upper deck had 9 booms used for surveys on both sides and the stern. The deck was equipped with 10 winches for cables and ropes ranging from 1,200 to 11,000 m, used for hydrological, physical, chemical, acoustic, geological, geomagnetic, and biological experiments, as well as trawling operations. In addition, the vessel had several laboratories and computer rooms for different disciplines, allowing for on-site experimentation, data collection, and analytical processing.

The vessel underwent thorough analysis and testing of its key technical elements to ensure optimal vessel performance. Its main design features include:

(1) Appropriate displacement: Considering that the vessel's load varies significantly during operation, with a notable difference in performance between departure and arrival states, it was deemed best to use the intermediate state as the design displacement. This involves taking into account the empty vessel

weight, personnel and luggage, and 50% to 60% of the variable load as the design displacement.

(2) Reasonable main dimensions: Ship length: Consideration was given to the general arrangement, resistance, and cost-saving in steel consumption. Ship breadth: Not only was the need for the helicopter pad, hangar, air inflation and deflation chamber, and fast work boats taken into account, but also the requirement for good initial stability when encountering Force 12 typhoons. Therefore, an appropriate ratio of ship breadth to draft was chosen. Draft: A deeper draft was adopted to improve seaworthiness, increase resistance to wind forces, allow for greater immersion of propellers and rudders, and prevent fuel deep tanks from protruding above the water surface. Molded depth: A smaller molded depth was adopted while ensuring sufficient unsinkability, so as to maintain a lower center of gravity in the empty vessel and a smaller wind-exposed area. Stability: To enhance stability at large angles of inclination and increase the volume of compartments, a long forecastle was employed.

(3) Consideration of resistance to wind and waves for profile design: The vessel's profile was designed with a V-shaped bow and stern, with the above-water section of the bow slightly elevated and a pear-shaped bulb bow underwater. The vessel's relatively large length and deep draft contributed to a relatively small block coefficient, resulting in comparable resistance to world-renowned vessels from the Taylor series.

(4) Wind stability capable of resisting Force 12 winds: At that time, China did not have a standardized method for calculating wind stability at Force 12 winds. After extensive research, calculation analysis, and studies of relevant domestic and international regulations and methodologies. Suitable calculated value of wind pressure and its calculation formula, as well as the correction to the initial heel angle were determined. This stood out as a pioneering work in the field of wind stability calculations at Force 12 winds.

(5) Enhanced vessel stability: To ensure the proper functioning of specialized equipment on the vessel, a high level of stability was required. To achieve this, a pair of anti-rolling fins was installed, with the control mechanism situated in the auxiliary engine room, and the control section and indicating instrument located in the electro-navigation instrument room and pilothouse, respectively. Furthermore, a passive anti-rolling tank was incorporated to improve stability at low speeds.

(6) Excellent unsinkability: The vessel featured nine watertight bulkheads that extended all the way to the upper deck. Even under full load conditions, when adjacent compartments were simultaneously flooded, the waterline on the subdivision deck would still be below 76 mm, fulfilling the requirements for unsinkability. With the exception of the main engine room, auxiliary engine room, and fore peak tank, the underwater section of the vessel consisted entirely of slop tanks. Through accounting for flooding in the main and auxiliary engine rooms, the vessel could maintain sufficient positive stability. In order to avoid excessive free surface correction and asymmetry in damaged conditions, the

lower liquid tanks of the vessel were interconnected on both sides to enhance the unsinkability.

(7) All-weather ocean communication system: In order to minimize the mutual interference of various signals during reception, especially the wideband RF interference caused by the strong RF magnetic field generated by high-power shortwave transmitters, it was necessary to ensure the smooth operation of communication signals during simultaneous reception and transmission. Electromagnetic compatibility on vessels posed a major technical challenge, thus effective measures were taken in the design of the vessel. These measures included shielding, setting antenna zones, locating receiving and transmitting antennas in shielded areas, and placing transmitting antennas as far away from living areas as possible. This was how the vessel guaranteed all-weather transoceanic communication during the "580" experiment.

(8) Helicopter landing system on desk: This system allowed for the long-term usage, storage, refueling, and maintenance of shipborne helicopters at sea. Through single and comprehensive practical experiments, it was demonstrated that the system was safe and reliable for extended deployment of a "Super Frelon" helicopter at sea. The successful development of this system filled a technological gap in this field in China.

(9) Emphasis of meteorological forecast and assurance system in design: Based on extensive and in-depth research and analysis through visits to meteorological stations, a comprehensive and effective meteorological center system was designed. This system was capable of detecting atmospheric elements at an altitude of 30,000 m and tracking the formation and movement of typhoons within a range of 500 km. It also enabled medium- and short-term weather forecasting and warnings of hazardous weather conditions, serving as an offshore center for meteorological forecast.

(10) High-power underwater acoustic power source: As per required, the vessel needed to equip with a high-power underwater acoustic experiment system in line with the requirements of long continuous experiment duration. Experiments were conducted using the "Shijian" survey vessel to seek a new power supply solution. Ultimately, it was decided to use diesel generator sets with low vibration and noise as a power source. This allowed for long-duration and continuous marine acoustic testing and sonar equipment experiments.

After being delivered and put into service, the "Xiangyanghong 10" oceangoing survey vessel participated in the "580" experimental mission. During the 62-day offshore joint training and oceangoing voyage, it demonstrated reliable meteorological and hydrological support, seamless communication and coordination, accurate navigation operations, proper functioning of electromechanical equipment, and safe handling of the vessel, earning full recognition from the maritime formation headquarters. The ship was honored with a collective second-class merit.

Subsequently, the vessel successfully completed significant special oceangoing investigation missions. It made multiple entries into the Pacific Ocean, completing tasks such as communication transponding and information transmission, gravity investigation and hydrographic survey, meteorological support, and investigation of the meteorological environment in the sea area. The vessel also sailed to Antarctica numerous times to conduct surveys and facilitate station construction. Antarctica is known for being the coldest, driest, windiest, and stormiest place on Earth. "Xiangyanghong 10" opened a new route to South America, which is the closest route from China and has been embraced by China's oceangoing shipping fleet. The ocean communication system established a shortwave communication network between Antarctica and Beijing, setting a new record for the longest distance shortwave communication in the history of Chinese telecommunications. It also provided initial insights into the communication mysteries in Antarctica and the Southern Ocean. The vessel also transported over 500 tons of supplies, equipment, tools, and numerous personnel for the construction of China's Antarctic "Great Wall" station.

During its inaugural scientific expedition to the Southern Ocean, the vessel crossed the Antarctic Circle, collecting tens of thousands of data and samples. It achieved breakthroughs in more than a dozen projects, some of which were unprecedented domestically. On January 26, 1985, while conducting oceanographic surveys near the Antarctic Circle, the vessel encountered a polar cyclone storm with wind forces exceeding 12. Enormous waves as high as 13 m surged, causing the vessel's body to experience substantial fluctuations, the propellers to emerge above the water surface, and main engine to experience nine times of propeller racing in the wild winds and huge waves. After more than ten challenging hours battling the furious winds and waves, "Xiangyanghong 10" finally broke free from the grip of the polar cyclone zone. The vessel withstood the test, ensuring the safety of both its property and the lives of the crew, earning praise from countries around the world. Afterwards, the vessel was awarded a collective first-class merit.

During the East China Sea continental shelf survey task, the vessel continued to operate without interruption in Force 9 winds, efficiently completing extensive survey activities. In this operation, the vessel obtained abundant scientific research data urgently needed by the nation, providing compelling evidence for foreign affairs.

"Xiangyanghong 10" received unanimous praise from units concerned for its appealing appearance, reasonable layout, and exceptional technical performance. Liu Huaqing, former Commander of the Navy, stated, "This vessel has truly passed the test." Chen Youming, the former Minister of Equipment Technology of the Navy, remarked during a visit to the vessel, "This vessel looks most beautiful." He further added, "The design of auxiliary ships, and even naval vessels, should draw inspiration from some of the positive aspects of 'Xiangyanghong 10'." Lu Wei, the former Deputy Director of the SOA, commented, "This vessel is the best within our Oceanic Administration. In the future, we plan to visit the US with this vessel and let them have a look at it." Comrades who participated in trial voyages of this vessel expressed, "We feel perfect in the trial voyages." Personnel who participated in the "580" experiment articulated, "We eagerly sought to be on this vessel. Work and living conditions there are convenient. The facilities are great, and we seldom get seasick." Zhang Zhiting, the former captain of the "Xiangyanghong 10", declared, "Even in Force 9 winds, we confidently deployed the winches for drifting survey work." Wu Heng, the former Director of the National South Pole Investigation Commission, stated at the ceremony welcoming the victorious return of the first Chinese Antarctic expedition formation, "Our domestically-built research vessel has withstood severe tests… All of this signifies a new level of advancement in China's science and technology." In reporting on the vessel triumphing over Force 12 winds, *Jiefang Daily* and *Wenhui Daily* wrote, "…This is unprecedented in the history of China's marine research."

The birth of the "Xiangyanghong 10" oceangoing survey vessel denotes that China's design and construction technology for oceangoing scientific survey vessels has reached a world-leading level. Its significance is profound and far-reaching.

The vessel received the First Prize for Major Scientific and Technological Achievements from the Commission for Science, Technology, and Industry for National Defense in 1979. In 1985, it was honored with the National Science and Technology Progress Special Award. In 2006, it was acknowledged as one of China's "Top Ten Famous Vessels".

In August 1998, "Xiangyanghong 10" was transformed into the space tracking ship "Yuanwang 4", primarily responsible for sea-based tracking, telemetry, communication, and control tasks related to missiles, satellites, and spacecraft. In October 1999, it embarked on its first voyage as the newly designated "Yuanwang 4", (see Fig. 6) traveling to the Indian Ocean to carry out measurement and communication tasks for the sea-based launch of the "Shenzhou 1" spacecraft. It has since undertaken multiple expeditions to the Pacific Ocean and the Indian Ocean, flawlessly fulfilling measurement and control tasks related to China's Fengyun satellites, manned spacecraft, and the "Chang'e 1" circumlunar satellite.

Fig. 6 Space tracking ship "Yuanwang 4"

2.6 Research Vessel "Kexue 1"

In the 1970s and 1980s, most of China's marine research activities were conducted using small vessels. These vessels, with a carrying capacity mostly in the range of a few hundred tons, allowed for loads of only up to several thousand tons. This fell far short of meeting the demands of the time. China was in urgent need of fully equipped research vessels with larger carrying capacities. In 1980, the birth of the first domestically designed and built specialized research vessel, "Kexue 1", (see Fig. 7) was of epoch-making significance for China's marine research at the time to some extent. The vessel embodied China's exploration capabilities in marine research.

"Kexue 1" was built by Hudong Shipyard in 1980 and classed by CCS. This vessel was designed with a total length of 104 m, a draft of 4.9 m, and a displacement of 3,324 tons. It was equipped with two diesel engines with a power of 5,280 horsepower each, allowing for a maximum speed of 19 knots, an economical speed of 15 knots, and a minimum speed of 4.5 knots. With endurance of 8,000 nautical miles, the vessel could accommodate 101 persons, including 38 crew members and 63 researchers. The vessel housed 10 laboratories that facilitated multipurpose surveys in geology, geophysics, physical oceanography, meteorology, chemistry, biology, and environmental protection among other disciplines. The total laboratory area measured 187.5 square meters, including a navigational positioning and data collection room of 24.5 square meters, a seismic room of 32.7 square meters, a geomorphology room of 22 square meters, a magnetic and meteorological room of 17.9 square meters, a wet laboratory of 20.5 square meters, a dry laboratory of 22 square meters, a gravity

Fig. 7 Research vessel "Kexue 1"

room of 22 square meters, an air-gun room of 8.4 square meters, an air compressor room of 8 square meters, and a data room of 9.5 square meters.

Moreover, the vessel's working deck was equipped with over 10 winches of various capacities, including a 3,000-m hydrographic winch, a 3,000-m CTD winch, a 6,000-m CTD winch, an 8,000-m hydraulic geological winch, and a 10,000-m hydraulic traction geological winch. The vessel also featured an A-frame and an inverted L-frame, 6 high-pressure air compressors, and 12 high-pressure gas cylinders, providing substantial support for specialized investigations.

Since its commissioning, "Kexue 1" has undertaken numerous research projects, such as the "863 Program", the "973 Program", National Key Projects, National Climbing Program, National Priority, international cooperation projects, defense science and technology projects, as well as investigation projects commissioned by the State Scientific and Technological Commission and the Chinese Academy of Sciences. It has ventured across the equator twice and involved in multiple expeditions to the Western Pacific. These voyages contributed tens of thousands of scientific research data and achievements of great value in scientific research. Notably, during the Pacific Ocean expeditions, scientists discovered the "Mindanao Undercurrent". This discovery challenged the traditional understanding of the dynamic structure of western boundary currents in the Pacific Ocean. It carried important implications for meridional heat transfer, balance, and climate in the oceans. Additionally, the vessel's surveys at the Chinese continental shelf led to the discovery of the "East China Sea Cold Eddy", advancing the study of the circulation in the Chinese continental shelf. Since 1980, "Kexue 1" has played a major role in China's scientific expeditions. It has navigated waters in the Bohai Sea, the Yellow Sea, the East China Sea, the South China Sea, and even reaching as far as Papua New Guinea, with a total distance traveled exceeding 600,000 nautical miles.

Throughout its service, "Kexue 1" cultivated a highly experienced and proficient team of marine research. As an integral part of the scientific research support system, "Kexue 1" had been dedicated to serving marine science and technology. With a spirit of exploration, bold reform, and innovation, it scored remarkable results in scientific expeditions, making significant contributions to China's ocean research undertakings.

On May 20, 2016, "Kexue 1" was decommissioned with honor in Qingdao. After 36 years of service, this esteemed research vessel was relocated to the Naval Park in the Military-Civilian Integration zone of the Qingdao West-Coast Economic New Area. At the park the vessel was transformed into a marine science popularization and education base, as well as a training base for marine science and technology professionals. The Qingdao Military-Civilian Integration Development Group Co., Ltd. at Qingdao West-Coast Economic New Area, a state-owned enterprise, invested in the operation of the "Kexue 1" research vessel in accordance with the basic concept of "embracing intelligence in reform, combining science with education". After being put into operation, it would showcase the development process of China's marine research. Meanwhile, as a popular science and education center, it would help teenagers enhance marine awareness, learn about marine knowledge, and understand marine research, becoming an interactive 116-type exhibition area that integrates popular science education, cultural tourism, and dining entertainment. Moreover, "Kexue 1" would also serve as a learning and practice platform for students majoring in marine science and media, aiming at nurturing more exceptional talents in marine science and technology.

2.7 Marine Geological Survey Vessel "Haiyang 4"

The marine geological survey vessel "Haiyang 4" (see Fig. 8) was completed by Hudong Shipyard in November 1980. In December 1986, it embarked on its maiden voyage to the Central Pacific, followed by over 10 ocean scientific investigation missions. In December 1990, it ventured to the Antarctic, earning the reputation as a hero vessel for scientific research.

"Haiyang 4" was fitted with equipment like large fuel tank, seawater desalting system, oily waste water treatment facility, winch (10,000-m), bow thruster, anticollision radar, and others. It was suitable for conducting long-distance ocean investigation missions. These features made it one of the most comprehensive and well-equipped marine geological and geophysical survey vessels in China at that time.

This vessel was designed with a length of 104.27 m, a molded breadth of 13.47 m, a molded depth of 7.8 m, a draft of 4.95 m, a displacement of 3,325 tons, a navigational speed of 15 knots, an endurance of 15,000 nautical miles, and a self-sustaining period of 60 days. It was equipped with two B&W 6K45GF diesel engines with a power of 3,533 kilowatts each.

"Haiyang 4" carries long standing reputation. Since its completion and initial service in 1980, it's courses have extended all over the seas of China and the Pacific Ocean. Moreover, it has made extensive voyages to Antarctica, undertaking a series of important tasks for China, covering surveys of marine basic geology, oil and gas resources, natural gas hydrate resources, and ocean mineral resources, among others. All in all, it has made important contributions to safeguarding China's maritime

Fig. 8 Marine geological survey vessel "Haiyang 4"

rights and interests and serving the construction of the marine economy. To further enhance its survey capabilities, the vessel required a refit, with design falling to the MARIC and construction to COSCO Shipyard, Guangdong. Since the start of the refit project, the institute conducted extensive work in preliminary research and analysis, feasibility studies, collaboration with the shipyard for bidding design, detailed design, and coordination during the construction process. Especially during the detailed design process, the design team closely collaborated with the ship user, the shipyard, the classification society, and the supervisory company, successfully overcoming various challenges such as old age of the vessel, lack of original vessel data, and tight schedule demand. With solid technical skills and a strong sense of responsibility, the team overcame multiple technical challenges such as shafting calculation and segment docking, laying the groundwork for the smooth execution of the entire project. The trial voyage results showed that the vessel has significantly improved in aft deck working area, hull vibration reduction, steerage, and hull stability. It was confirmed that all technical properties of the vessel met the design requirements, with all technical indexes achieving the expected targets. Through refit, "Haiyang 4" expanded its geological survey approaches, enhanced the technological content of field acquisition, and promoted the development of underwater survey equipment towards mechatronics, multi-functionalization, and intelligence.

The main survey equipment equipped on the vessel were the global positioning system, integrated navigation system, multi-beam sounding system, marine gravimeter, magnetometer, sub-bottom profiler, single-channel seismometer, digital and analog single-channel seismic system, deep-sea camera system, CTD measurement system, Aanderaa current meter, myriameter depth sounder, large-scale gravity piston corer, cable gravity corer, ocean-50 cable grab sampler, box corer, multi-corer sampler, cable-free self-return grab, crust trawl, crust and surface biological trawl. The on-site chemical laboratory was also equipped with X-ray fluorescence analyzer, atomic spectrophotometer, and pH meter. Physical, chemical, and biological on-site analysis instruments such as sediment shear apparatus and polarizing microscope were also available. The vessel's advanced survey techniques includes:

The first is the multi-beam full-coverage sounding technique, which enables the reproduction of deep-sea seabed topography and geomorphology. This technique is one of the primary survey methods used for the exploration of international seabed mineral resources in China. By collecting data using the multi-beam sounding system and processing it into images, marine geologists can reproduce the topography and geomorphology of the seabed at depths of several kilometers. This important survey method is crucial for identifying the distribution and enrichment regions of polymetallic nodules and cobalt-rich crust resources on the seabed.

"Haiyang 4" first introduced the SEABEN M2110 multi-beam sounding system in the mid-1990s. Based on this foundation, the Guangzhou Marine Geological Survey conducted further research and development. They integrated the sounding system with differential GPS navigation positioning technology, refined post-processing, and electronic digital automatic mapping technology. The outcome was an array of precise techniques for multi-beam full-coverage sounding, enabling China to reach the international level in terms of seabed topography and geomorphology sounding technology at that time.

Starting from 1997, "Haiyang 4" has been using this vital survey method on multiple voyages. It has completed multi-beam measurements of 4,363 km in crust areas, covering an area of 30,000 square kilometers. Compared to the single-beam sounding system used in the past, the multi-beam sounding system has shown significant improvements in terms of collected data volume, work efficiency, and sounding accuracy.

The second is the seabed color camera system, which serves as a "telescope" for marine geologists. "Haiyang 4" was equipped with a domestically developed seabed color camera system that could reach depths of up to 6,000 m. The seabed camera system acts as the eyes of marine geologists, revealing the mysterious underwater world to the world. This system consists of two parts, the deck and the underwater device, which are connected and transmit communication signals through a 7,000-m-long coaxial armored cable. The underwater camera captures close-up shots of the seabed, while control personnel in the instrument compartment and winch operators on the vessel's deck can monitor and observe the underwater camera unit in real time through monitoring images and height data.

In 1998, this system was first deployed and successfully obtained clear seabed records at four stations in the East Pacific. Through underwater cameras, it clearly observed the "red clay", as well as activities and traces of fishes, mollusks, and benthic organisms, in the deep-sea plain at a depth of over 5,000 m. More importantly, it directly observed potato-sized black-brown manganese nodules scattered on the seabed partially covered by the "red clay". Subsequently, this system was quickly put into trial production and achieved color camera recordings in the complex seabed terrain conditions of the cobalt-rich crust target area in the Central Pacific. It was also used in the natural gas hydrate resource survey in China in 2001. During this voyage, "Haiyang 4" conducted seabed camera recordings at 23 stations in the crustal and nodule contract areas of the Magellan Seamount region and the Central Pacific Seamount region, obtaining firsthand visual documentation of the surveyed areas.

In 1999, China became the fifth country in the world to register with the United Nations International Seabed Authority, obtaining exclusive exploration rights and mining priority for polymetallic nodule contract areas in a 75,000-square-kilometer international seabed region in the East Pacific. During the "10th Five-Year Plan" period, the survey of cobalt-rich crust resources in international seabed areas was listed as a key task under the unified organization of the China Ocean Association (China Ocean Mineral Resources R&D Association, COMRA). Efforts were made to conduct surveys in this resource area. The endeavors included identifying promising cobalt-rich crust mining areas for timely application for China's contract mining area with the International Seabed Authority.

"Haiyang 4" carried out over 10 oceanic surveys and scientific investigations in the Antarctic and South Pacific. It is currently the earliest and most successful scientific research vessel in China to participate in oceanographic survey missions, with each research mission being a great success. The Chairman of the COMRA once highly praised "Haiyang 4" for setting a benchmark for China's oceanic scientific investigations.

On October 30, 2002, Guangzhou Whampoa Military Academy Pier was bustling with people, and colorful flags were fluttering in the air. "Haiyang 4" oceangoing research vessel of the Guangzhou Marine Geological Survey returned to Guangzhou after completing the scientific investigation tasks of the DY10513 voyage for *China's "10th Five-Year Plan" for the International Seabed Area Research and Development Program*. This 174-day scientific expedition, covering a distance of over 24,000 nautical miles, achieved significant results, discovering two potential cobalt-rich crust areas with exploration prospects. It further clarified the resource situation in key target areas, confirming the resource prospects of three crustal investigation areas. China has taken the initiative in the upcoming new round of international competition for seabed mineral resources. It reflects the important strategic adjustment of China's oceanic exploration and development activities, expanding from a single polymetallic nodule resource to various strategic resources in international seabeds.

In late May 2003, "Haiyang 4" set sail from Guangzhou, embarking on a long journey to the Pacific. After 164 days of offshore scientific research and investigation, covering a distance of over 20,000 nautical miles, it completed the scientific investigation task of *China's "10th Five-Year Plan" for the International Seabed Area Research and Development Program* (codenamed DY10515 voyage) and returned to Guangzhou in early November.

The DY10515 voyage of "Haiyang 4" is an important voyage that represents the strategic adjustment of China's oceanic exploration and development activities from a single polymetallic nodule resource to various strategic resources in international seabeds. It conducted surveys of cobalt-rich crust resources and exploration of polymetallic nodule contract areas in the Pacific International Seabed Area, achieving significant investigative results. Cobalt-rich crust is one of the most important solid mineral resources in the deep sea. This voyage preliminarily identified candidate areas for cobalt-rich crust applications, laying a solid foundation for upcoming cobalt-rich crust mining area applications. For the previously identified key seamount areas in the Magellan Seamount region and the Central Pacific, "Haiyang 4" contributed to a clearer understanding of the distribution patterns and resource conditions of cobalt-rich crusts through intensive trawl sampling and seabed video recording, further identifying the cobalt-rich crust resources in key areas. Additionally, this expedition further revealed the small-scale distribution patterns of polymetallic nodules in the eastern contract area of China. These findings provided a basis for the design of future mining equipment and the selection of mining plans.

2.8 Multipurpose Oceangoing Research Vessel "Dayang 1"

In order to coordinate the strengths of various marine sectors in China, carry out work in international sea areas, and safeguard China's maritime rights and interests, the China Ocean Mineral Resources R&D Association (China Ocean Association, or COMRA) was established under the leadership of the SOA in April 1990.

In response to the demand for ocean mineral resource surveys, the COMRA purchased an marine geology and geophysics research vessel from the Far Eastern Branch Russian Academy of Sciences in July 1994. The vessel, originally named "Pyotr Andropov" and built in 1984, was renamed "Dayang 1" (see Fig. 9) and managed by the South China Sea Branch of the SOA on behalf of the COMRA.

This vessel was built in accordance with Soviet ship classification and construction specifications. Its hull was made of steel and strengthened for operations in icy areas, equivalent to China's B3-class ice-resistant capability for non-restricted waters.

The vessel was designed with a total length of 104.5 m, a molded breadth of 16 m, a molded depth of 10.2 m, a draft of 5.7 m at full load, a maximum speed of 16.2 knots, a self-sustaining period of 60 days, a capacity of 75 persons, and a displacement of 5,600 tons at full load.

Based on the usage situation, the COMRA commissioned the China Ship Development and Design Center to carry out the refit designs for three consecutive times.

The first refit was conducted in Guangzhou Wenchong Shipyard in 1995. The main modifications included: removing the stern single-boom crane, poop and related winches; moving the foremast forward and relocating the aftermast to the middle between the two funnels; installing a bow thruster, a multi-beam sounder, a deep water (10,000-m) towing winch and its hydraulic station, an underwater robot and a cable winch, two telescopic cranes, and a folding arm crane; and setting up several laboratories. Following this refit, the vessel became a versatile work platform for multidisciplinary research in marine geology, geophysics, marine chemistry, marine biology, marine acoustics, and other fields. It was capable of conducting surveys and

Fig. 9 Multipurpose oceangoing research vessel "Dayang 1"

experiments in areas such as seabed topography, gravity and magnetism, geology and structure, comprehensive marine environment, marine engineering, and deep-sea technology and equipment.

After several years of ocean geological surveys following this refit, it was revealed that there were certain issues with the ship system and the survey system, making it challenging to meet the demands of extensive survey tasks. For instance, the vessel's roll period exceeded the original design value, resulting in poor seaworthiness. The ballast system was damaged, leading to poor flotation and difficulties in adjustment. The main propulsion system had defects, causing the actual speed to be lower than the design speed. The main generator ran with problems, resulting in insufficient power supply capacity for the entire vessel. The height of the aft deck working area from the water surface was too high, leading to limited functionality of lifting equipment and difficulties in launching and recovering survey equipment, posing safety risks. The number and effective area of the laboratories were inadequate. Communication and navigation equipment were subpar. There was no dynamic positioning system, preventing the execution of more complex survey projects. There was no ship-wide computer network system, hindering ship-wide data sharing capabilities.

In 2001, to align with China's strategic shift in ocean resources survey from single polymetallic structural resources to multiple resources, and to better fulfill the country's ocean resources survey tasks, a second refit was carried out at Hudong Shipyard. Additional facilities were added, including a multi-beam and sub-bottom profiling laboratory, gravity and ADCP laboratory, magnetic laboratory, seismic laboratory, comprehensive electronic laboratory, geological laboratory, biological gene laboratory, deep towing and ultra-short baseline (USBL) laboratory, as well as various survey equipment. Among the advanced equipment independently developed in China were the modern ship network system, 6,000-m deep towing optical system, 4,000-m sounding side-scan acoustic towage control system, 3,000-m sub-bottom rock core drill, 3,000-m TV grab, 3,000-m underwater camera continuous observing system, shipborne deep-sea barophile continuous cultivation system, as well as various deep-sea sampling equipment such as multiple corers, box corers, and trawls. Imported advanced instruments and equipment included multi-beam sounding system, high-precision dual-frequency swath-type sounding system, bottom profiling system, full-ocean broadband nonlinear difference frequency sub-bottom profiling system, 6,000-m (long-range) ultra-short baseline positioning system, optical fiber motion reference unit system, 6,000-m deep-towed acoustic system, 6,000-m sounding/side-scan and multi-beam system, 6,000-m deep-sea photographic system, and 6,000-m acoustic release system, among others.

This refit transformed "Dayang 1" into a modern, multipurpose, and close-to-international-advanced-level oceangoing research vessel. It was capable of meeting China's resource development needs at international seabeds, being open for domestic and foreign oceanography with integration of oceanographic surveys and deep-sea equipment experiments. Since 2003, "Dayang 1" has successfully carried out numerous demanding marine survey tasks, making significant contributions to

China's oceanic scientific research during the "10th Five-Year Plan", "11th Five-Year Plan", and "12th Five-Year Plan" periods.

This vessel has successfully completed seven ocean survey voyages for the China's ocean mineral resource R&D special project, as well as numerous survey voyages for continental shelf exploration. It conducted seabed nodule resource surveys in the East Pacific and West Pacific, as well as hydrothermal sulfide resource surveys in the Southwest Indian Ocean. These efforts laid a solid foundation for the development of China's oceanic endeavor.

From April 2005 to January 2006, "Dayang 1" completed China's first-ever global scientific expedition across the three major oceans, achieving four significant outcomes:

(1) It obtained samples of sulfides, rocks, sediments, organisms, and other materials near the hydrothermal fluid vents in target areas of the three major oceans, providing valuable data for further research.
(2) It conducted preliminary surveys on the distribution of hydrothermal sulfide "black smokers" in certain regions.
(3) It explored cutting-edge and focal issues within the scientific community, driving the advancement of oceanic scientific research.
(4) It stimulated the development of related marine technology and equipment.

From July 2009 to March 2011, the vessel undertook two more global expedition missions. The second global scientific expedition, which took place from July 2009 to May 2010, achieved significant historic breakthroughs in various aspects. The scientists discovered polymetallic sulfides in hydrothermal fluid vents in the Atlantic Ocean, making China a country capable of independently discovering hydrothermal fluid areas in all three major oceans. Moreover, this expedition marked the inaugural use of a domestically developed deep-sea unmanned tethered submersible for the observation and recovery of "black smokers" groups. The 22nd voyage of the research vessel, which set sail in December 2010, marked its third global scientific expedition. It covered a total distance of 45,000 nautical miles and lasted for 376 days, making it one of the longest ocean scientific expeditions undertaken by China at that time.

All these three global scientific expeditions conducted by "Dayang 1" shared an essential mission closely related to future economic development, which entailed the acquisition of hydrothermal sulfides from "black smokers".

"Black smokers" refer to volcanic vents located at hydrothermal zones, where the continental and oceanic plates meet. Resembling chimneys, these structures are rich in various rare metals and serve as valuable mineral resources that attract considerable international attention.

The search for seabed "black smokers" holds significant importance for any nation. These dark substances comprise various resources vital to industrial societies, including crucial metals such as copper, nickel, and zinc, as well as indium used in the production of flat-screen displays. Even precious metals like gold can be found within "black smokers". Furthermore, the vicinity of these structures gives rise to a unique ecosystem known as the "dark biological community", which thrives without reliance on sunlight and photosynthesis. These organisms contribute greatly to the future development of advancements in biomedicine.

Up to now, China has discovered 17 hydrothermal zones, accounting for approximately one-tenth of the hydrothermal zones discovered worldwide in the past three decades within the three major oceans.

From November 2014 to June 2015, the vessel completed its 34th voyage, covering a distance of 28,125 nautical miles over 215 days. This venture achieved six significant accomplishments:

(1) It successfully fulfilled China's Southwest Indian Ocean polymetallic sulfide exploration contract, identifying multiple mineralized anomaly zones within the contract area and gaining new insights into the distribution range and structural characteristics of hydrothermal zones.
(2) It implemented engineering exploration of seabed polymetallic sulfide resources, carried out exploration for methods such as sediment geochemical exploration and near-bottom magnetometry, and formed a set of detection methods for identification of mineralized anomaly zones.
(3) It achieved breakthroughs in the application of key equipment for sulfide exploration, including China's self-developed "Jinquzhe (Enterpriser)" deep-hole rock core sampling drill and electric method-based detector.
(4) It conducted environmental baseline surveys within the polymetallic sulfide contract area, obtaining annual environmental baseline data and biological samples as mandated by the International Seabed Authority.
(5) It collected a comprehensive range of geological, geophysical, hydrological, and environmental data, as well as samples from certain blocks within the polymetallic sulfide contract area. These findings provided crucial supporting data for scientists investigating the diversity characteristics of sulfides within the area.
(6) It made the first discovery of large-scale rare metal-rich sediments in the Central Indian Ocean Basin, which were subsequently subjected to on-site testing and analysis. This analysis revealed high concentrations of rare earth elements, indicating suitable conditions for mineralization.

To address concerns arising from long-term usage and aging equipment, as well as to ensure safe operation and meet the requirements of international seabed resource surveys in a new era, "Dayang 1" underwent its third repair and refit by Wuchang Shipbuilding Industry Group Co., Ltd. in December 2016. This refit involved the installation of internationally advanced ROV winches manufactured in Denmark and 12-ton folding arm cranes produced in Italy. Additionally, the main engine and corresponding auxiliary equipment were replaced. Besides, the bridge and control console underwent modifications, achieving full integration of navigation equipment, ship handling, and dynamic positioning. The refitted "Dayang 1" experienced a substantial enhancement of its overall performance, with significant improvements in noise reduction, ship information intelligence, and crew living facilities, among other aspects. It is expected to remain in service for approximately 15 more years.

2.9 Antarctic Research Icebreaker "Xuelong"

To provide stronger support for scientific research activities in the Antarctic regions, the SOA decided to build an Antarctic research icebreaker based on the marine environment of the Antarctic ice region. In 1990, the MARIC was commissioned by the SOA's Antarctic Office to conduct preliminary research and provide the feasibility report. After thorough research and analysis, the pre-research team proposed the SOA for both a construction plan of a new icebreaker or the option to refit the vessel "Xiangyanghong 10", considering the national financial situation and the shipbuilding cycle at that time. Later, the team learned about a new vessel under construction in Ukraine that was to be abandoned due to lack of funds for payment as the Soviet Union collapsed. The vessel had a total length of 167.0 m, a molded breadth of 22.6 m, a molded depth of 13.5 m, a draft of 9.0 m when fully loaded, a displacement of 21,025 tons, a maximum speed of 18 knots, and an endurance of 20,000 nautical miles. It was classified as a B1 icebreaker and could break ice continuously at a speed of 1.5 knots with ice thickness up to 1.1 m, including 0.2 m of snow. After analyzing the vessel's data, the pre-research team concluded that the overall performance, electromechanical equipment, and icebreaking capabilities of the vessel met the requirements for navigation in Antarctica. With appropriate refits, it could become a multipurpose vessel for scientific research in the Antarctic regions. Based on this, the SOA decided to purchase the vessel. In April 1993, the purchase succeeded and multiple refits were carried out as needed.

Key points of the refits are as follows:

To meet the needs of polar scientific investigations, the first refit took place in 1994, with a focus on refitting the first cargo hold. The cargo hold covers on the upper and lower decks were completely removed, and the upper and lower deck spaces were sealed. The strength of the sealed structure met the requirements for carrying rolling vehicles. The rolling cargo holds between the upper and lower decks were retained, and facilities for securing and fastening cargo were added. Four additional levels were built on the upper deck of the first cargo hold. Laboratories and survey equipment were installed on the upper deck, while a restaurant and living compartments were arranged on the forecastle deck. Two large laboratories, three sample processing rooms, one clean room, one data center, one sample library, and one breeding room were added, with a total laboratory area of approximately 200 square meters. The newly added laboratories, workrooms, living compartments, and public areas were all equipped with air conditioning systems. A fish finder, an acoustic Doppler current profiler, a CTD winch, and a 1,200-m biological winch were also installed. A freshwater tank with a capacity of 400 tons and a light diesel oil tank with a capacity of 600 tons for Antarctic station use were installed between the inner bottom and lower deck.

After the refit, the vessel was named "Xuelong" (Snow Loong) (see Fig. 10) based on the proposal by Wu Heng, who was among the pioneers and organizers of China's Antarctic exploration. "Loong" represents China, while "Snow" signifies the icy world of Antarctica.

Fig. 10 Antarctic research icebreaker "Xuelong"

After nearly a decade of operation, "Xuelong" no longer met the requirements for safe polar navigation. Another upgrade and refit was carried out in 2004. This refit was conducted in two phases, with the first phase taking place from 2004 to 2007, completed by the Shanghai Merchant Ship Design and Research Institute together with the Shanghai Shipyard. The second phase took place from 2012 to 2013, completed by the MARIC together with Jiangnan Shipyard. Although the project was carried out in two phases, under the effective planning and management of the shipowner, the COMRA, the two phases had clear goals and were planned in a reasonable sequence, making them highly complementary. This upgrade turned "Xuelong" into a pioneer among international first-class polar exploration icebreaker, fulfilling the goal of being "safe, advanced, and applicable".

After the upgrade, all equipment and facilities on the main deck of "Xuelong" were updated. The laboratory area onboard was expanded from over 200 square meters to 580 square meters, and all 126 laboratory equipment items were replaced. The upgraded "Xuelong" possessed an advanced navigation and positioning system and automatic steering system. It was equipped with an advanced communication system, a helicopter landing platform capable of accommodating two helicopters, a hangar, and supporting equipment. The vessel housed a center for processing atmospheric, hydrological, biological, and computer data, a meteorological analysis and forecast center, as well as laboratories for marine physics, marine chemistry, biology, geology, meteorology and cleanness analysis. Upgrades were also made to the fish finder used to explore krill and other aquatic animals in polar regions, an acoustic Doppler current profiler for measuring seawater flow speed and direction during navigation, and a wide range of advanced instruments for measuring seawater temperature, salinity, and depth (CTD).

After being refitted, "Xuelong" could accommodate up to 130 persons. The living compartments for the expedition team members were generally double rooms, with each room being 10 square meters in size. Each room was equipped with a central air conditioning system, 24-hour hot water supply, as well as other amenities such as a refrigerator, wardrobe, and writing desk. Additionally, there were internet ports available for accessing the internet and sending emails. "Xuelong" also featured various facilities for daily living, including academic lecture hall, post office, hospital, swimming pool, library, gym, sauna room, indoor basketball court, internet cafe, and laundry room.

The "Xuelong" research vessel was the largest polar research vessel in China at that time and the only vessel capable of continuously navigating through polar ice. Equipped with adjustable pitch propellers, it allowed for flexible maneuvering, particularly beneficial for icebreaking. The vessel's hull was made of E-class steel plates to ensure excellent performance even in the extreme cold climate of minus 40 degrees Celsius. It had the capacity to transport general cargo, large and heavy goods, various vehicles (with roll-on/roll-off capability), refrigerated goods, valuable items, explosives, minerals, standard containers, and various types of fuel.

Since its first Antarctic expedition and logistical resupply mission in October 1994 until 2018, "Xuelong" had completed a total of 34 voyages to the Antarctic and nine voyages to the Arctic, traversing all five oceans. These expeditions set numerous records in Chinese maritime history.

In July to September 1999, China organized its inaugural large-scale multipurpose scientific expedition to the Arctic. "Xuelong" embarked on its first journey to the Arctic, carrying 124 expedition members. The expedition lasted for 71 days, covering a distance of 14,180 nautical miles. In this expedition, various surveys were conducted to study the ocean, atmosphere, biology, geology, fisheries, ecological environment, and other aspects in the Arctic region.

In July 2003, the Chinese government organized the second scientific expedition to the Arctic. "Xuelong" carried 109 expedition members, piercing through the ice and reaching the North Pole at 80 degrees north latitude. This expedition lasted for 74 days, covering 12,600 nautical miles. The expedition team conducted multidisciplinary surveys on the ocean, atmosphere, sea ice, and biochemistry, among others. Advanced technologies such as underwater robots were utilized to enhance the understanding of the interaction between the Arctic's ocean, sea ice, and atmosphere.

In 2007, "Xuelong" conducted a multipurpose survey of the ice dome A region in Antarctica, which included studying ice cap typical sections, ice core drilling of the ice dome A region, geophysical exploration, and astronomical observation. The ice dome A region, known as the "uninhabitable extreme", is the highest part of the Antarctic ice cap, characterized by extremely harsh climate conditions.

This scientific expedition achieved several groundbreaking accomplishments, including the production of the first 1:50,000 topographic map of a 450-square-kilometer area within the ice dome A region, the first seismic observations in the inland of Antarctica, and the first optical astronomical experimental observations. Astronomers selected a site for an astronomical observatory in the ice dome A region. They planned to build an automated observation station and establish a massive array composed of hundreds of telescopes to conduct research on variable stars and dark matter in the universe. Ice core drilling was also carried out. Ice cores with complete ice caps were obtained from a depth of approximately 3,200 m and the existence of ancient ice dating back over 1.5 million years in the ice dome A region was checked.

Another crucial mission of this expedition was to conduct surveys within a 6,000-square-kilometer area surrounding the base camp in the ice dome A region. The goal was to determine the site and environmental parameters of the Antarctic inland establishing China's third Antarctic research station.

In March 2012, China conducted the 28th Antarctic expedition. "Xuelong" successfully crossed the "roaring forties" at 45 degrees south latitude. This marked the fourth crossing of the "roaring forties" during the expedition and was the most challenging one. The voyage was greatly affected by cyclones and lasted the longest compared to the previous three crossings. On the evening of the 6th, "Xuelong" encountered a cyclone shortly after departing from Prydz Bay, where the Zhongshan Station was located. This cyclone was caused by the southward movement of a cyclone in the westerlies and, under the combined influence of the cyclone and the polar continental high-pressure system, the sea winds and waves gradually intensified. From the daytime of the 7th until the evening of the 11th, "Xuelong" experienced severe turbulence and wobbling, continuously navigating northward amidst the sways and jolts. Given the maximum sway angle of the vessel reached 20 degrees during this period, large objects within the compartments were securely fastened, and anti-slip mats were placed on the tables to prevent them from sliding. Despite these precautions, with each significant sway, the sound of items falling from elevated surfaces and rolling back and forth on the floor could be heard from the compartments. Finally, on the evening of the 11th, "Xuelong" managed to traverse the "roaring forties". Throughout this Antarctic expedition, "Xuelong" completed a "one vessel, three stations" mission, crossing the "roaring forties", where strong winds and waves prevail all year round, four times, with a total distance reaching 28,000 nautical miles. The vessel demonstrated its resilience, validating the success of its refit.

2.10 *Multipurpose Maritime Practice and Research Platform "Dongfanghong 2"*

Ocean University of China is the cradle for nurturing marine science professionals in China. Throughout the process of talent development, the university has relied on its marine training survey vessel, the "Dongfanghong", which has played a significant role since the 1960s. As the "Dongfanghong" neared its decommissioning after nearly 30 years of service, the Ocean University of China recognized the need for a new vessel to take its place. With approval, the State Development Planning Commission, the Ministry of Finance, the Ministry of Education, and local authorities, collectively raised funds to construct a multipurpose oceanographic survey vessel for the university. Designed by the MARIC and constructed by Zhonghua Shipyard, the new vessel, named "Dongfanghong 2" (see Fig. 11), was delivered on December 28, 1995. "Dongfanghong 2" effectively met the diverse needs of marine disciplines at various levels in affiliated higher education institutions, serving as a distinctive platform for multipurpose maritime practice and research that aligns with China's national conditions. It was capable of conducting comprehensive observations, research, and student navigating training in multiple disciplines of marine science, including hydrology, meteorology, physics, chemistry, geology, biology, and geophysics, in both coastal waters and the open oceans. In addition, it could also undertake international collaborative projects for global marine research, provide theoretical and technical support for the development and utilization of marine resources in the ocean industry sector, facilitate marine education, and enhance public awareness of the ocean.

During the design phase, the team thoroughly researched the utilization of "Dongfanghong" and carefully considered the university's requirements while staying within the allocated shipbuilding budget. Through meticulous design and testing, "Dongfanghong 2" had significant improvements in main dimensions, seagoing qualities, operational capabilities, equipment and instrument configuration and other aspects, compared to the "Dongfanghong" vessel, which earned the vessel extensive praise from the users.

The vessel was equipped with 15 different types of laboratories to support comprehensive and specialized surveys in marine physics, marine meteorology, marine chemistry, marine biology, marine geology, and marine geophysics, as well as analysis and research work. Its design layout allows for interdisciplinary research, and

Fig. 11 Multipurpose maritime practice and research platform "Dongfanghong 2"

the navigation support equipment ensures secure and reliable operations, simple management, and easy maintenance, with low cost and high maneuverability.

The vessel was designed with a length of 96.0 m, a molded breadth of 15.0 m, a molded depth of 8.8 m, a draft of 5.0 m, and a displacement of 3,400 tons. It boasted a maximum speed of 18 knots when the wind is no stronger than Beaufort scale 3 at a designed draft of 5.0 m, along with an economic speed of 14 knots. Propelled by adjustable pitch propellers, the vessel offers seamless speed variation between 0.5 and 18 knots. Its endurance was set at 12,000 nautical miles at economic speed, and it could accommodate 196 persons with a self-sustaining period of 60 days.

Equipped with a long forecastle and an asymmetric layout on the upper deck, the vessel provided ample workspace for marine researches. Its bulb bow enhanced its wave resistance, while the low ship's rail at the stern could facilitate marine research operations.

Key configurations of the vessel:

(1) Main Equipment and Instruments

The vessel was equipped with a self-contained CTD, direct-reading CTD, shipborne phased array ADCP, sub-bottom profiler, shallow-water multi-beam system, deep-water multi-beam system, high-precision satellite differential global positioning system (DGPS), optical fiber ship attitude measurement instrument, conventional meteorological observation instrument, sound velocity profiler, alkalinity meter, and self-contained wave height gauge.

(2) Main Winches and Booms

The vessel was outfitted with various winches and hangers, including a 6,000-m hydro-hydraulic winch, 3,000-m bio-hydraulic winch, 1,300-m hydraulic winch, 2,500-m CTD hydraulic winch, 8,000-m CTD hydraulic winch, hydrological sampling boom, biological sampling boom, large and small gantry booms.

(3) Laboratories

The navigation deck was equipped with a weather room and a remote sensing room. The boat deck housed a navigating training room and a meteorological sounding room. The upper deck was equipped with a central computer station, a gravimeter room, and five general laboratories. These facilities could be used in dry, wet, or semi-wet conditions, depending on specific marine survey tasks. They provided an ideal platform for conducting oceanographic surveys and scientific research in hydrology, physics, chemistry, biology, and geology.

The vessel addressed nine main design challenges. The first was the determination of the vessel's tonnage. Considering the status and development trends of global multipurpose oceanographic survey vessels at that time, vessels of around 5,000 tons are considered ideal for such purposes. However, taking into account the initial investment, future maintenance costs and other economic considerations, the tonnage was preferred to be controlled below 3,500 tons. Through detailed analysis and calculations, under the premise of meeting the requirements of use, the final tonnage was determined at 3,400 tons, which was approved by the authorities. Second, to obtain the optimal main dimensions and form coefficient of the vessel, the design team collected data from 11 existing oceanographic survey vessels built between 1965 and 1989. After analyzing and comparing the data, the best design solution was selected, resulting in a favorable admiralty constant of 304 and good speed performance. Third, to improve the vessel's seakeeping capability and adaptability to sea conditions, in addition to overall design considerations, the vessel adopted an pneumatic anti-rolling tank device, which achieved a reduction of approximately 50% in rolling motion. The fourth was to improve the maneuverability at low speeds. Traditional single-propeller, single-rudder ships typically have poor maneuverability during low-speed maneuvers. Particularly when dealing with strong winds and rough seas which result in severe stalling, it can be challenging to make timely adjustments to the heading. Considering the high cost of employing a dynamic positioning system, a cost-effective and generally effective bow thruster installation solution was adopted after studying, which effectively addressed the issue of poor heading adjustments. Fifth, the structural design of the vessel must comply with the regulations stipulated in the *China Classification Society's Code for Classification and Construction of Steel Sea-Going Vessels* published in 1989, as well as the *Notification of Modification of Steel Ship Classification and Construction Code* issued in 1991. To enhance the safety of navigation in ice zones and meet the requirements for class B ice strengthening, the steel used for the outer plating in the ice-strengthened areas was upgraded to Grade D steel, which was one level higher than the code requirements. Sixth, in addition to the essential various systems, the vessel was equipped with the IBS-200 integrated navigation system. This system could provide real-time panoramic display of various navigation parameters on the central control display, such as track, route plan, and actual ship navigation conditions. Through calculations of various parameters, it could control the fully digital autopilot, which ensures optimal automatic tracking of the route, contributing to the safety and reliability of the vessel. Seventh, a twin-engine tandem single-shaft adjustable pitch propeller propulsion system was employed. This system offers flexibility and can meet the demands of various tasks. With both engines operating at full power, the vessel can achieve a speed of 18 knots, while a single engine can achieve a speed of 14 knots. The system allows for low-speed navigation during survey activities, optimizing propulsion efficiency and fuel economy. Additionally, the twin engines could serve as backup for each other to enhance the vessel's reliability. Eighth, network lines and interfaces were configured. The vessel's central computer station was equipped with network lines and interfaces, facilitating the internet connection of laboratory computers with the navigation system. This enables the sharing and real-time transmission and processing of survey data among different disciplines. Laboratories could access real-time positioning and navigation information through the network, such as latitude, longitude, speed, heading, water depth, wind speed, wind direction, compass heading, and other data, which can be recorded and replayed in real time. Ninth, a global maritime distress and safety system (GMDSS) compliant communication system was configured. The vessel was equipped with a very high frequency (VHF) telephone, single side band (SSB) radio, digital selective calling (DSC), satellite C station, and satellite B station. The Danish satellite B system could enable communication between the vessel and other vessels through satellite connectivity, as wells as direct communication between the vessel and the land, allowing for real-time transmission of information between land and sea via voice calls, telegrams, fax, and sending and receiving emails.

As of August 2011, over the course of 16 years since its delivery, the multipurpose oceanographic survey vessel "Dongfanghong 2" had completed more than 300,000 nautical miles of scientific research journeys, equivalent to circling the Earth's equator 14 times. Its routes covered the Bohai Sea, the Yellow Sea, the East China Sea, the South China Sea, and the Northwestern Pacific Ocean. The vessel has undertaken more than 100 comprehensive maritime research missions for national-level major research projects such as the "973" Program, "863" Program, and National Natural Science Major Foundation projects. It has involved more than 4,000 scientific researchers and collaborated extensively with over 30 domestic and international marine research institutions and academies.

In October 2016, "Dongfanghong 2" successfully completed a scientific research mission on deep-sea (10,000 m) exploration in the Mariana Trench. Through the use of self-developed marine instruments and equipment, valuable oceanographic data was obtained, filling gaps in various areas of marine research. This expedition marked a national oceangoing scientific research development-sharing voyage, guided by strategic missions and led by the National Laboratory for Marine Science and Technology. It signified the entry of the deep and far sea scientific research platform for the Laboratory into its substantive operational phase.

The "10,000-Meter Deep-sea Operation" program is a sub-task of the "Transparent Ocean" major strategic mission of the National Laboratory for Marine Science and Technology. It is equivalent to placing "human eyes" at a depth of 10,000 m in the ocean, driving a significant transformation in China's deep-sea technological innovation from "tracking international advanced levels" to "leading global deep-sea scientific innovation".

This voyage primarily focused on accomplishing three major scientific research tasks. First, it involved retrieving the internationally recognized first-ever Mariana Trench marine science comprehensive observation network, which was established by the Ocean Dynamic Processes and Climate Change Functional Laboratory of the National Laboratory for Marine Science and Technology. Second, it involved conducting systematic sea trials of various independently developed marine instruments and equipment such as the deep-sea (4,000-m) intelligent buoy, remote underwater vehicle, wave glider, deep-sea real-time self-contained high-definition camera, deep water (10,000-m) sampling device, and deep-sea (10,000-m) gravity sediment sampler. These trials aimed to gather technical data for the development of domestically-made deep-sea instrument and equipment products, promote the development of innovative deep-sea technologies in China, and break free from the passive state of relying on imported deep-sea instruments and equipment. Third, it involved conducting all-depth sea environment parameter measurements, large-volume water sampling, and deep-sea seabed sediment sampling along a major transect of the Mariana Trench. The scientific research personnel conducted in-depth studies on the germplasm resources, enzyme resources, and medicinal resources of deep-sea bacteria, filling the gaps in China's research on deep-sea microorganisms. These studies also provided valuable data, samples, and information for the study of the origin and evolution of deep-sea life, deep-sea carbon cycling, geobiochemical circulation, and deep-sea marine medicinal resources.

On April 22, 2017, "Dongfanghong 2" successfully completed a 20-day scientific expedition mission for the spring season debut of the national key research and develoopment program project "Biogeochemical Processes and Climate Effects of Biogenic Active Gases in the Eastern China Shelf Sea". This project marked China's first major basic research project to systematically study the non-carbon dioxide biogenic active gases in the ocean. Seventy-two cientists from six participating organizations, including Ocean University of China, Xiamen University, Shandong University, Fudan University, Chinese Academy of Meteorological Sciences, and Tianjin University of Science and Technology, joined the scientific expedition. This voyage closely focused on the research objectives and key scientific questions of the project. Multipurpose surveys across multiple disciplines were conducted, covering marine chemistry, marine biology, marine physics, marine sedimentology, and atmospheric environments, among others, with the East China Sea and the Southern Yellow Sea as research areas. Over 20,000 samples were collected, providing important data support for the study of the migration and transformation processes of biogenic active gases in different interfaces.

2.11 Research Vessel "Shiyan 3"

The research vessel "Shiyan 3" (see Fig. 12) serves as a multipurpose oceanographic survey vessel. In order to cater to the requirements of various marine research projects, "Shiyan 3" was developed in a modular direction and has become a multipurpose offshore experimental platform. By equipping different experimental devices for different voyages, the vessel can carry out marine research tasks in marine geology, hydrology, chemistry, biology, acoustics, sounding, and meteorology. The primary purpose of this vessel was to conduct marine research missions in the waters of the Nansha Islands. Built in 1979 at the Hudong Shipyard and delivered in 1981, it is owned by the South China Sea Institute of Oceanology, Chinese Academy of Sciences.

The vessel was designed with a length of 104.2 m, a molded breadth of 13.7 m, a draft of 4.95 m, a displacement of 3,243 tons, a navigational speed of 19.5 knots, an endurance of 9,000 nautical miles, and a self-sustaining period of 40 days. It can accommodate a total of 94 personnel, including 35 crew members and 59 scientific researchers. It has two main engines, each with 4,800 horsepower. Onboard, there are 14 laboratories with a total area of 230 square meters.

"Shiyan 3" was equipped with advanced navigation and positioning system, collision avoidance equipment, CTD (conductivity temperature depth) detection system, towed body system, rossette water sampling system, marine optical multi-parameter measurement instrument, polarograph, myriameter depth sounder, 956 direction

Fig. 12 Research vessel "Shiyan 3"

wave buoy, wave rider, plankton collector, underwater TV system, and benthic organism trawl for conducting multipurpose oceanographic surveys. The vessel was also equipped with auxiliary facilities, including a 3-ton crane, a 10-ton hydraulic gantry hanger, a 6,000-m geological winch, and an 8,000-m biological winch.

In addition, it was equipped with a large-scale observation system, capable of continuously and uninterruptedly measuring the water temperature, salinity, and depth in the navigated sea areas, with relevant data observed and recorded automatically.

Since its commissioning, "Shiyan 3" has successfully completed various multipurpose oceanographic surveys, such as the *Comprehensive Scientific Investigation and Research Report on the South China Sea Islands and Its Adjacent Sea Areas*, a national science and technology project during the "7th Five-Year Plan" to the "9th Five-Year Plan" period stated in *Zengmu Ansha: Comprehensive Investigation and Research Report on China's Southern Frontier*. It has also undertaken the marine survey tasks of the national 908 Program on "Comprehensive Investigation and Evaluation of China's Offshore Waters" as well as the "Open Voyage Scientific Expedition of the Northern Part of the South China Sea" task.

In July 1984, June 1985, and April 1986, the research vessel "Shiyan 3" conducted three marine scientific expeditions in the waters of the Nansha Islands respectively. These expeditions lasted for over 140 days, covered a distance of 12,000 km, and involved the sampling and observation at 167 stations. Special research was also carried out in the Zengmu Ansha sea area, yielding a vast amount of first-hand data on seabed topography, geomorphology, sediment, marine hydrometeorology, seawater optical properties, marine chemistry, and marine biology, among other aspects.

In March 1987, the State Council approved a multipurpose survey led by the Chinese Academy of Sciences, with the participation of relevant ministries and commissions, for a period of five years (1987–1991) in the Nansha Islands and its adjacent waters. This expedition was named the *Multipurpose Scientific Research in the South China Sea Islands and Its Adjacent Sea Areas.* From 1987 to 1990, "Shiyan 3" conducted four marine expeditions, yielding significant findings in the fields of territorial resources, ecology, environment, and maritime rights and interests in the Nansha Islands. Since 2004, the vessel has completed four open-voyage marine surveys, and in 2009, it embarked on its fifth open-voyage survey. On May 25, 2010, the vessel, owned by the South China Sea Institute of Oceanology of the Chinese Academy of Sciences, successfully returned to Guangzhou Xinzhou Pier after completing the multidisciplinary integrated voyage in the South China Sea for the year 2010, funded by the National Natural Science Foundation of China. The scientific objectives of this particular voyage were to study marine environmental issues in the South China Sea, gain insights into the regional impacts of ocean dynamics, environmental, and ecological processes in the context of global climate change, as well as the influence of natural and human activities on the South China Sea ecosystem. The intention was to provide scientific evidence to support the coordinated and sustainable development and decision-making in the region's economy and society.

In response to an invitation from the National Institute of Oceanography, Pakistan, the research vessel "Shiyan 3" embarked on a journey from Guangzhou on December 30, 2017. It sailed through the Strait of Malacca and reached the Makran Trench in the offshore waters of Pakistan on January 14, 2018, to conduct a multidisciplinary survey encompassing marine geology, physical oceanography, marine biology, and microbiology. This marked the first time that a Chinese research vessel had reached the Makran sea area. During the expedition, Chinese and Pakistani scientists obtained first-hand research data and samples in the target area, promoting scientific research of significant importance in the Makran sea area. This expedition enriched the experience of maritime cooperation between the two countries and laid a solid foundation for further promoting China-Pakistan cooperation in marine science and technology. After successfully completing the mission, "Shiyan 3" departed from Karachi, Pakistan, on February 4 and returned to China. On the vast expanse of the ocean, the crew and expedition members spent an unforgettable Chinese New Year at sea. Led by the South China Sea Institute of Oceanology, Chinese Academy of Sciences, the joint expedition involved more than 70 personnel from over 10 Chinese research institutes and the National Institute of Oceanography, Pakistan. The expedition focused on the study of large-scale geological structures in the Makran Trench and geological hazards such as earthquakes and tsunamis in the adjacent areas, with the aim of providing scientific basis for Pakistan's maritime security and disaster risk reduction while serving the construction of the Belt and Road Initiative. After a voyage of 12,230 nautical miles, "Shiyan 3" successfully completed the inaugural joint China-Pakistan expedition in the northern Indian Ocean and returned to Guangzhou on February 22, 2018.

On April 15, 2021, the research vessel "Shiyan 3" embarked on the East Indian Ocean Shared Voyage, funded by the National Natural Science Foundation of China for the year 2020. The voyage involved 34 scientific expedition members from various units, including the South China Sea Institute of Oceanology of the Chinese Academy of Sciences, Institute of Oceanology of the Chinese Academy of Sciences, Xiamen University, Tianjin University, Tianjin University of Science and Technology, Ocean University of China, China University of Geosciences, Chinese Academy of Meteorological Sciences, among other institutes. The survey primarily covered the waters of the East Indian Ocean, carrying out survey design of six sections and 103 large-section CTD stations. The expedition focused on hydro-meteorological observations, marine sediment sampling, biochemistry, atmospheric aerosols, and atmospheric electromagnetic wave surveys, involving multiple disciplines such as marine biology, marine physics, marine chemistry, and marine geology. A significant amount of field observation data and samples were collected. In particular, researchers conducted continuous 25-h observations in the special marine dynamic areas of the low-oxygen zone and the ridge in the Bay of Bengal. Repeat observations were also conducted in the ridge area during the first and second halves of the voyage. This voyage collected information on the characteristics of mesoscale oceanic dynamic processes in the ridge area of the East Indian Ocean during different time periods before and after the outbreak of the Indian Ocean monsoon, creating conditions for studying the unique phenomena such as biogeochemical processes that are influenced by the mesoscale features in this region.

The multipurpose research voyage to the East Indian Ocean was one of the shared voyages funded by the National Natural Science Foundation of China. Its aim was to deepen the understanding of the ocean and atmosphere processes through a multidisciplinary and interdisciplinary investigation in the tropical East Indian Ocean, as well as the understanding of the seasonal and intra-seasonal characteristics of the circulation in this sea area. Meanwhile, it also aimed at elucidating the cycling processes of key biogeochemical factors and the contributions of the circulation system to these processes, as well as revealing the geological features of the area, so as to advance the research in Indian Ocean sedimentology and pale-ocean environmentology.

Chapter 5
Scaling New Heights and Striding Towards the Forefront

1 Overview

In the early twenty-first century, China entered the third phase of research vessel development. Under the leadership of the Communist Party of China, the diligent, courageous, and wise Chinese people had been steadily building a well-off society. During this phase, the country's economic strength gradually increased, and its cultural and scientific undertakings continued to flourish.

Self-reliance is the cornerstone of China's position among the nations of the world, while independent innovation paves the way for reaching new technological heights. The development of research vessels follows this same philosophy. Among special vessels, research vessels stand out as an overall high-performance, representative, and technologically advanced type. Over the past 70 years, in the research and development of scientific research vessels in China, researchers have strengthened their research efforts to address the integrated, coordinated, practical, and exploratory design features of scientific research vessels. Recognizing that key technologies cannot be acquired through external means, they have dedicated themselves to mastering essential capabilities that form the foundation for the development and advancement of China's research vessels.

Since the 18th National Congress of the Communist Party of China, with General Secretary Xi Jinping at its core, the Party Central Committee launched a new era of socialism with Chinese characteristics, and the whole nation embarked on a new journey towards building a modern socialist country. On the one hand, for China, a major maritime nation, achieving modernization is closely linked to exploring, developing, and governing the oceans. On the other hand, human society, in the process of developing and utilizing marine resources, has often lacked planning and coordination, leading to individualistic and profit-driven activities. In addition, the lack of effective regulation and the unpredictable factors of nature have resulted in severe pollution of the oceans. This includes the annual leakage of thousands of tons of oil into the sea, forming marine pollution events known as "Kuroshio Current".

L. Han et al., *History of the Research and Development of Chinese Marine Research Vessels*, https://doi.org/10.1007/978-981-96-8414-4_5

River pollution, wastewater from fish and shrimp farms, petroleum waste, and other factors have triggered red tide events. Additionally, climate change, rising sea levels, and the formation of hurricanes from ocean storms pose significant threats to human safety. To address these challenges, research efforts need to comprehensively improve the quality of studies on marine hydrometeorology, marine ecology, marine chemistry, marine physics, and seabed sediment. As a responsible major country in the world, China not only needs to consider its own interests but also must strengthen and promote scientific research and surveys of the oceans from the perspective of a shared future for humanity. This brings greater responsibilities and missions to the marine scientific community, marine industry, and education sector. First, the construction of a marine power requires the development of marine science and the accelerated exploration of marine resources, which is the foundation of building a strong maritime nation. Second, further understanding the changes in marine ecological environment is essential. Third, it is necessary to promote deep-sea scientific research, which allows for strategic scientific investigations with a broader vision. In striving to fulfill this formidable task, China has proposed to develop a batch of high-grade, precision and advanced multipurpose or specialized scientific research vessels with distinctive characteristics, as long as the national financial capability allows. Simultaneously, China's specialized equipment for marine research has also undergone significant advances, enhancing research capabilities and setting new requirements for research vessel development. The competent national marine management authorities and industrial product quality management departments attached great significance to marine research and surveys. They established several certification systems specifically for qualifications, including certificates for marine geological surveys, laboratory testing, ocean surveying and mapping, and laboratory accreditation by the China National Accreditation Service for Conformity Assessment. This implies that newly developed research vessels not only need to be designed according to the relevant regulations of classification societies and numerous international conventions and rules but also need to meet the requirements of these certification systems. Ships must attain the relevant certificates and qualifications to engage in oceanographic survey activities.

The Chinese shipbuilding industry was brave in rising to this new challenge. Science and technology advance the nation, and strong science and technology lead to a strong country. The ability of a nation to innovate in science and technology fundamentally shapes its future and destiny. This holds true for the development of industries, including the research vessels. As a result, the government attached great importance to research vessel development and continuously set new requirements and indexes for this field. As humans entered the twenty-first century, China has become the world's leading shipbuilding nation, with a research, development and innovation capacity ranking among the best globally. Furthermore, its construction capabilities were sufficient to meet the needs for building various types of new research vessels. China has developed world-class research vessels that fully cater to the needs of the marine scientific community, the marine industry, and the education sector.

The research vessel "Haiyang 6", successfully developed in 2009, is China's first multipurpose survey vessel for natural gas hydrates (combustible ice). It has conducted surveys of natural gas hydrate resources in the South China Sea, laying the foundation for the subsequent development of semi-submersible platforms.

The delivery of the small waterline area twin-hull multipurpose research vessel "Shiyan 1" in 2009 marked the beginning of a new era in the development of scientific research and survey vessels.

The Multipurpose Oceanographic Research Vessel "Kexue" is a significant technological infrastructure project during China's "11th Five-Year Plan" period. It featured innovative design concepts that aligned with international trends in research vessel development, laying the foundation for the development of a new generation of multipurpose research vessels in China.

The multipurpose survey vessel "Dayang 1" serves as the core foundational equipment for the China Ocean Mineral Resources R&D Association's initiative of "conducting deep-sea exploration, developing deep-sea technologies, and establishing the deep-sea industry". It has significantly improved China's multipurpose survey capabilities, detection capabilities, and research levels for deep sea resources and environments.

The abyssal trench research vessel "Zhang Jian" serves as the mother ship for China's manned submersible capable of diving to depths of over 10,000 m. It has enabled China to realize the dream of "capturing turtles in the five oceans" in the field of marine science.

The construction of polar research icebreaker "Xuelong 2" is an important project of the "Two Loongs ("Xuelong and Xuelong 2") Exploring Polar Region" program during China's "13th Five-Year Plan" period. With a gross tonnage of 12,769 tons, it had an icebreaking capability of 1.5 m of ice plus 0.2 m of snow and allows for bidirectional icebreaking sailing in polar waters, making it a polar research icebreaker with dual-class certification from the CCS and Lloyd's Register.

In addition, more than 40 research vessels have been developed over the past 20 years, including the "Xiangyanghong" series, the "Haiyang Dizhi" series, and the "Haiyang" series. The designers of research vessels in China, driven by a strong sense of responsibility and scientific spirit, upheld a deep passion for the ocean and the mission they are entrusted with. They fearlessly embraced innovation during the development process, resulting in great breakthroughts in the quality of research vessels and propelling the industry to the forefront of the world.

The typical features of this batch of research vessels are that they were not only in line with international standards but also tailor-made for the practical needs of marine research in China, exhibiting innovative ship forms and advanced design concepts. From design capabilities and construction quality to ship performance and functionality, they have met and even surpassed the advanced levels of international research vessels.

In terms of main dimensions, the ship forms evolved from slender and elongated shapes to shorter and wider shapes to enhance stability and resistance to harsh sea conditions during navigation and scientific investigations. In this way, the transverse working area on the deck has been increased and the ship breadth has been broadened.

In addition, the superstructure floors have been increased, providing a larger available working deck area, improved visibility from the pilothouse. Besides, more living compartments have been relocated above the main deck, providing the crew with more comfortable living conditions.

The emergence of new technologies and equipment have brought about a significant transformation in design concepts. Notably, the introduction of full electric propulsion technology and all-direction propulsion system has played a pivotal role. "Zhongguo Haijian 83", built and delivered by Jiangnan Shipyard in August 2005, became the first Chinese marine surveillance vessel to employ full electric propulsion technology and all-direction podded propulsion system. With their superior performance, low noise, and ability to meet the special requirements of research vessels, the full electric propulsion technology and all-direction podded propulsion system began to be employed extensively in research vessels. As a result, electric propulsion technology became a standard configuration in modern Chinese research vessels.

Advanced technologies such as dynamic positioning, communication navigation, acoustic detection, and vibration and noise control have facilitated a leap forward in the operational and exploratory capabilities of 21st-century research vessels.

The integrated design of multi-beam anti-bubble interference is widely recognized as a key technology in research vessel hull design. Research vessels primarily serve marine research, where underwater detection capabilities are of utmost importance. Accordingly, various types of deep-sea acoustic equipment are commonly installed on both domestic and international research vessels. The hull shape of research vessels features a pointed and slim bow, a flat bottom, and narrow side lines. In addition, a bow thruster is typically fitted. Acoustic equipment mounted at the bottom of the bow, such as deep-water multi-beam equipment, shallow-water multi-beam equipment, and myriameter depth sounder, often necessitates a low-noise work environment. Thus, during the initial hull design, careful consideration needs to be given to the dimensions and installation methods of such acoustic equipment. The research vessel development team from the MARIC took the lead in tackling this issue. Through computational fluid dynamics analysis and evaluation and innovative ship model bubble release observation experiments, they repeatedly optimized the bow lines of the vessel. Simultaneously, they conducted comprehensive comparisons of transducer detection accuracy and focused on studying the bow lines with embedded installation, aiming to improve the economic efficiency of ship navigation while considering the effectiveness of acoustic equipment. The success of this optimized hull shape has been validated through physical verification tests on actual vessels.

The research vessels designed and built during this period generally possess integrated capabilities for multidisciplinary, multifunctional, and multitechnology surveys, allowing for synchronous investigations and research activities across various disciplines. For instance, the multipurpose research vessel "Xiangyanghong 3" can meet the interdisciplinary research needs of deep-sea marine studies, encompassing detection and experiment functions for marine dynamic environments, geological environments, ecological environments, submarine resources and energy sources. It supports deep-water exploration, atmospheric detection, seabed exploration, deep-sea extreme environment survey, remote sensing information field validation, and onboard test, featuring cutting-edge onboard scientific technologies such as onboard network. Moreover, it possesses the capability to conduct high-precision and long-term comprehensive marine observations and detection of ocean dynamics, marine geology, marine ecology, atmospheric environments, as well as the capability of truth-preserving sampling and field analysis.

During this period, new ship forms, such as energy-efficient ship types with smaller length-to-breadth ratios, multi-beam anti-bubble interference bulb bow ship types, and bow and stern bidirectional icebreaker types, emerged. A novel ship model test method for validating the drag reduction effect of the bow's bubble layer was introduced. The research vessels designed and built during this period exhibit the following features:

(1) Improved ship forms: To expand the working area on the deck, ship forms have transitioned from pointed and slender to fuller and larger forms, featuring smaller length-to-breadth ratios. This allows for larger deck operation and laboratory areas with smaller main dimensions and gross tonnages, thereby enhancing the living conditions of crew members.
(2) Controlled weight and center of gravity: Special programs are developed to control weight and center of gravity. The entire ship weight and center of gravity are calculated, dynamically evaluated, and timely predicted at critical stages of ship design and construction.
(3) Comfort-oriented designs: Wind resistance stability and capabilities are improved, and additional living facilities are incorporated to enhance living conditions, ensuring comfort for crew members and scientific researchers.
(4) Automated designs: Next-generation research vessels place greater emphasis on automation after achieving "unmanned engine room and single person navigation". With the advancement of electronic technology, research vessels are equipped with automatic navigation systems, which utilize satellite positioning and computer systems to achieve route automation, navigation automation, and collision avoidance automation.
(5) Reduced vibration and noise: Vibration and noise reduction in ship design have transitioned from former passive to present active noise reduction as the main approach, with passive noise reduction as a supplementary technique. The concept of vibration and noise reduction has been integrated throughout the entire process of ship design and construction. Special programs for forecasting vibration and noise are developed to guide targeted vibration and noise reduction measures.
(6) Application of new propeller technology: Full electric propulsion technology and all-direction propellers have been widely utilized. These devices feature compact structure, minimal space occupation, flexible maneuverability, minimal vibrations and noise, effectively satisfying the varying operational conditions of research vessels.

(7) Widespread adoption of dynamic positioning (DP) technology: DP-1 systems have become standard configurations, and a few research vessels are equipped with DP-2 systems.
(8) Computer networking: Newly constructed research vessels are networked, facilitating the sharing of system software and hardware resources, database information, and survey data exchange. The vessels are equipped with various sensors, enabling omnidirectional intelligent sensing, acquisition, exchange, and display of vessel data, significantly enhancing the level of ship automation.
(9) Modular construction: With survey projects becoming more diverse and unpredictable, modular laboratories are widely adopted to minimize the time required for instruments and equipment replacements and improve the efficiency of oceanographic surveys. Container laboratories can be equipped with different functional container modules based on different voyage requirements, allowing for the completion of surveys of various types.

2 Classical Research Vessels

2.1 Research Vessel "Kexue 3"

"Kexue 3" is a research vessel constructed by Wuchang Shipyard in 2005 (see Fig. 1). It is primarily used for comprehensive research and experiments in physical oceanography, marine geology, marine ecology and environmental sciences, and marine chemistry in China's offshore areas. The construction of the vessel began in July 2005. It was launched in February 2006, and delivered in July 2006. The vessel is managed and used by the Institute of Oceanology, Chinese Academy of Sciences.

This vessel was designed with a total length of 73.3 m, a molded breadth of 10.2 m, a molded depth of 4.6 m, a maximum draft of 3.4 m, a full-load displacement of 1,224

Fig. 1 Research vessel "Kexue 3"

tons, an economic speed of 14 knots, a maximum speed of 16 knots, an endurance of 5,000 nautical miles, a self-sustaining period of 30 days, and a capacity of 48 persons (30 scientists and 18 crew members). The power system consists of two B&W 6K45GF diesel engines, each with a power of 3,882 kilowatts. The vessel has a total laboratory area of 124.6 square meters, including a wet laboratory of 43.3 square meters, a general laboratory of 48 square meters, an analysis room of 16 square meters, and an information processing center of 17.3 square meters. Additionally, the vessel can also carry a 20-foot standard container laboratory, furnished with interfaces for water, electricity, network, and communication. Wiring holes are present between the laboratories and the aft deck working area. The deck offers a working area of 330 square meters (150 square meters on the fore deck and 180 square meters on the aft deck). The laboratories are equipped with network terminal, network interface, compass interface, 6,000-m depth sounder, freezer, drying oven, fume hood, and water tank.

The vessel's aft deck features an 8-ton A-frame, a 2-ton folding arm crane, a 1-ton telescopic crane (hydrological), a 1-ton inverted L-frame (hydrological), a 6,000-m geological winch, and a 3,000-m hydrological winch. On the fore deck, there is a 2,000-m hydrological winch and two booms, with all winch and boom operations controlled from the console.

On June 8, 2010, "Kexue 3" embarked on the 2010 East China Sea (Yangtze River Estuary) Marine Science Shared Voyage Mission. This voyage was one of the five pilot projects funded by the National Natural Science Foundation of China, titled "Marine Science Research Vessel Time and Fee Shared Voyage", initiated in 2009. Organized and implemented by the Institute of Oceanography, this voyage involved 18 national natural science foundation projects. The expedition conducted investigations for 46 sampling stations across 7 scientific survey sections within the Yangtze River Estuary and the East China Sea continental shelf. The expedition counted 25 experts from the Institute of Oceanography, Xiamen University, Ocean University of China, Nanjing University, and East China Normal University as participants.

In December 2010, "Kexue 3" departed from the Qingdao Maritime Bureau Wharf to venture into the Yellow Sea and the East China Sea, undertaking the autumn 2010 open voyage mission. The anticipated duration of this voyage was around 15 days, covering an approximate distance of round 1,800 nautical miles and involving a participation of 48 researchers for the scientific expedition. This voyage included survey activities such as marine hydrological observation, ocean currents and meteorological observation, marine biology and ecological chemistry observation, and marine geology observation. The survey region covered the offshore areas of the Yellow Sea and the East China Sea, between 31 degrees north latitude and 39 degrees north latitude. A total of 49 stations were established, consisting of 49 CTD measurement points, as well as 25 stations each for plankton trawl and benthic organism trawl.

On June 12, 2012, "Kexue 3" departed from the Qingdao Olympic Sailing Base Pier, embarking on the 2012 marine science shared voyage mission for the "East China Sea Scientific Expedition and Experimental Research" funded by the National Natural Science Foundation of China.

This expedition focused on the interaction between land and sea in the East China Sea, revolving around the key scientific question of the "Evolution of the Marine Environment and Ecosystem Response in the Yangtze River Estuary and Adjacent Continental Shelf". Centered on the principles of efficiency and versatility, the researchers conducted multipurpose surveys pertaining to the hydrodynamic environment, marine organisms and ecology, as well as water and sediment environment within the region. The aim was to explore the structure, function, and temporal-spatial evolvement rule of the marine ecosystem in the East China Sea, elucidate the impact of physical, chemical, and biological processes on the evolution of the marine environment, and investigate the response and feedback mechanisms of the ecosystem to these changes. The expedition sought to further enhance the understanding in the above-mentioned aspects within the East China Sea continental shelf under the context of global changes.

More than 30 members from seven research institutes and universities including Ocean University of China, Xiamen University, Tongji University, Jinan University, Liaocheng University, East China Sea Fisheries Research Institute of Chinese Academy of Fishery Sciences, and Institute of Oceanology of Chinese Academy of Sciences participated in the scientific expedition along with this voyage.

On December 2, 2019, "Kexue 3" was assigned to conduct the "Healthy Ocean" autumn joint voyage in the Southern Yellow Sea of the Center for Ocean Mega-Science, Chinese Academy of Sciences, successfully completed its scientific expedition mission and returned to the pier at the West Coast campus of the Institute of Oceanology, Chinese Academy of Sciences. This expedition lasted a total of 19 days and completed interdisciplinary investigations, including physical oceanography, marine chemistry, plankton, and marine biology in 7 sections with a total of 49 stations in the Southern Yellow Sea.

During the expedition, scientists measured various oceanographic parameters such as temperature, salinity, and nutrients at different stations and depths, which hold guiding significance for studying changes in the marine ecological environment. Valuable samples of benthic organisms and shellfish were obtained through box coring and Agassiz trawling operations, playing a vital role in studying the biodiversity and migration patterns of marine benthic organisms. Real-time atmospheric and ocean current parameters were obtained using onboard automatic weather stations and other equipment, enabling three-dimensional ocean observations and providing a solid data foundation for continuous and in-depth research in marine science. The expedition was organized by the CAS Institute of Oceanology, and involved a team of 21 scientists from the Institute, CAS Yantai Institute of Coastal Zone Research, CAS Nanjing Institute of Soil Science, Peking University, Shandong University, and other institutions.

On July 5, 2018, the two advanced research vessels, "Kexue" and "Kexue 3", successfully passed the temporary audit certification of the vessel's Safety Management System (SMS) by the CCS Qingdao Branch at the Qingdao home port, receiving temporary SMC certificates on site. Following the research vessels "Xiangyanghong 1" and "Xiangyanghong 18", managed by Qingdao Huayang, Qingdao has once again obtained SMC certificates for these two scientific survey vessels, opening a new chapter in the standardized management of Chinese scientific survey vessels.

The certification of the "Kexue" and "Kexue 3" has strengthened the fleet of voluntary compliance with the *International Safety Management Code* (ISM Code) for research vessels and public service ships. This laid a solid foundation for the scientific and standardized management of future research vessels, the continuous improvement and enhancement of ship safety management capacities and safe navigation, as well as the support of ocean development cause. It played a leading role in the audit and certification of the safety management systems for Chinese research vessels and public service ships.

2.2 Marine Petroleum Geophysical Vessel "Haiyang Shiyou 718"

The marine petroleum geophysical vessel "Haiyang Shiyou 718" (see Fig. 2) is a six-cable dual-seismic source deep-sea work geophysical vessel, refitted from the "Binhai 256" triple-purpose work ship built by Jinling Shipyard in July 2001. Shanghai Chengxi Shipyard Co., Ltd. began the refit project in March 2005 and completed it in December 2010. The vessel was then delivered to China Oilfield Services Limited (COSL).

The vessel was designed with a length of 78 m, a length between perpendiculars of 70.2 m, a molded breadth of 18 m, a molded depth of 7.4 m, a draft of 6.06 m, a navigational speed of 13.5 knots, and a gross tonnage of 5,082 tons.

After the refit, the hull was equipped with a three-level deck configuration, consisting of a seismic source deck, a cable deck, and a helicopter deck. The key

Fig. 2 Marine petroleum geophysical vessel "Haiyang Shiyou 718"

geophysical equipment on the vessel includes hydraulic system, high-pressure air pipeline system, cable oil system, and industrial TV camera surveillance system. All the geophysical equipment is installed and debugged in compliance with the specified requirements at one time. In addition, the seismic exploration equipment used on the vessel was of world-class quality. For instance, the mechanical equipment employed on the vessel, such as the deck cable LARS and the seismic source LARS, was supplied by ODIM, a Norwegian company. These systems exhibited advanced technological features such as high automation, fast deployment speed, and safety protection for the cable and seismic source systems as well as the operating personnel. High-power, high-pressure air compressor units were employed to drive wide and long array air guns to generate and release high-energy seismic waves. These waves, capable of penetrating seabed layers ranging from 6,000 m to 9,000 m, could be received by the multichannel detection cables released by the geophysical vessel. The vessel is capable of towing six 5,000-m long streamer cables and can conduct 3D seismic acquisition operations in various seas around the world. It is considered the most powerful 3D seismic vessel in terms of operational capability in the Asia region. On December 27, 2010, the refitting work of the vessel was completed, injecting new energy into the development of the Geophysical Exploration Division of the company. After being put into operation, the annual 3D seismic acquisition capability of China Oilfield Services Limited (COSL) would increase by 60%.

The seismic acquisition operation in Block 53/30 of the South China Sea represents COSL's largest single-block deep-sea 3D seismic acquisition project. Since its initiation, the "Haiyang Shiyou 718" geophysical vessel has overcome adverse factors such as harsh sea conditions, significant fishing impacts, and short operation survey lines of work areas, and meanwhile strictly adhered o Chevron's high standards of safety and quality management for the project. Through close communication and collaboration among the various units within the fleet, along with the implementation of a spate of measures and rigorous requirements, the vessel successfully completed Chevron's 3D seismic acquisition task in Block 53/30 of the South China Sea in August 2011. This achievement further enhanced the competitiveness of COSL and laid a solid foundation for expanding its presence in the international geophysical market.

When operating in the Enping work area of Shenzhen Branch of China National Offshore Oil Corporation (CNOOC), "Haiyang Shiyou 718" experienced a prolonged duration that needed to dock for refueling and replenishing food supplies based on fuel consumption. However, the normal procedure of replenishing at the dock would take an estimated total of 128 h (approximately 5.3 days), including cable launching and recovering, dock replenishment, and round-trip navigation. In order to save time and enhance efficiency, China Oilfield Services Limited (COSL), after discussions with Shenzhen Branch of CNOOC (China), decided to carry out offshore refueling. With the valuable assistance and close coordination of the "Nanhai 219", "Haiyang Shiyou 718" accomplished the task of refueling and replenishment in a berthed state. This marked the first time that refueling was carried out during the offshore multi-cable seismic operations in the South China Sea, and it was a resounding success. This refueling operation only took five hours, whereas normally, the process of launching and recovering a cable and an extender would take around 8 h, totaling 13 h. Consequently, the offshore refueling option saved a significant amount of time, approximately 115 h (about 4.8 days), effectively conserving precious operational time and maximizing operational efficiency, and generating substantial economic benefits, both direct and indirect. This experience contributed to future offshore refueling during multi-cable seismic operations.

Toward the end of 2015, "Haiyang Shiyou 718" successfully secured a bid via international tender, winning a contract to conduct 2D seismic acquisition operations for Shell in the waters of New Zealand. Shell was renowned for its stringent requirements concerning safety, environmental protection, and fleet management. Throughout the entire tendering process, the Chinese team meticulously prepared, focusing closely on the client's demands and operational requirements. As a result, the team won the bid for its extensive operational experience, complete operational plan, and outstanding team collaboration.

On December 15, 2015, "Haiyang Shiyou 718" arrived at New Plymouth Port, New Zealand. Upon arrival, the vessel immediately commenced the operations in full swing and successfully fulfilled the contracted task.

This marked the first collaboration between China Oilfield Services Limited (COSL) and Shell in international waters, playing a positive role in enhancing the competitive edge and reputation of COSL in the international market for its geophysical exploration service.

2.3 Oceangoing Teaching and Training Vessel "Yukun"

Dalian Maritime University (DMU), a comprehensive university of science and technology featuring shipping majors, is directly affiliated with the Ministry of Transport. Looking back on its past, DMU has gone through challenges and hardships. However, it has also accumulated a profound cultural heritage and achieved numerous brilliant accomplishments over the course of a century. As we step into the twenty-first century, which is full of uncertainties and rapid changes, we, carrying the weight of history and aspirations of our nation, are steadily progressing towards greater heights. The rapidly changing global landscape and the constant advancements in science and technology are presenting us with ever-increasing challenges. More importantly, the future of the oceans signifies not only markets and wealth, but also the key to the progress of our entire nation. In this critical period, it is clear that relying solely on our existing technologies, management methods, and systems is insufficient. As DMU, which dedicates to cultivate shipping elites, it once had oceangoing training vessels including the "Hongzhuan", "Yuhong", "Yuying", "Zheng He", and "Yulong", among others. However, only the "Yulong" is currently in service. Depending on just one training vessel, the "Yulong", has become increasingly unsustainable. With the development of higher education in maritime affairs, DMU's planned enrollment has been increasing year by year. In order to improve the quality of maritime education and cultivate more outstanding sailors for our country, the Ministry of Transport approved the construction of a new oceangoing teaching and training vessel for DMU in 2002.

DMU entrusted the Shanghai Merchant Ship Design and Research Institute (SDARI) with the design, and the Wuchang Shipyard with the construction of the oceangoing teaching and training vessel. The vessel was launched on November 15, 2006, and delivered for use in April 2008. It was named "Yukun" (see Fig. 3).

This vessel was designed with a length of 116 m, a molded breadth of 18 m, a molded depth of 8.35 m, a draft of 5.4 m, a displacement of 5,878.8 tons, a navigational speed of 16.7 knots, and an endurance of 10,000 nautical miles. It can accommodate 236 persons, including 40 crew members, and 196 teachers, apprentices, and trainees.

"Yukun" is a midship engine vessel with a single engine, single propeller, shaft generator, forward-inclined stem column, bulb bow, and transom stern. The vessel has two longitudinal decks, a long poop, and four-deck rooms. It is equipped with a MAN B&W 6S35MC diesel engine (MCP 4,440 kW), a variable pitch propeller, and a bow thruster.

The vessel has excellent seagoing qualities and safety performance. It is equipped with integrated piloting system, AUTO-level engine room monitoring system, adjustable pitch propeller, shaft generator, diesel generator set, bow thruster, and retractable anti-rolling fin. In addition to complying with regulatory requirements, ample margins are reserved for the vessel's performance in stability, subdivision, fire protection, lifesaving, and escape passage, among other aspects. The vessel is capable of worldwide navigation, and its innovative lifeboats can dive up to 60 m. Advanced testing instruments, such as radar wave tester and ship attitude tester, are equipped. A computer Local Area Network (LAN) system is configured to transmit electrical and video data from the vessel's machinery to the work areas, as well as to

Fig. 3 Oceangoing teaching and training vessel "Yukun"

the classrooms and laboratories, providing comprehensive real-time information and data for teaching experiments and marine research purposes. This vessel serves not only as a modern specialized vessel for maritime teaching and training but can also be utilized for ship trials and scientific research in the fields of traffic engineering, maritime engineering, and marine engineering. Each voyage can accommodate 196 students for onboard training. Since the vessel is devoid of cargo holds and dedicated solely to student training, it increases the practical operational space for students and ensures the quality of their training. "Yukun" not only serves maritime students at DMU but also serves as a base for ideological and moral education for counselors in higher education institutions in Liaoning Province. "Yukun" is mainly used for the cultivation of maritime professionals, scientific research and ship trials and tests in maritime disciplines, and the promotion of maritime scientific culture. Moreover, it is also used to participate in external exchanges on behalf of China's higher maritime institutions. The construction of this vessel fills a gap in the dedicated experimental teaching vessel used in China's maritime institutions of higher education, signifying a milestone in China's higher maritime education. Compared to similar teaching and training vessels around the world, "Yukun" is designed specifically for the training of maritime students, with the rational arrangement of various teaching, training, and research facilities, high-standard compartments, and a well-designed functional layout.

From July 9 to 12, 2008, "Yukun" was invited to participate in the celebration of the Maritime Day of China. The vessel berthed at the container terminal in Taicang Port, Jiangsu Province on the afternoon of July 9. On the morning of July 10, Taicang Municipal Party Committee and Government held a grand welcoming ceremony, during which the captain of "Yukun" exchanged commemorative gifts with the two offices. In the afternoon of July 10, leaders from Association for Relations Across the Taiwan Straits (ARATS), including the President, inspected the vessel. On July 11, "Yukun" participated in the celebration activities held at the Taicang Port branch venue for the celebration of the Maritime Day of China. Through live broadcasting, the scenes at the main venue and the branch venue were displayed on large screens, creating a lively and spectacular atmosphere. The immaculate and advanced "Yukun" in festive attire, together with the energetic teachers and students, received an inspection and high praise from central leaders, attending officials, and guests at the celebration. During the participation in the celebration activities, "Yukun" was interviewed several times by Suzhou TV and Taicang TV on different channels. The teachers and students of "Yukun" also had exchange with the crew members of the "Nanhai Rescue 101" vessel, the "Xin Suzhou" vessel, and the Taiwan wooden vessel "Zheng He 1". A reception was held at the Zheng He Park, where they commemorated the great achievements of the legendary seven voyages of Zheng He to the Western seas.

On September 2, 2008, a delegation representing Chinese maritime institutions, including "Yukun" of Dalian Maritime University (DMU), set sail from Dalian Port for a ten-day exchange visit to South Korea. This marked the first overseas visit of dedicated teaching and training vessels from Chinese maritime institutions. Hosted by the Ministry of Transport, the event delegation mainly consisted of students and teachers from maritime majors at DMU, Shanghai Maritime University, Wuhan

University of Technology, Jimei University, and Qingdao Ocean Shipping Mariners College, as well as representatives from various maritime bureaus, port and shipping companies, and other relevant institutions, totaling 130 persons. During the visit, "Yukun" made stops at Busan Port and Mokpo Port in South Korea, where the delegation visited the Korea Maritime & Ocean University and Mokpo National Maritime University. The delegation also participated in cultural, sports, and maritime skills exchange activities with students and teachers from the two universities, who were also invited to visit the vessel. The delegation acted as envoys for Sino-Korean maritime cultural exchanges, showcasing the excellence of Chinese maritime education, fostering friendship between the people of China and Korea, and promoting exchanges and collaborations between maritime institutions in both countries, thus making a positive contribution to the development of maritime education in China.

With over 100 years of history, DMU is recognized as a world-class maritime institution. Each year, the university produces more than 1,000 maritime graduates who are sought after by shipping industries worldwide. In comparison, National Taiwan Ocean University (NTOU), a similar institution in Taiwan, only graduates 100 maritime students annually. As a result, Taiwan has a strong demand for maritime graduates to serve in their shipping industry. In recent years, there has been an increase in Taiwanese students expressing interest in studying at DMU. To further expand cooperation between the two sides, DMU accepted the invitation of the presidents of NTOU and National Kaohsiung Marine University (NKMU). With the approval of relevant authorities, they decided to visit these two higher education institutions in Taiwan on their oceangoing teaching and training vessel. The purpose of this visit was to showcase the strength and accomplishments of DMU and its teachers and students. Moreover, the two sides engaged in extensive and in-depth discussions on further strengthening inter-school cooperation and exchange through the exchange of students, consolidating the direct "three links" – trade, transport and postal services – between the two sides. In May 2009, "Yukun" set sail from Dalian Port to Taiwan, embarking on a 7-day exchange visit to the two higher education institutions. The visiting delegation consisted of over 100 teachers, students, and staff members from DMU. This marked the first visit of an inland oceangoing teaching and training vessel to Taiwan, ushering in a new era of exchanges and cooperation in maritime education across the Taiwan Strait. During the visit, the "Yukun" delegation visited NTOU and NKMU, extended invitations for students and teachers from the two universities to visit the vessel, and organized events such as inter-school basketball games and receptions. Additionally, DMU and NTOU signed a memorandum of cooperation, which includes visits by scholars and students from both universities, academic exchanges on research papers and academic monographs, organization of academic seminars and specialized research lectures, as well as collaborations and exchanges in scientific research between the two institutions.

On June 20, 2012, "Yukun" was officially designated as the "Maritime Science Popularization Education Base" by the China Institute of Navigation. As an important platform and carrier for providing maritime science popularization and education services to the public, "Yukun" fulfilled its role as a non-profit maritime science popularization and science education infrastructure. Combining its own strengths and leveraging its professional expertise and ship-based uniqueness, it fully implemented the relevant tasks outlined in the *Outline of National Scientific Literacy* in line with the *Law on Popularization of Science and Technology* and the *Law on Progress of Science and Technology* under the guidance of the China Institute of Navigation. It has been dedicated to building a unique and high-level maritime science education platform, making significant contributions to this field.

In addition to fulfilling practical teaching tasks, "Yukun" also engaged in numerous international exchange activities with maritime institutions in South Korea, the Philippines, Vietnam, Singapore, Russia, Malaysia, Japan and other countries. This played a crucial role in facilitating China's maritime exchange and cooperation, and fostering an international mindset and vision of Chinese maritime talents. From May 14 to June 6, 2013, "Yukun" completed its visit to Malaysia. During this visit, the "Marine Petroleum Hydrocarbon Pollution Investigation and Research" team from the College of Environmental Sciences and Engineering of DMU, along with onboard teachers and students, conducted a summer comprehensive pollution investigation voyage along the Dalian-Sanya-Malaysia Port Klang route from May 14 to 23. They conducted analysis on the collected seawater samples, measured the content of petroleum hydrocarbons in the seawater, and assessed the ecological risks to the marine environment based on the current pollution situation.

2.4 Small-Waterline Twin-Hulled Multipurpose Research Vessel "Shiyan 1"

"Shiyan 1" (see Fig. 4), delivered to the Chinese Academy of Sciences in 2009, is a new type of multipurpose research vessel. It is the first large-scale (over 2,000 tons) small-waterline twin-hulled vessel in China and the country's only small-waterline twin-hulled multipurpose research vessel at that time.

In January 2002, with approval from the Chinese Academy of Sciences, "Shiyan 1" was designed by the China Ship Scientific Research Center and built by Bohai Shipbuilding Heavy Industry. The vessel was delivered in April 2009.

The vessel was designed with a total length of 60 m, a molded breadth of 26 m, a mold depth of 10.5 m, a draft of 6.5 m, a full-load displacement of 2,560 tons, a maximum speed of 15 knots, an endurance of 8,000 nautical miles, a self-sustaining period of 40 days, and a capacity of 72 persons (including 27 crew members and 45 researchers), supporting normal operation under sea conditions up to grade 6.

The main hull of "Shiyan 1" is made of steel with a fully welded structure. It is propelled by a frequency control electric propulsion system, featuring twin engines and twin propellers. It adopts a small-waterline twin-hulled ship form with a double-deck continuous deck. The right hull's bow is equipped with a side thruster, and both the inside bows and sterns of the left and right hulls are equipped with a pair of bow fins and stern fins.

Fig. 4 Small-waterline twin-hulled multipurpose research vessel "Shiyan 1"

Design features of the vessel:

The vessel is equipped with ship-wide vibration and noise reduction facilities. It has advanced features such as full automation and dynamic positioning, enabling it to meet the needs of multipurpose surveys in multiple disciplines and cross disciplines in offshore and ocean areas, including acoustics, marine physics, marine biology, marine chemistry, marine geology, and marine atmospheric environment. The vessel is also capable of deploying, observing, controlling, and monitoring large-scale and wide-ranging observation networks. It can be used for studying real-time, three-dimensional marine environment monitoring systems and integrated information systems. Additionally, it can support comprehensive experimental research on underwater robot at sea.

The vessel is equipped with 11 laboratories that operate under networked and automated management. The scientific data obtained in these laboratories can be transmitted in real time to the onshore experimental center via satellite communication.

The design and construction of "Shiyan 1" faced high requirements and great challenges. Thanks to its unique small-waterline twin-hulled ship form, an excellent seakeeping and stability are ensured. This contributes to its capability to navigate normally in Force 11 winds. Even under adverse sea conditions, the vessel can maintain excellent stability without significant pitching and rolling motions, playing a leading role in the domestic ship design and construction industry. With its numerous features tailored to marine research, such as excellent seakeeping capability, low noise, good maneuverability, superior dynamic positioning, spacious laboratory and deck spaces, as well as good economics, the vessel became one of the most advanced multipurpose research vessels in China at that time. It served as an important platform supporting the multipurpose surveys in disciplines, such as acoustics and oceanography, and interdisciplinary studies in offshore and ocean areas.

In April 2010, "Shiyan 1" made its maiden voyage to the Indian Ocean to conduct a scientific expedition mission. This voyage lasted for 48 days and covered a distance of 7,900 nautical miles. The main task of the expedition is to conduct surveys in the Indian Ocean and the western regions of the South China Sea. This expedition yielded two significant results:

(1) Comprehensive oceanographic data were collected from the Bay of Bengal to the eastern equatorial Indian Ocean. By combining these data with the accumulated data from the South China Sea, researchers were able to conduct systematic scientific research and explore in-depth the relationship between the South China Sea and its surrounding sea areas.
(2) Dynamic and static observations were carried out for marine physics, marine ecology, marine geology and marine meteorology. This provided a preliminary understanding of regional oceanography in the tropical Indian Ocean, laying the foundation for future marine investigations.

In March 2017, the vessel was deployed to the South China Sea to undertake the 2017 Comprehensive Scientific Expedition Voyage of the South China Sea Central Basin, funded by the National Natural Science Foundation of China. Lasting 44 days and covering a distance of approximately 4,000 nautical miles, the expedition involved 70 observation stations and over 40 researchers from institutions such as the South China Sea Institute of Oceanology of the Chinese Academy of Sciences, Guangzhou Institute of Geochemistry, National Space Science Center, Second Institute of Oceanography of the Ministry of Natural Resources, Xiamen University, Ocean University of China, and Tianjin University.

The deep-water observations in the central basin of the South China Sea this time mainly encompassed marine hydrological observations, ocean currents and meteorological observations, investigations into marine biology, ecology, and chemistry, and the collection of marine sediments. These efforts covered disciplines such as marine physics, marine chemistry, marine ecology, marine environment, and marine geology. The data collected provided a scientific basis for a deeper understanding of the deep-sea dynamic, biochemical, and geophysical processes in the central basin of the South China Sea.

2.5 Geophysical Exploration Vessel "Haiyang Shiyou 720"

Geophysical exploration is the first step in oil and gas exploration. 3D seismic is considered the "eyes" and "ears" of exploration scientists in their search for oil and gas. It provides detailed geological research data, effectively increasing the success rate and reducing the risks of exploration, especially in deep-water exploration. In the vast South China Sea, there are abundant oil and gas resources, accounting for one-third of China's oil and gas reserves, with 70% located in deep-water areas. Due to the lack of deep-water equipment, the development of offshore oil and gas fields in China is generally limited to water depths below 300 m, and the development of deep-water areas greater than 300 m is still in its infancy. Good tools are prerequisite to the successful execution of a job. To overcome the obstacles in deep-water exploration and development, it is necessary to build a deep-water geophysical vessel.

During geophysical explorations, the geophysical vessel uses shipborne artificial seismic sources and signal receiving cables towed at the stern to receive the data of multiple subsurface vertical planes or cross-sections beneath the seabed during transmission and reflection of seismic waves. This forms a three-dimensional representation, consisting of numerous continuous slices or cross-sections. The precise position of the equipment (seismic source "air gun array" and signal cable internal detector) towed by the geophysical vessel must be determined using an offshore integrated navigation and positioning survey system. Its role is to guide the geophysical vessel's direction, determine the vessel's position, measure the position of the seismic source and the orientation of the cable internal detector, control the firing and reception in real time, and provide information on changes in the seabed terrain. By completing thousands of complex and accurate position calculations within seconds, seismic data can be collected in a complete and real-time manner.

The 12-cable geophysical exploration vessel is one of the core equipment for China's offshore oil and gas development during the "11th Five-Year Plan" period. It is included in the Science and Technology Major Project "Large-scale Oil and Gas Field and Coalbed Gas Development". The MARIC undertook the sub-project "Design and Research of 12-Cable Geophysical Vessel" of this major project. Concurrently, they also undertook the research project "High-tech Ship Scientific Research Project - Autonomous Development of High-performance Geophysical Vessels".

China Oilfield Services Limited (COSL) commissioned Skipsteknisk, a Norwegian design company, for the basic design of the vessel. In September 2008, COSL entrusted the MARIC to conduct the detailed design. The contract clearly stated that the MARIC would take full technical responsibility for the design proposed by Skipsteknisk.

The design team of the MARIC, with the guidance of experienced designers serving as technical advisors, drawing upon the technical reserves from two national-level geophysical vessel research projects and leveraging the rich expertise of the shipowner in geophysical exploration, successfully completed the design work for the vessel with full support and cooperation from the shipyard. Based on the completion of various model trials and comparisons, the optimization of general arrangement and structural forms, the adjustment in weight and center of gravity, the selection of major electromechanical equipment and geophysical equipment, as well as a thorough analysis of the design scheme, each speciality proposed over a total of 300 technical optimization measures for the basic design provided by Skipsteknisk. The strengthened detailed design not only met the underwater noise and towing force indexes of geophysical vessel, but also complied with the latest regulations from CCS (2009) in areas such as stability, Clean mark, emergency towing, ballast tank coating protection, ballast water treatment and management plan, and the comfort of compartments.

The vessel has a total length of 107.4 m, a molded breadth of 24.0 m, a maximum breadth of 28.0 m, a molded depth of 9.60 m, a designed draft of 7.50 m, a gross tonnage of 12,797, and a capacity of 75 persons. It is equipped with four main engines, each with a power of 4,320 kilowatts, alongside a speed of 750 revolutions per minute. It also has two propulsion motors with a power of 4,500 kilowatts each, a bow thruster with a power of 1,200 kilowatts, and a telescopic all-direction propeller at the stern with a power of 1,500 kilowatts. The helicopter deck has a diameter of 22 m and can accommodate helicopters with a maximum takeoff weight of 12.8 tons. The vessel incorporates a controlled passive anti-rolling tank.

The vessel was constructed at the Chongming Base of Shanghai Shipyard and was delivered on April 22, 2011. It was named "Haiyang Shiyou 720" (see Fig. 5), and owned by COSL.

The geophysical vessel is mainly used to create a scientific and reasonable space for the use, maintenance, and storage of 12 cables and associated geophysical equipment while meeting the latest safety and environmental regulations set by CCS. It also provides a low-noise and low-vibration working environment, ensures excellent towing capability (solutions for normal towing under harsh sea conditions), precise navigation and positioning, excellent endurance, and comfortable living conditions for long-term offshore operating personnel. Additionally, the vessel addresses fuel consumption reduction challenges during different operating conditions for shipping and seismic operations.

Fig. 5 Geophysical exploration vessel "Haiyang Shiyou 720"

The vessel is equipped with a dual-source air gun array. It also features three air compressors for the exhaust gun array, each rated at 3,000 psi[1] and a flow rate of 51 m^3/min. It is also equipped with five duplex gun winches for the seismic source winches, each with a drum diameter of 4,200 mm, and six duplex cable winches for the 12-cable 8,000-m cable winches, each with a drum diameter of 5,600 mm.

"Haiyang Shiyou 720" is one of the "Six Vessels in Five Types" used for deep-sea oil and gas development by COSL, boasting the first and latest generation of 3D seismic geophysical vessel in Asia. The vessel, also known as the 12-cable 3D geophysical vessel, is the first large-scale deep-water geophysical vessel designed and built domestically in compliance with the Performance Standard for Protective Coatings (PSPC) standards. As a 12-cable dual-seismic source large-scale geophysical vessel, it is driven by a diesel-electric propulsion system, supporting non-restricted waters navigation. Design features of the vessel:

1. Optimization of the Basic Design Provided by Skipsteknisk

The design team of the MARIC, in accordance with the established technical roadmap and the requirements of the user, carefully studied and digested the foreign drawings, proposed modification plans, and obtained the approval of CCS and the user. This created the conditions and paved the way for detailed design.

1) Optimization of the Basic Design

(1) Based on the basic design provided by Skipsteknisk, the arrangement of the living compartment areas was reconfigured, the interior layout of instrument compartments was optimized, the seismic source duty room was adjusted, additional offices were added for the marine engine and deck departments, the number and placement of laundry rooms were adjusted, and the layout of the kitchen and cold storage was optimized. The detailed design by the MARIC fully met the safety requirements of the regulations for the vessel, and the use function requirements for the vessel. It incorporated necessary equipment and systems, enabling a reasonable passageway width, a shortened passageway overall length, and an improved compartment space utilization. The interior layout was neat and orderly, making it align more with Chinese culture, living, and working characteristics.

(2) The angle of the stern shaft bracket was adjusted, reducing local eddies and lowering resistance. At the same time, it prevented the propeller from inhaling the bubbles generated from the root of the shaft bracket. According to the estimation through Computational Fluid Dynamics (CFD) analysis, the adjustment resulted in a decrease in resistance and a slight increase in navigational speed.

(3) The two model towing trials for the project demonstrated that the vessel achieves a towing force of over 108 tons at a draft of 6.5 m and a speed of 5 knots. This meets the requirement of a towing force greater than 100 tons under similar conditions. The actual ship mooring force test conducted at

[1] Unit of pressure, 1 pound per square inch = 6894.757 Pa.

the shipyard under corresponding test conditions also achieved satisfactory results.

(4) The capacity of the fuel and freshwater tanks was increased, resulting in an improved endurance of the vessel.

(5) The passive anti-rolling tank was upgraded to a controllable passive anti-rolling tank, improving its anti-rolling performance and expanding its capability to adapt to various sea conditions.

2) Design according to the Latest Regulations

(1) The vessel was modified to comply with the new *Special Purpose Ship Safety Rules* (SPS, 2008) instead of the previous compliance with the International Maritime Organization's (IMO) A534 (13) requirements for special purpose ships. This change includes modifications in damaged stability calculation, as well as updates in fire protection and life-saving requirements.

(2) The vessel was modified to meet the new edition of the *International Convention for the Safety of Life at Sea* (SOLAS, 2009) instead of the previous compliance with the SOLAS (2004) edition. Thus, for the irregular double bottom structure, additional damaged calculations were included. Individual liquid tanks were adjusted to avoid any unreasonable subdivision, while also simplifying the damaged calculations.

(3) CCS has special requirements for the use of buoyancy blocks, requiring a need of calculation of departure and arrival conditions for ice loads. In the detailed design process, adjustments were made to ensure compliance with CCS requirements for buoyancy blocks and calculation for ice loads. Additionally, actual values were adopted for assessing for the free surface of the anti-rolling tank after discussions with CCS.

3) Weight Control

Weight control was a crucial aspect of the vessel design. The basic design provided by Skipsteknisk estimated the empty weight of the vessel to be 6,750 tons. At the beginning of the detailed design phase, the design team closely tracked the weight control and asked for weight updates on an average monthly basis. The initial estimation of the weight was around 7,500 tons for the first version. In the second version, it was adjusted to approximately 7,400 tons. However, during the construction process, the feedback on weight data was not ideal, and at one point, it reached an unexpectedly high value of 7,873 tons. The following measures were taken to determine the true weight:

(1) Ensured consistency between weight breakdowns from the shipyard and the MARIC.

(2) Calculated and verified each item multiple times. It was recommend that the user and the shipyard adhere to the design requirements of the MARIC during construction, minimizing the use of alternative or additional unnecessary materials.

(3) Performed calculations and estimates for the extreme weight and center of gravity, with contingency measures prepared for worst-case scenarios. Upon

verification, it was found that the weight breakdown for piping systems and cable systems at the shipyard was overestimated by 200–300 tons. It was also discovered that the shipyard had a higher factor of safety for certain weight breakdown items.

The two preliminary inclining tests conducted before the vessel's launch and delivery verified the accuracy of the initial estimations by the MARIC. The first inclining test result before launching (December 20, 2010) indicated an empty weight of 7,530 tons, while the second inclining test result before trial voyage (April 10, 2011) confirmed an empty weight of 7,379 tons. The weight of the vessel was effectively controlled.

2. Main Dimensions of the Hull and Arrangement of Geophysical Equipment System

   1) Analysis of Main Dimensions of the Hull

      The characteristics of main dimensions of geophysical vessels present in being short, wide, and tall. The arrangement of the geophysical equipment system plays a crucial role in determining the main dimensions, operational performance, and development success of the vessel. Statistical data shows that the *L/B* ratio (length-to-breadth ratio) of geophysical vessels with 8 or more cables ranges from 3 to 4.2. Whether to increase the ship breadth is determined by the spacing required for the arrangement of unit-type cables. As the number of cables increases, so does the breadth. Therefore, the *L/B* ratio for this vessel was set at 4.025.

      In order to minimize the negative effects of increased breadth on resistance while ensuring stability, the breadth of multi-cable geophysical vessels often exceeds the waterline width above the main deck. This design reduces resistance at the waterline during navigation. The breadth above the main deck of this vessel was 24.0 m, and the maximum breadth at the stern was increased to 28 m, which was the result after optimizing the general arrangement.

      The choice of maximum draft should consider ensuring the immersion depth of the propeller in adverse sea conditions, so as to ensure that the propeller can provide sufficient towing force. An outwardly protruding design of the panoramic pilothouse on both the larboard and starboard is more advantageous for operational observation.

   2) Geophysical Equipment System

      The geophysical equipment system typically consists of a seismic source equipment system, a cable equipment system, and an integrated navigation and positioning and data post-processing system. Each component has different technical requirements. The seismic source equipment system requires sufficient air compressor capacity and a sufficient number of air gun arrays. Hence, it has specific requirements for vibration isolation, hold capacity, spare parts, and operating space. For cable equipment system, in addition to the requirements for equipment, operating space, and storage of supporting equipment, the number and placement of cables are the most important factors influencing the maximum breadth of the vessel (B_{max}) and the windward area. The geophysical machinery room and instrument room where the shipborne equipment for the integrated navigation and positioning and data post-processing system are located have strict requirements for temperature control, ventilation, low noise, low vibration, spaciousness, and convenient communication to ensure the normal operation of equipment, computers, and software.

3. General Arrangement

   One of the solutions is to establish a geophysical operation area, which involves the following equipment:

   (1) A dedicated seismic source deck with sufficient length is installed to accommodate the seismic sources (i.e., air gun arrays). Dual seismic sources typically consist of two sets of air gun arrays, with each set including three to four rows. The air compressor compartment is ideally located beneath the main deck, far from the navigation equipment area and living compartments.
   (2) A dedicated cable deck is installed to reduce the deck's breadth. According to statistical data, the average width of individual cable on a 3D geophysical vessel ranges from 1.88 to 2.3 m per cable (2.0 m per cable for this vessel). Under the premise that stability is ensured, the individual cable width affects the breadth of the vessel, and the breadth directly impacts the vessel's resistance, speed, and stability.
   (3) The navigation and data processing system instrument room (i.e., the geophysical machinery room) is arranged keeping off the top area of the generator compartment. It also needs to take into account the design for vibration and noise reduction and installation of independent air conditioning units.
   (4) The main functions of the geophysical equipment are closely related to the space where the geophysical equipment system is located. Therefore, the geophysical functions are better to be arranged in the following six different deck areas: navigation functions are arranged in the pilot and compass deck area; the deployment and storage of the crane and extender in the aft area of the helicopter landing platform deck; cable winches and associated equipment in the cable deck area; air gun arrays in the seismic source deck area; the geophysical core computer data processing center in the front instrument compartment area of the seismic source deck; the air compressor compartment in the aft area of the lower deck.
   (5) 3D geophysical vessels operate with their unique principles and methods at sea. The arrangement of the geophysical area is one of the key design considerations. The contradiction between arrangement and sailing resistance can be resolved by increasing the maximum vessel breadth (B_{max}) at the stern deck. It is effective to address issues posed by structural weight, stability, control of underwater radiation noise, and air noise through employing a

double-hull design below the main deck, optimizing the local structure in the geophysical area, and enhancing damping and vibration reduction treatments in the seismic source equipment area. It is required to minimize circuit harmonic interference, reduced vibration and noise, maintain a lower and consistent engine room temperature, and ensure the normal operation of advanced shipborne integrated navigation and positioning system. It is also necessary to achieve international advanced levels in safety, environmental protection, and energy efficiency.

4. Maximum Towing Force and Towing Force Maintaining, Maximum Speed, Underwater Radiation Noise

   1) Towing Force Requirements and Challenges Involved

      The functionality of the vessel necessitated a larger block coefficient, smaller *L/B* ratio, and dual-duct adjustable pitch propellers, resulting in higher underwater appendage resistance, lower hull efficiency (η), and lower propeller efficiency, making the vessel being of inherent deficiency. However, the vessel was required to realize a maximum speed not less than 16 knots, a maximum underwater radiation noise not greater than 168 decibels at a speed of 5 knots during towing operations, and a maximum towing force not less than 100 tons. Additionally, it needed to ensure emergency cable and gun array recovery under adverse sea conditions. Therefore, it was necessary to simultaneously address the key technological challenges related to underwater low radiation noise, maximum speed, maximum towing force, and towing force maintaining in adverse sea conditions.

   2) Solutions for Propeller, Towing Force and Hydrodynamic Performance

      Combining ship model trials, computational fluid dynamics (CFD) calculations and propeller design with the proper selection of propulsion unit power, type, quantity and location is an integrated approach to address issues in the maximum speed, maximum towing force, and towing force maintaining. The specific technical measures include:

      (1) Two sets of adjustable pitch 5-blade inclined-duct propellers, with a diameter of 4.20 m, were used.
      (2) Maximum pitch ratio was designed for the propeller to achieve maximum speed. The maximum towing force at 5 knots was ensured by using an appropriate pitch ratio, and an optimal pitch ratio was designed to meet the requirements of various operational conditions, cable variations, and towing operations at different speeds.
      (3) The hull efficiency varied by less than 1% between shaft bossing and shaft bracket solutions. Further referencing the ship model trial results and conducting CFD optimization calculations, the vessel addressed problems identified during the trial, such as local bubbles and turbulent flow, by reasonably adjusting the forward tilt angle and transverse angle of the shaft bracket, which also simplified the construction process.
      (4) By utilizing towing force estimation methods, the vessel tackled safety concerns during towing operations at 2 knots in the event of single

propeller failure or adverse sea conditions, by utilizing telescopic all-direction rudder propellers in conjunction with duct-adjustable pitch propellers. The vessel's towing force primarily relies on empirical formulas and ship model trial, whereas the towing force required for operations depends on equipment specifications, statistical data, and comprehensive analyses. According to the actual ship application results, the towing force achieved the index and surpassed the anticipated value.

3) Solutions for Underwater Radiation Noise from the Propeller and Hull

This vessel had specific limitations for underwater radiation noise during towing operations at 5 knots. Therefore, for the overall vessel design, conducting preliminary noise assessments and proposing optimization measures were necessary for noise control. This included using finite element method to evaluate the impact of hull vibrations on underwater radiation noise, assessing the influence of onboard power equipment noise on underwater radiation noise through calculations, and determining the impact of propeller noise on underwater radiation noise through ship model basin test. Based on this, the design was optimized, including modifications for noise reduction in the unloading of propeller blade tips, and measures to reinforce local weak structures while maintaining the overall double-hull design of the vessel.

4) Geophysical Functional Index

The vessel fully met the requirements of the design specification in terms of underwater radiation noise, speed, and towing force (see Table 1).

The combination of duct-adjustable pitch propellers and variable frequency electric propulsion broke the conventional mode of electric propulsion. This innovative approach takes into account the interrelationship between functionality, environmental protection, and energy efficiency. It has significantly reduced the noise levels in various operating conditions, variable cable numbers, and variable speed cable towing operations. It has proven particularly effective in addressing key technological challenges such as low radiation noise, maximum speed, maximum towing force, and maintaining towing force under harsh sea conditions. This technological breakthrough become one of the highlights of the vessel's design.

Table 1 Comparison of design indicators and actual indicators for underwater radiation noise, speed, and towing force

S/N	Indicator	1/3 octave band maximum noise/decibel	Underwater noise total/decibel	Maximum speed/knot	Maximum towing force/ton
1	Design indicator	168	176	≥ 16	≥ 100
2	Actual indicator	Complied	Complied	16.3	110

5. Structural Strength, Overall Ship Weight, and Center of Gravity Control
   1) Weight's Impact on Stability, Effective Vibration and Noise Reduction, and Functional Use

      The weight of the structural components in geophysical vessels often constitutes a significant proportion of their overall weight. As the first large geophysical vessel designed and built domestically, this vessel faced critical design challenges in achieving a well-occupied superstructure and effective vibration and noise reduction, while well controlling its weight and center of gravity. These considerations also directly impact the vessel's stability and its geophysical functionality. With an *L/B* ratio (length breadth ratio) of only 4.025 and the need to comply with ice strengthening regulations, the conventional structural calculation method alone could not fully conform to CCS specifications.
   2) Solution
      (1) Hull structure solution based on finite element calculation: For the vessel with an *L/B* ratio less than 4.5, utilizing finite element calculation based on the North Atlantic wave distribution chart was deemed one of the effective approaches to address these challenges. By comparing results obtained from both traditional and finite element calculations, the dimensions of the vessel's steel structural components could be determined. The pillars were used to replace steel bulkheads, and an alignment strategy for the vessel's longitudinal and transverse structures was adopted. This method was effective in reducing structural weight, mitigating vibrations, and ensuring hull strength.
      (2) Calculation of local structural natural frequency (F_c): Analysis revealed that the primary sources of excitation in the instrument room and geophysical machinery room were the nearby double-bottom diesel generator sets. Through finite element modeling, calculations were made to compare the natural frequency of critical areas, such as the geophysical instrument room (and engine room) and living compartments, against the excitation frequency generated by the ship's diesel generator sets. It was discovered that the natural frequency of the instrument room and machinery room were effectively offset from the excitation sources created by the vessel's diesel generator sets. This ensured that the structural design of the geophysical instrument room (and engine room) and other relevant areas complied with the requirements for low vibration and noise without increasing weight.
      (3) Comprehensive approach: Following a detailed analysis of ice strengthening, control of underwater radiation noise, displacement reserve, stability margin, and the weight of electromechanical equipment and outfitting, a series of methods were implemented. These included increasing the thickness of the vessel's shell below the main deck, optimizing the local structural form of the geophysical vessel, integrating lightweight interior outfitting materials, enhancing damping and

vibration reduction measures in the seismic source equipment area, and reducing the use of damping materials in the upper superstructure. Through these strategies, weight was effectively managed, and vibration and noise levels were reduced, resulting in an extended service life for the vessel's shell. The actual empty weight of the vessel was found to be 0.43% lower than the estimated value, with the center of gravity height measuring 5.0% lower.

6. Integrated Navigation and Positioning Survey System

The purpose of the integrated navigation and positioning survey system is to guide the geophysical vessel, determine its position, measure the real-time location of the seismic source and the orientation of the cable internal detectors, and control the firing and reception in real time. It also dynamically controls the position of the towed cable to achieve automatic, visual, control, and predictive capabilities for the entire offshore exploration operation. In recent years, the DGPS positioning system has been widely adopted in the integrated navigation and positioning survey system of offshore oil geophysical exploration.

In addition to utilizing high-performance marine navigation positioning equipment, ensuring a stable and compliant working environment for shipborne positioning devices is also crucial for the proper functioning of the advanced integrated navigation and positioning system. With this in mind, the vessel's design incorporated several considerations: the vessel's electrical design ensures that the harmonic interference of the control circuit remains below 5%; low levels of vibration and noise are required (the instrument room and surrounding compartments maintain noise levels below 60 decibels); and through extensive collaboration between the shipowner, design team, and domestic equipment manufacturers, the vessel incorporates domestically-produced controllable passive anti-rolling tanks, which are protected by three patented inventions. These tanks considerably expand the range of sea conditions in which the vessel's anti-rolling effectiveness remains intact, resulting in excellent anti-rolling performance.

The vessel incorporated multiple new technologies in its design and construction, and has set several operational records since its delivery.

(1) Breakthroughs in key technologies and application of new technologies are the trend of modern geophysical vessel development. The vessel met various technical requirements, such as controlling whole vessel noise, maintaining sufficient towing force at a navigational speed of 5 knots, achieving a maximum navigational speed of 16 knots, complying with the Clean Mark standards, fulfilling the safety regulations for Special Purpose Ships (SPS2008), and adhering to the Emergency Towing Guide (implemented since 2010). It also featured increased fuel tank and freshwater tank capacities, as well as the integration of low-sulfur oil tanks and ballast and bilge water treatment systems, resulting in energy-saving measures. Additionally, measures were implemented to control harmonics in the electric propulsion system. Its design incorporated a double-hull and double-bottom structure, with strengthened specialized geophysical equipment and consideration for local fatigue calculations, overall longitudinal

strength, and processes. Consequently, it achieved advancement in the design technology of geophysical exploration concepts.

(2) During the construction process of this vessel, Shanghai Shipyard fully demonstrated its humanized management, efficient execution, and team cohesion. It provided comprehensive guarantees for safety, project duration, and quality, ensuring a construction process that is accident-free across 3.8 million work hours. Through close coordination among specialized teams responsible for safety, quality, design, procurement, planning, construction, and commissioning, the "Haiyang Shiyou 720" construction team overcame several technical challenges, including segment pre-installation of the dome, control of overall weight, laying of a 400-km cable, and fine-tuning of medium-pressure generator, among others. Shanghai Shipyard also capitalized on the resources of renowned domestic institutions, applying a combination of production, education, research, and application to tackle technical difficulties related to the installation and commissioning of geophysical equipment and the composite controllable passive anti-rolling tank, among other difficulties. Consequently, they acquired key technologies for geophysical vessel construction and successfully completed the equipment selection, installation, and commissioning of "Haiyang Shiyou 720". During the design and construction process, the research and development team achieved leading status in various technological aspects, including geophysical exploration control and layout, hydrodynamic performance and ship form optimization, 3D design and virtual reality application, and underwater noise analysis and control.

(3) The construction of the "Haiyang Shiyou 720" fully embodied the characteristics of the new generation of high-performance 3D geophysical vessels: safety, tranquility, accuracy, clarity, high capacity, long endurance, environmental friendliness, and low energy consumption. Several technological innovations were realized through the design of the vessel, resulting in overall technical performance that is at the forefront of the global industry. With a daily acquisition area reaching as high as 96 square kilometers, the vessel substantially surpassed foreign ship types of similar capacity, while exhibiting significantly lower line-filling rates and failure rates. These achievements garnered praise from the users. Within one month of being put into operation, the vessel had generated over USD 10 million in value, which would continue to contribute to China's geophysical exploration and offshore oil and gas development. The successful development of this vessel has bridged a domestic gap, strengthened the nation's capacity for independently designing and building such ships, and broken foreign monopolies in this field. China's capabilities in marine engineering exploration were greatly enhanced, with deep-sea exploration capacities expanding from a water depth of 500 m to 3,000 m. This significant improvement has brought about profound and enormous social benefits, marking China's successful entry into the top tier of deep-sea exploration equipment for marine engineering. In 2013, the vessel received the first prize of the Science and Technology Progress Award from the Chinese Society of Naval Architects and Marine Engineers and the same honor from the China State Shipbuilding Corporation. In 2014, it was

awarded the second prize of the Science and Technology Progress Award from the Chinese Society of Oceanography and the first prize of the Shanghai Science and Technology Progress Award.

(4) In June 2011, this vessel set a record for the highest daily acquisition area in China, reaching 75.93 km, during the Liwan 43/11 exploration operation. In July of the same year, during the Dongsha 25 working area operation, the single-day acquisition record was further increased to 96.495 km. In August, it achieved an outstanding monthly acquisition area of 1,607.79 km, repeatedly surpassing the acquisition operation records of China Oilfield Services Limited. In the winter of 2011, it conducted unconventional 3D operations for the first time in Chinese waters, setting a precedent for the geophysical vessel fleet to conduct all-weather operations in Chinese waters.

2.6 Deep-Water Engineering Exploration Vessel "Haiyang Shiyou 708"

On December 16, 2011, a grand delivery ceremony for the first domestic 3,000-m deep-water engineering exploration vessel, "Haiyang Shiyou 708" (see Fig. 6), was held at the Longxue plant of CSSC Guangzhou Huangpu Shipbuilding Co., Ltd. This vessel marked the target ship of the research project "Deep-Water Engineering Exploration Vessel and Supporting Technology" under the National Science and Technology Major Project "Major Equipment and Supporting Engineering Technology for Deep-Water Engineering in the Ocean". This vessel is the world's first integrated engineering exploration vessel incorporating functions such as lifting, exploration, and drilling. It can operate at a depth of 3,000 m, with a drilling depth capability of up to 600 m below the seabed. The development of this vessel has filled the gap in China's deep-sea exploration equipment for marine engineering. As a crucial equipment for deep-sea oil and gas field development, the successful development of the "Haiyang Shiyou 708" has played a key role in the exploration and development of oil and gas fields in the South China Sea. It has greatly enhanced China's capacity for deep-sea marine resource exploration and development, as well as the core competitiveness of China's marine engineering. This achievement marks China's entry into the top field of deep-sea exploration equipment for marine engineering, as it signifies the improvement of China's marine engineering exploration capabilities from a depth of 300 m to 3,000 m.

"Haiyang Shiyou 708" was conceptually and basically designed by VIK-SANDVIK, a renowned Norwegian company, while the detailed design was carried out by the Shanghai Merchant Ship Design and Research Institute. The vessel was owned by China Oilfield Services Limited and built by CSSC Guangzhou Huangpu Shipbuilding Co., Ltd.

This vessel was designed with a total length of 105 m, a molded breadth of 23.4 m, a molded depth of 9.6 m, an endurance of 16,000 nautical miles, a self-sustaining period of 70 days, and a capacity of 90 persons. It possesses a displacement of

Fig. 6 Deep-water engineering exploration vessel "Haiyang Shiyou 708"

approximately 11,600 tons for non-restricted waters Designed to reach a maximum speed of 14.5 knots and equipped to withstand winds up to Force 12, it ensures safe navigation in sea conditions up to grade 9 in hurricanes (12 on the Beaufort scale).

Featuring an aft deck with a spacious working area of 846 square meters, the vessel is equipped with a 7-m by 7.2-m moonpool. It is classed by both the CCS and DNV.[2] The vessel has a short forecastle, a bulb bow, class B3 ice strengthening, a continuous freeboard deck, and a helicopter landing platform. The vessel features two 3,500-kilowatt all-direction fixed-pitch rudder propeller devices, two 1,500-kilowatt variable frequency bow thrusters, two 1,500-kilowatt variable frequency telescopic all-direction propellers, which are powered by an electric propulsion system consisting of four 3,360-kilowatt diesel engines. In addition, a 700-kilowatt emergency diesel generator is configured to empower DP-2 dynamic positioning capability. There is a reverse osmosis desalination system onboard that can produce 40 tons of freshwater per day. Additionally, a deep-water drilling and sampling system is equipped, boasting a derrick height of 34.5 m, and a capability to drill up to 600 m under the maximum operating water depth of 3,000 m with heave compensation functionality. It is also equipped with a 150-ton large-scale marine engineering crane, capable of operating at a water depth of 3,000 m with a maximum reach of 31 m. Moreover, the vessel features a 3,000-m AUV[3] capable of navigating at a maximum speed of 5 knots for up to 20 h. It is further equipped with high-precision multi-beam and single-beam depth sounders, accompanied by a 250-ton A-frame, a 5,000-m deep-water sampling winch, and a 23.5-m ultra-long insulation and pressure-holding water sampler. These systems enable the vessel to conduct engineering geophysical surveys, engineering geological drilling and sampling operations in water depths of up to 3,000 m. The

[2] Det Norske Veritas.

[3] Autonomous underwater vehicle.

employment of "Haiyang Shiyou 708" marks a huge leap in China's marine engineering exploration capability, which is elevated from operating in a water depth of 300 m to 3,000 m.

The development of "Haiyang Shiyou 708" gave rise to several technological innovations, including the upgrade from towing operation to self-propelling operation for engineering geophysical surveys, the utilization of AUVs to carry exploration instrument for data collection of water depth geomorphology and sub-bottom in deep-water areas, and the implementation of underwater precise positioning. The deep-water high-fidelity sampling technique enabled the collection of 23.5-meter-long samples of deep-water hydrates with heat and pressure preserved, which could be tested and stored on site. The vessel is equipped with a 150-ton/10-m deck crane, which incorporates heave compensation functionality that ensures stable operation of the crane. Additional configurations include one eight-ton crane and three two-ton cranes. Besides, it supports ROV operations.

"Haiyang Shiyou 708" represents a highly integrated engineering exploration vessel, capable of conducting multiple operations such as engineering explorations, seismic surveys, engineering geological hole and well drilling, deep-sea coring, seabed surface sampling, in-situ static cone penetration tests (CPT), and large-scale marine engineering lifting. It employs electric propulsion and DP-2 dynamic positioning system, offering functionalities including engineering geophysical surveys, digital seismic surveys, 3,000-m water depth engineering geological drilling, ROV operations, seabed surface sampling, in-situ static cone penetration tests (CPT), offshore engineering support services, and large-scale marine engineering lifting.

Design features of the vessel:

Firstly, it is capable of unmanned operation. A single person can handle the responsibilities of five people including captain, sailor, operator, communicator, and commander. It can also achieve unmanned autonomous navigation.

Secondly, it can easily achieve operations in the air and underwater. The vessel is designed to withstand winds up to Force 12 and impacts from ice blocks, ensuring safe navigation even in sea conditions of up to grade 9. Robots, small submarines, and helicopters are equipped, allowing for smooth operations both in the air and underwater. The AUV can dive to around 500 m above the seabed for exploration, operating autonomously and automatically returning after completing its mission. The robots can accompany scientists to conduct designated tasks down to the seabed. The helicopter landing platform is located at the top of the superstructure, enabling charging, takeoff, and landing of the "Puma" helicopters for transportation of supplies and personnel. Given the frequent winds and waves in the South China Sea, reliable ship positioning is crucial for exploration work. The vessel's positioning system can handle wave heights of up to 2.5 m, ensuring that drilling operations can proceed even in strong winds and rough waves. Dynamic positioning and compensated positioning systems are configured. When hit by winds and waves, all five propellers on the vessel operate simultaneously to maintain its position. In facing a wave of 2.5 m high, the vertical displacement of the underwater drill bit does not exceed 5 cm, allowing it to operate normally. The vessel

is designed to have a wind resistance capability of no less than Force 12 and is capable of safe navigation in sea conditions up to grade 9, allowing for operating about 280 days at sea per year. In addition, the vessel can withstand ice blocks of about 20 cm thickness and continue to navigate even in icy conditions at sea.
Third, it serves as a five-star hotel on the sea. It is the first vessel in China to pass the Norwegian C3V3 comfort certification. The tables and chairs on the vessel are no longer fixed to the floor with hooks but can be arranged freely, resembling the decoration style of a land-based room. No matter how rough the winds and waves encountered are, the vessel remains stable. Stepping into the cabins is like entering a five-star hotel, as there are beautifully decorated meeting rooms, restaurants, luxurious lounges, compartments with bathrooms and LCD TVs, as well as saunas. Crew members from different countries in the East and West can choose between Chinese and Western cuisine according to their dietary preferences. During their breaks from work, they can also enjoy a cup of coffee, indulging in leisure at sea.

2.7 *Multipurpose Natural Gas Hydrate Survey Vessel "Haiyang 6"*

Natural gas hydrates, or "combustible ice", are ice-like crystalline substances formed by the combination of natural gas and water under high pressure and low temperature conditions. These substances are a new type of clean energy that ignite upon contact with fire. They are primarily found in seabeds and land permafrost regions, with reserves surpassing those of oil and natural gas. Since the 1960s, Chinese scientists have been engaged in research on this resource. However, extracting seabed combustible ice poses technical challenges far beyond those of oil and gas exploitation, which has resulted in it remaining in the research and investigation phase for an extended period. Since the beginning of the twenty-first century, with global oil and gas resources gradually depleting, numerous countries worldwide have turned their attention towards the extraction of offshore combustible ice, leading to a new wave of interest. While conducting multiple scientific surveys and investigations, China has also formulated a comprehensive and precise plan to investigate combustible ice in its national waters.

In 2002, the Ministry of Land and Resources granted approval for the construction of a hydrate survey vessel to the Guangzhou Marine Geological Survey. Following a bidding process, the design task for the vessel was awarded to the MARIC. Detailed design work was completed in December 2007, and construction was carried out by the Wuchang Shipbuilding Industry Group Co., Ltd., with the vessel being delivered in October 2009 and christened as "Haiyang 6" (see Fig. 7). The vessel boasts a total length of 106.00 m, a designed draft of 5.5 m, a displacement of 4,600 tons, a capacity of 65 persons (including 39 scientific researchers), an endurance of 15,000 nautical miles, a self-sustaining period of 60 days, and a trial sailing speed of 15.2 knots. It features two-compartment floodability, and capability to navigate in non-restricted waters and withstand Force 12 winds and class B3 ice.

To effectively fulfill the vessel's principal scientific investigation tasks while optimizing operational efficiency, the design team, in collaboration with experts from the Guangzhou Marine Geological Survey, conducted meticulous analysis based on the characteristics of offshore scientific investigation operations. As a result, they designed the vessel to be a multipurpose survey vessel that integrates seismic operations, geological survey operations, and other multipurpose operations. The vessel is equipped with various devices, including the 4,000-m deep-sea underwater robot "Sea Lion", deep-water multi-beam sounding system, deep-water sub-bottom profiling system, and seismic acquisition system of long spread, high capacity, and high resolution. The entire vessel is divided into three working areas, namely, geological survey area, seismic survey area, and acoustic equipment transducer area. Its survey equipment comprises four major categories, namely, geophysical survey, geological sampling survey, deep-sea underwater remote sensing, and hydrological survey. In actual operation, the vessel can be modified with additional or alternative equipment based on specific operational requirements, allowing for a multipurpose survey capacity that encompasses various functionalities.

In addition to holding a CCS certificate, the ship also possesses certificates including marine geological survey class A certificate, experimental testing class A certificate, marine surveying and mapping class A certificate, qualification accreditation and metrological certification certificate, laboratory accreditation certificate from the China National Accreditation Service for Conformity Assessment (CNAS), and document of compliance (DOC) and safety management certificate (SMC) from

Fig. 7 Multipurpose natural gas hydrate survey vessel "Haiyang 6"

the CCS. These credentials guarantee the capabilities of the vessel to carry out investigations of marine land resources and environmental surveys, the strategic exploration and evaluation of mineral resources in international waters, while also facilitating marine geological science and other related activities supported by high-tech technology.

This vessel addressed the following technical challenges in order to meet the requirements for use:

1) Interference from the Bulb Bow and Channel-type Side Thrust Device Hole on Deep-sea Multi-beam Operations

   The poor interference caused by the bulb bow and side thrust device hole in newly built research vessels, both domestically and internationally, significantly impacts the operation of deep-sea multi-beam sounders. Inevitably, this vessel was also equipped with a bulb bow and a channel-type side thruster at the bow to meet the navigational speed, port docking, and dynamic positioning needs. To overcome the issue, extensive numerical analysis and experimental research were conducted to optimize the hull design of this vessel, contributing to the final proper operation of the deep-sea multi-beam sounder.
2) Control of Weight and Center of Gravity

   Weight and center of gravity are critical considerations for research vessels. In the ship type evaluation process, careful attention was given to the local layout of major equipment for optimization. Meanwhile, careful evaluations was conducted in advance to assess the impacts of optimization on changes of weight and center of gravity, followed by certain sensitivity analysis. After taking effective control measures and close collaboration with the shipyard, the desired target was achieved.
3) Wind Stability and Seaworthiness

   Research vessels require high levels of stability, while comfort and safety (stability) requirements may be in conflict with each other. In designing this vessel, many efforts were made to optimize the vessel's roll characteristics, such as rolling period, roll angle, and roll acceleration, while ensuring the highest resistance to wind stability. After numerous model trials and hull form optimization, this conflict was successfully addressed.
4) Power System

   This vessel was equipped with an electric propulsion system, featuring three large (1,900 kW each) and one small (760 kW) diesel generator sets. These generator sets enable long-duration parallel operation and offer flexible power configurations, minimizing energy waste. Additionally, advanced technologies and equipment such as dynamic positioning and all-direction rudder propellers were employed for this vessel. Instead of traditional rudders and propellers, the vessel utilizes electric motors to drive short shafts, which in turn drive the propellers through a gear transmission. This integrated design combines the functions of both rudder and propeller, allowing the ship to achieve 360-degree steering and continuously variable frequency and variable speed. This design provides the vessel with high maneuverability and excellent low-speed navigation characteristics, giving it a significant advantage for research vessels in navigating in variable navigation conditions and complex areas.

5) Vibration and Noise Reduction

Measures were implemented to reduce vibration and noise levels. These include the use of floating rafts for diesel generator sets, the installation of silencers in the exhaust pipeline of the generator sets to reduce noise by 25 decibels, the adoption of elastic mounting for fans and air compressor units, the employment of streamline-shaped inlets for fans, the coating of sound absorption materials for fans' inner sides, the use of flexible braces in the exhaust pipes, the application of damping materials on the generator set base surface, on both sides of the engine room, and on the top plates of the front and rear bulkheads, as well as isolation and sound absorption measures for fans and air ducts. According to measurements, both the vibration and noise levels were found to be low.

6) Fire Safety

Given the large number of personnel on the research vessel, excluding the crew members, research investigators temporarily working onboard may be unfamiliar with the vessel. To enhance the fire safety of the vessel, a fire zone was designed, with fire barriers being set. The accessibility of staircases and passageways was improved, and measures for fire detection and fire protection were strengthened.

7) Collision Avoidance

The *International Regulations for Preventing Collisions at Sea* has stipulated specific requirements for the placement and height of navigation lights on ships. According to the regulations and requirements, it was necessary to install a taller light mast in the midship or aft section of the vessel. However, this mast may cause an impact on the use of cranes on board. To address this issue, efforts were made to optimize the placement and form of the main mast to fully comply with the regulations and requirements for the positioning of navigation lights while avoiding hindrance to crane operations.

8) Living Environment

This vessel fully embodied humanized design principles. All living compartments were equipped with bathrooms, showers, televisions, and internet access. In addition to the mentioned amenities, the vessel also has a gym, a library, a sauna, and other facilities, creating a comfortable living environment for the crew and personnel on board. The design of the vessel strictly separated living areas from mechanical areas. Moreover, the water generating system incorporated a two-stage filtration process to produce water of purified water level. In addition, a UV-disinfected direct drinking water device was arranged on each deck.

"Haiyang 6" is China's first domestically developed and successful multipurpose survey vessel dedicated to natural gas hydrate exploration. It is also the country's first specialized geological survey vessel with independent intellectual property rights. The successful development of this vessel signifies that China has entered the forefront of international marine geological survey equipment.

As a survey vessel with multi-functional scientific investigation capabilities, "Haiyang 6" has accomplished numerous significant oceanographic survey missions since its delivery.

In May 2012, the vessel embarked on a precise survey of combustible ice resources in the northern region of the South China Sea. The survey covered areas such as the Qiongdong waters, Xisha waters, and Dongsha waters, preparing for future more precise exploratory work. Notably, successful scientific exploration activities have already been conducted in multiple areas of the South China Sea, contributing to China's goal of extracting hundreds of tons of combustible ice from the region by the 2020s. Going forward, the vessel will continue to play a pivotal role as the main vessel in the overall survey of combustible ice resources in the targeted sea areas.

From June to November 2011, "Haiyang 6" carried out the 23rd voyage mission for scientific investigations for the China Ocean Mineral Resources R&D Association (COMRA). The aim of this oceangoing expedition was to safeguard China's maritime rights and fulfill the obligations included in the exploration contract with the International Seabed Authority (ISA). The survey area encompassed the CC Zone[4] in the Pacific Ocean and the seamount region in the Western Pacific. The main objectives included the investigation of polymetallic nodules and cobalt-rich nodules, as well as environmental and biological surveys within the area and associated scientific researches. Throughout the expedition, which lasted for over 20 days, the vessel successfully accomplished the following four tasks:

(1) Providing surveillance and support for the 5,000-m sea experiments of China's manned submersible, "Jiaolong". In this task, it conducted CTD measurements and multi-beam submarine topography and geomorphology surveys in the deep diving experimental area.
(2) Conducting marine environmental and biological surveys, as well as geological sampling, in polymetallic nodule areas. Thus, it successfully fulfilled the "exploration contract" tasks for polymetallic nodule resources signed between the COMRA and the ISA.
(3) Investigating cobalt-rich nodule resources and environmental baselines in seamount areas.
(4) Surveying areas for the deeper-sea trials of the "Jiaolong" submersible.

Upon its return from the voyage, the COMRA presented the "Haiyang 6" with a prestigious golden plaque recognizing its "outstanding achievements as the pioneer in deep-sea oceanic scientific expeditions". This accolade confirmed the vessel's status as a formidable new force in China's oceanic scientific expeditions. It affirmed that the vessel is fully capable of undertaking major oceanic scientific expedition tasks, thereby making greater contributions to the safeguarding of China's maritime rights and interests as well as the peaceful use of marine resources. Moving forward, the vessel will greatly drive the advancement of China's marine geological undertakings.

[4] A famous area rich in mineral deposits of polymetallic nodules.

In July 2012, during the mission to provide support and surveillance for "Jiaolong's" 7,000-m sea experiments, "Haiyang 6" conducted an precedent high-precision multi-beam survey of the "Challenger Deep" at the southern end of the Mariana Trench: the world's deepest trench. This survey significantly filled a research gap for China in this field.

2.8 New-Generation Multipurpose Oceanographic Research Vessel "Kexue"

By the successful completion of China's "10th Five-Year Plan", various marine research institutions in the country had nearly 40 types of research vessels. With the help of these vessels, effective scientific surveys have been conducted in coastal and international waters, as well as in the Arctic and Antarctic regions. Since the beginning of the twenty-first century, there has been a growing international competition for ocean rights and an increased interest in scientific investigation of marine resources and disasters. China has attached great importance to this and has developed high-level marine research projects in the "973" and "863" programs. Simultaneously, institutions responsible for the development of marine oil and gas and mineral resources also urgently required high-precision marine mineral information. Therefore, the Institute of Oceanography of the Chinese Academy of Sciences, as the main force in China's marine science, planed to construct a 4,000-ton multipurpose oceanographic research vessel that meets the requirements of navigation in non-restricted waters, based on the major science and technology infrastructure projects of the "11th Five-Year Plan". The aim was to provide a marine mobile laboratory and experimental platform for fundamental research of marine science and R&D of high and new technology in the deep sea and ocean areas. The project was officially approved on July 26, 2007.

In May 2009, following an open bidding process, the vessel design work was awarded to the MARIC, and the construction work to the Wuchang Shipbuilding Industry Group Co., Ltd. Construction of this vessel began on October 28, 2010. Afterwards, the vessel was launched on October 30, 2011, and delivered on June 30, 2012, named "Kexue" (see Fig. 8).

The vessel was designed with a total length of 96.60 m, a molded breadth of 17.8 m, a molded depth of 8.9 m, a draft of 5.6 m, and a displacement of 4,600 tons, a service speed of 12 knots, an endurance of 15,000 nautical miles, a self-sustaining period of 60 days, and a capacity of 80 persons (including 49 scientific survey personnel). It was classed by the CCS.

It is propelled by a twin-pod electric propulsion system with a power of 1900 kilowatts each. In addition, the vessel features class B3 ice strengthening capability and a helicopter landing platform.

Fig. 8 Multipurpose oceanographic research vessel "Kexue"

The design team fully recognized that the vessel would undertake multipurpose deep-sea survey missions in major ocean regions, including marine dynamic environment survey, geological environment survey, ecological environment survey, seabed resources exploration, comprehensive energy detection, comprehensive observation and experiment of military oceanography. These activities aimed at providing technical support for resolving critical issues related to marine resources, energy, national defense security, and mitigation of the impact of natural disasters at that time. At the same time, it was believed that the vessel would serve as an advanced platform for Chinese scientists to participate in major international marine research programs and further to enhance China's influence in the international marine research community, becoming one of the main vessels for China's multipurpose oceangoing surveys. Therefore, the design team incorporated the development trends of international multipurpose oceanographic research vessels. They formulated a design concept that aimed at achieving an overall international advanced level and ensuring the ship's performance remained advanced for the next 20 years.

1) Design of New Ship Type for Deep and Far-Sea Multipurpose Research Vessel

The vessel was designed with smaller main dimensions and a more rational tonnage, properly satisfying the needs of comprehensive scientific research. It outperformed foreign counterparts of the same period in terms of capacity, endurance, oil tank volume, deck working area, and laboratory space allocation.

The vessel's demonstration of main dimensions and analysis of economic viability focused on repeatedly demonstrating and optimizing the vessel according to various technical indexes of ships. By repeatedly comparing displacement index, speed (resistance) index, endurance index, scientific research operational area index, port depth, power station scale and configuration, and stability calculations, the most optimal main dimensions and form coefficients can be determined. Additionally, while serving as a platform for a diverse array of scientific research equipment and measurement instruments, the research vessel also accommodates a large number of scientists and crew members. Therefore, apart from speed, the profile design of research vessels should also take into account the matching of the optimization of the bow lines and the acoustic equipment. Meanwhile, the seakeeping capacity and seaworthiness of the vessel should also be considered. The ship type analysis and profile design are interconnected. Different ship type main dimensions and lines determine various resistance characteristics, seakeeping capacities, and other hull performance aspects. They also influence the operational effectiveness of underwater acoustic equipment, the selection of thrusters, and the configuration of the main engine, further impacting the overall efficiency, functionality, and seaworthiness of this ship type. Therefore, overall optimization is necessary in ship type analysis and profile design.

Based on the analysis of several excellent hull forms, the design team proposed multiple main dimension schemes. Through comparative analysis and experimental optimization between bottom multi-beam protruding mounting and hull embedded multi-beam mounting, they took into consideration three aspects, including scientific research efficiency, navigational economy, and control of bubble interference. Ultimately, the final hull form was worked out after the preliminary comparison and selection of CFD schemes and optimization by ship model trials. "Kexue" was designed to be a 4,000-ton deep and far-sea multipurpose research vessel. Through full demonstration of its speed, seakeeping capability, and economy, it was redesigned based on previous research ship types. The new design features a smaller length-to-breadth ratio and a larger block coefficient, resulting in improved capabilities of corresponding indexes compared to similar international ship types under the design conditions. Thanks to the successful development of this profile, "Kexue" is able to meet the demanding economic requirements of operating a single generator at a speed of 12 knots. It consumes approximately 10 tons of fuel per day, fulfilling the energy-saving needs envisioned by the user. Model trials and operational performance after delivery confirmed that the vessel's speed and economy indexes exceed those of comparable international research vessels, earning high praise from the user. This is the first domestically designed and energy-saving new ship type that successfully applies a small length-to-breadth ratio in recent years in China's research vessel industry. The key technologies employed in the vessel's hull analysis and profile design have gradually been incorporated into the subsequent designs of research vessels.

The vessel exceeds the international standards in wind resistance. It has been proven to be safe in harsh sea conditions through its operation records.

2) Overall Arrangement of Scientific Research System

The overall layout of the scientific research system combined the development trends of mainstream international research vessels. It features a left placement of chimney and a starboard L-shaped mainstream layout, while incorporating a comprehensive arrangement of dry/wet general laboratories. The stern working deck is designed with a large low freeboard working space, which provides benefits for scientific research operations and future space expansion. Additionally, the deck is equipped with embedded fasteners to facilitate the installation of future mobile container onboard. The sheltered working deck space at the stern not only provides a conducive working environment but also offers excellent conditions for truth-preserving sampling. The vessel also features a bow science mast, midship lifting fins, and a centralized control room for underwater operations on the starboard side. These features ensure efficient scientific research operations.

The design team optimized the installation of various equipment. They strategically planned and arranged the limited compartment space for reasonable arrangement, utilizing integrated and movable designs to ensure reasonable spatial layout and equipment functionality. This approach enables the shared use of winches, lifting equipment, and other similar devices with similar load-bearing and operational conditions. They also conducted comprehensive planning for survey equipment in terms of navigation segments and time allocations. Based on different operation periods and voyage schedules, they implemented an overall plan that incorporates modular designs as much as possible. This allows for a unified module interface on the vessel, facilitating future upgrades and modifications to the equipment system while saving a significant amount of cost on upgrade and maintenance.

Due to the reasonable arrangement of the scientific research system, the scientific research laboratory area on this vessel reaches 665 square meters, with a unit displacement area parameter of 0.142 square meters per ton. The scientific research working deck area is 647 square meters, with a unit displacement area parameter of 0.139 square meters per ton. These parameters are all superior to those of similar international ship types.

3) Development and Installation of Scientific Research Support Equipment

The lifting fin system on the "Kexue" vessel is the first of its kind to be manufactured and installed domestically. This system consists of multiple devices with complex mechanisms and requires high installation accuracy. Welding cannot meet the positioning and installation precision required. Therefore, a template marking method was used. During equipment positioning and installation, a template was first positioned on the vessel using an angular measurement. Other equipment was then positioned using the template as a reference by employing a perpendicular line method. This ensured the accuracy of equipment positioning and installation.

Due to the large size of the multi-beam transducer on the vessel's bottom, the installation accuracy of the base is also demanding, with a requirement for the horizontal levelness error of the lower surface of the base to be not greater than one millimeter. Due to the influence of the segmented construction precision of the hull, it is difficult to guarantee the required levelness with welding alone. Through technological breakthroughs, the factory adopted an embedded installation method to ensure the installation accuracy of the multi-beam transducer module. This installation process has been patented.

4) Design of Pod Integrated Power System

Research vessels usually operate under complex conditions with large load variations, requiring high maneuverability and positioning performance. Conventional diesel propulsion systems could no longer fully meet these operational requirements. Therefore, variable frequency electric motor propulsion was employed. This technology offers advantages such as high flexibility in arrangement, good maneuverability, low vibration, low noise, and high redundancy. It shows clear advantages, particularly in special-purpose vessels, and has a promising application outlook in research vessels.

The configuration of the power station on this vessel was proved the best through analysis and demonstration of different power loads and fuel economy considerations. A notable design highlight is the power station configuration of three large generator sets (2,600 kW each) and one small generator (758 kW). During the trial operation, the flexibility, economy, and reliability of this configuration were fully demonstrated under various operating conditions, particularly during economical sailing, drift operations, and seismic towing. This configuration garnered recognition and appreciation from the users.

5) Vibration and Noise Analysis and Control

The vessel is equipped with five major systems for water detection, atmospheric detection, seabed detection, deep-sea extreme environment survey, and field verification of remote sensing data. It carries various equipment, precision instruments, and instruments for detection, all of which have high requirements for the working environment. The requirements for environmental temperature, vibration, and noise levels are also quite stringent.

To tackle this, research, analysis, and demonstration of the vessel's own vibration and noise levels were carried out during the design and construction phase. Starting from the design stage, vibration and noise prediction analysis was conducted, and specific measures to control vibration and noise were proposed. Overall control was achieved through a combination of hull structure, power system, compartment layout, selection of mechanical equipment, and design of air-conditioning ventilation system. Through these measures, the vessel achieved satisfactory results.

6) Control of Empty Vessel Weight and Center of Gravity

The weight and center of gravity of a research vessel are not only related to stability indexes, floatation characteristics, load capacity indexes (primarily fuel, freshwater, and other consumables, which affect self-sustaining period), but also affect critical factors such as bending moments and shear forces on structural components, as well as the seakeeping performance. In this way, it further impacts the effectiveness of anti-rolling devices, lifeboat launching/recovering devices, helicopter landing, and other performances.

The control of the empty weight and center of gravity of the vessel is crucial throughout the design and construction process, from the initial design control to the control during the construction process, and finally precise control before delivery.

Estimating the empty vessel weight and center of gravity during the design phase is complex and tedious, along with limited accuracy. Therefore, based on rough estimates from the detailed design phase and multiple calculations using production design models, the floating condition of the vessel underwater, and the movement of heavy outfitting pieces, the vessel's weight and center of gravity were monitored as early as possible.

The "Kexue" vessel was proven to be safe, efficient, comfortable, energy-saving, environmentally friendly, and maneuverable, along with advanced control systems and reliable network and communication equipment. The vessel was proven to meet and even exceed the design requirements in terms of its scientific investigation operation performance indexes, ensuring the collection of accurate and reliable data that fully meets the demands of marine research.

Since its inception, "Kexue" has successfully completed numerous scientific research missions, including those for the Strategic Leading Science and Technology Special Projects of the Chinese Academy of Sciences, the "973" Program, the "863" Program, major projects of the China National Offshore Oil Corporation, and the National Natural Science Foundation of China. As of 2015, the vessel has undertaken over 510 days of navigation, covering a distance of 63,783 nautical miles. Notable achievements include the autonomous observation of active hydrothermal vents in the Okinawa Trough, the acquisition of high-resolution seabed topographic maps, and the first-ever deployment of the 18 sets of large-scale deep-sea subsurface buoy arrays in the Western Pacific's main current area. In addition, the vessel conducted the first large-area geophysical survey in the south-central region of the South China Sea.

On April 8, 2014, "Kexue" set sail for the Western Pacific to undertake scientific research tasks related to the Strategic Leading Science and Technology Special Project of the Chinese Academy of Sciences, focusing on the "Material and Energy Exchange in the Tropical Western Pacific Ocean System".

On January 19, 2015, during the scientific research tasks in the Yap Seamount area of the Western Pacific, "Kexue" deployed seven seabed seismometers in the designated area. This marked China's first-ever deployment of such instruments in this region, and all the seabed seismometers successfully indicated normal operational status.

On February 9, 2015, "Kexue" deployed heat flow probes in the vicinity of the Yap Trench in the Western Pacific to obtain information on seabed heat flow. This process was likened by the researchers to taking the temperature of the seabed.

The instrument utilized for this purpose is a 7.5-meter-long, 965-kilogram heat flow probe. It is lowered into the seabed from the aft deck of "Kexue" using a steel cable. With the addition of an extra 500 kg of weight, the probe can be vertically inserted into the shallow layer of the oceanic crust, composed of seabed sediment formed over millions of years.

When the heat flow probe is properly positioned, the 22 temperature sensors on the probe are instantly heated by the built-in power supply, recording temperature readings every second during the cooling process. Upon recovery, the distribution of submarine thermal gradient can be deduced based on the in-situ temperature gradient and thermal conductivity of the seabed sediment obtained from the sensors.

The submarine thermal gradient refers to the diffusion of heat from the Earth's interior to the surface of the seabed. By combining seabed heat flow values with deep seismic and other geophysical data, researchers can explore the characteristics of oceanic crust subduction. As planned, "Kexue" aimed to deploy heat flow probes at 14 stations during this voyage, spanning the Yap Trench area with water depths ranging from 2,000 km to 5,000 km.

The most active geological regions of Western Pacific focus on the plate boundaries. The Western Pacific serves as a "natural laboratory" for studying plate subduction processes and has significant influence on the marine environment near China.

Chinese researchers plan to conduct continuous research in the Western Pacific over the next few years. By employing various methods such as heat flow probing and rock sampling, it aims to gain a deeper understanding of the structural characteristics of the deep-sea lithosphere, further to facilitate the understanding of the Pacific plate structure and seamount evolution.

Oceanic real-time observation data has traditionally relied on satellite remote sensing and buoys. However, subsurface buoys used for underwater and deep-sea data collection can only be retrieved once a year, making it impossible to acquire real-time information like satellite remote sensing and buoys. This is due to the fact that the top buoy of the subsurface buoys is a few hundred meters (400–600 m) below the sea surface, preventing data transmission to satellites. Hence, in this voyage, the scientific research team placed a floating body for real-time data transmission on the water surface, establishing both wireless and wired connections with the subsurface buoys. The subsurface buoys transmitted data to the floating body, which in turn sent the data to satellites for relay back to land-based laboratories.

The challenge of this technology lies in establishing a stable connection between the floating body and the subsurface buoys, while offshore construction operations also present various difficulties. In this voyage, two sets of floating buoys were employed for real-time data transmission, enabling both wireless and wired connections. It demonstrated the feasibility of both the two solutions developed by Chinese scientists. Real-time transmission of deep-sea data could provide critical technical support for marine environmental and global climate research, with the data enhancing the accuracy of ocean climate and environmental forecasting. This voyage successfully overcome the worldwide challenge of real-time transmission for deep-sea subsurface buoy observation data.

On February 22, 2015, the "Kexue" research vessel, which was conducting operations in the vicinity of the Yap Trench in the Western Pacific, concluded its seismic data collection work amidst strong monsoon winds. The voyage collected a total of 19,839 data feedbacks from successful shots fired by air guns, with a combined survey line length of approximately 1,000 km.

During this voyage, “Kexue” carried out seismic operations along three survey lines in the area (two perpendicular to the trench and one diagonal). The vessel traveled at a speed of 5 knots, firing one shot every 50 m with an interval of approximately 20 s.

Multi-channel seismic operation refers to the process of generating seismic waves from an artificial seismic source, which propagate to the seabed and deeper depths before reflecting back. These reflected waves are then received and analyzed. Multi-channel seismic data provides the basis for further study of geological structures such as the direction, inclination, and faults of strata. These data can ultimately be used to interpret marine geological phenomena such as the morphology of oceanic subduction zones.

The introduction of the “Kexue” research vessel has significantly enhanced China’s capability in marine research, laying the foundation for breakthroughs and leap-forward developments in ocean research and exploration in international frontiers. The vessel has also contributed to safeguarding maritime defense security and rights and interests, as well as the development of marine science and technology. It has played a significant role in the development of marine resources, the protection of marine environments, and the sustainable use of marine resources. With remarkable scientific value and societal benefits, the vessel is instrumental in facilitating China’s transformation from a major maritime country into a strong maritime nation.

As a new generation of advanced research vessel in China, “Kexue” is a modern integrated marine research platform with independent intellectual property rights. It incorporates multiple disciplines, functions, and technological capabilities, capable of meeting the diverse research needs of multidisciplinary marine science. With its global navigation capability, “Kexue” stands on par with newly built and under-construction multipurpose research vessels of other leading maritime nations in terms of technological prowess and research capability. The deployment of this vessel will provide an advanced maritime mobile laboratory and research platform for national basic research in marine science and development of high-tech marine technologies. It will particularly facilitate research in fundamental marine science in deep-sea and oceanic areas. In the next ten to twenty years, this type of ship is expected to become one of the mainstay types of vessels for China’s oceangoing scientific multipurpose surveys. It will complement and cooperate with existing and planned research vessels and specialized research vessels, gradually establishing a complete system for marine research in China.

This project has already generated nine research papers and obtained three utility model patents. “Kexue” has received many awards, including the first prize of Science and Technology Progress Award from the China State Shipbuilding Corporation in 2014, the first prize of Science and Technology Progress Award from the Chinese Society of Naval Architects and Marine Engineers and the first prize of Shanghai Science and Technology Progress Award in 2015.

Fig. 9 3D High-performance multi-cable geophysical vessel "Faxian 6"

2.9 3D High-Performance Multi-Cable Geophysical Vessel "Faxian 6"

In recent years, the ocean has become the primary battleground for global oil and gas exploration. More investments are being put into offshore oil and gas exploration and development, particularly in enhancing the support for offshore oil and gas engineering operations and improving equipment standards, aiming to take a favorable position in future offshore oil and gas exploration and development. It has become a crucial direction for global energy companies to seek sustainable development. Sinopec[5] holds a significant number of exploration-promising sea areas and blocks both domestically and internationally, making the ocean a vital front for securing oil and gas resources of the company. Constructing a multi-cable geophysical vessel serves as one of Sinopec's initiatives to drive breakthroughs in offshore oil and gas exploration.

Sinopec's Shanghai Offshore Oil Bureau commissioned Rolls-Royce plc. of the UK to design and Shanghai Shipyard to construct a geophysical vessel. The construction of the vessel, named "Faxian 6" (see Fig. 9), began in September 2011 and was delivered in September 2013.

The vessel was designed with a length of 100.1 m, a molded breadth of 24 m, a draft of 6.5 m, a displacement of 10,882 tons, a speed of 16 knots, and a total active power of 8,000 kilowatts (two generator sets, 4,000 kW of each).

The "Faxian 6" is a 3D high-performance multi-cable geophysical vessel. It is capable of conducting large-scale 3D seismic operations in global waters, satisfying the highest environmental and waste discharge standards of most countries

[5] China Petrochemical Corporation.

worldwide. This successful delivery would significantly elevate Sinopec's deep-water exploration capacity. Not only would it symbolize Sinopec's collaboration with China State Shipbuilding Corporation (CSSC) to secure a leading position in high-end offshore engineering equipment, but it would also testify to China's ever-improving construction capabilities in the high-end deep-water equipment sector. Shanghai Shipyard assembled a team of highly skilled professionals from various fields in the production department to form the construction team for the "Faxian 6" project. They meticulously organized the design and construction to ensure a smooth completion and delivery of the vessel. The undertaking of the "Faxian 6" marked a solid step for the shipyard towards product structure adjustment, transformation, and upgrade. Throughout the vessel's design, construction, and commissioning process, Shanghai Shipyard tackled multiple challenges related to 3D modeling, weight and center of gravity control, progress control, vibration and noise control, and installation and commissioning of specialized geophysical equipment. 3D high-performance deep-water geophysical vessels are a crucial link in the industrial chain of offshore oil and gas exploration, development, and utilization. However, there was a limited number of such vessels in China, and the nation's 3D geophysical capabilities remained limited at that time. The "Faxian 6" geophysical vessel represented the latest international geophysical exploration technology. It is capable of navigating in global non-restricted waters and operating at water depths of up to 3,000 m. It can conduct 3D seismic acquisition operation and on-site data processing, equipped with a high-speed data transmission system. Furthermore, the vessel can collect seismic data even under sea conditions of up to grade 5 and a sea current speed of 3 knots. The vessel is equipped with 14 solid cables, each measuring 10,000 m in length. It also features a dual-seismic source air gun array consisting of 8 rows and a seismic integrated navigation system. At the time of its construction, it was considered one of the most advanced geophysical vessels in Asia.

As China's most advanced multi-cable geophysical vessel, "Faxian 6" features a double-hull design that meets high-level environmental standards. It complies with the clean design standards set by the Det Norske Veritas and the latest international and industry regulations, as well as the safety regulations for special purpose ships. The vessel has a more flexible crew configuration, a spacious working area, a towing capability of up to 14 cables, and a high level of automation.

During its 3D seismic exploration operations in the South China Sea, "Faxian 6" achieved a record-breaking daily production rate, marking a notable milestone of over 70 square kilometers covered in a single day, reaching 72.975 square kilometers. In the "Continuously Tackle Hard Issues and Create Greater Benefits" campaign, the vessel employed a collaborative approach across multiple departments to elevate operational efficiency. To minimize interference with fishing vessels and efficiently carry out operations, the deck team installed fishery walkie-talkies to enhance communication with fishing vessels. In areas with a high density of fishing vessels, dedicated boats were deployed to warn and persuade them to leave, thus ensuring the continuity of operations. The engine team intensified their inspection of the main engines on both sides to ensure reliable power support throughout the entire operation. The gas explosion team strengthened maintenance through regular inspections of key equipment such as the main bearer ropes, in-batch recovery of the air gun arrays, and replacement of any unstable or damaged components to ensure the reliable performance of the air gun arrays. The instrument team took advantage of turnaround periods to replace damaged hydrophones, remove fishing nets and fishing lines entangled in the cables, and thus to ensure smooth data collection. The navigation team utilized turnaround periods to replace damaged rudders, acoustic birds, and poorly performing trailers, and thus to ensure reliable functioning of underwater navigation equipment. Hard work paid off. With the concerted efforts of all departments, the vessel successfully achieved a daily survey area of over 70 square kilometers in 3D seismic exploration operations.

Over the years, "Faxian 6" has successfully completed multiple international bidding projects in offshore oil exploration. For example, in 2018, it undertook the South China Sea SK project, followed by the Husky project in 2019. In 2020, it participated in the Sino-ROK Cooperative South China Sea SK 3D Seismic Exploration Project.

2.10 *Multipurpose Oceanographic Research Vessel New "Xiangyanghong 10"*

To meet the demands of deep-sea marine research, the Second Institute of Oceanography of the Ministry of Natural Resources required a research vessel that could serve as a scientific research platform on the sea. It was planned to commission Zhejiang Taihe Shipping Co., Ltd. to raise funds and construct the required research vessel. In February 2011, the MARIC was entrusted by Zhejiang Taihe Shipping Co., Ltd. to design the research vessel.

Relying on their technical expertise, the MARIC collected, organized, assimilated, and digested cutting-edge technologies and design concepts from around the world for 4,000-ton-around research vessels. Leveraging their experience in similar ship type design projects, they formed a technical proposal that not only embodied the current development trends and advanced technologies of research vessels worldwide, but also possessed unique characteristics tailored to China's needs. Furthermore, they took into full consideration the characteristics and requirements of the Chinese users and incorporated their suggestions. Through meticulous demonstration, comprehensive optimization, innovative exploration, and independent design, they created a research vessel with complete intellectual property rights, distinctive features, and catered to China's national conditions.

This vessel is a 4,000-ton modern research vessel designed to meet the requirements of non-restricted waters navigation and global travel. It integrates multi-disciplines, multi-functions, and multi-technologies to cater to the cross-disciplinary deep-sea research. With these capabilities, it serves as a marine mobile experimental platform for basic marine research and research of high and new technologies.

The main scientific objectives that the vessel can achieve include: a. oceanic circulation system and climate change; b. ocean dynamic processes and hazards; c. deep-sea mineral products, organisms, gene resources, and biodiversity; d. oceanic ecosystem and carbon cycle; and e. mid-ocean ridges, hydrothermal systems at continental margins, and deep Earth processes.

As one of the main research vessels for China's multipurpose oceangoing research endeavors, the vessel undertakes multipurpose deep-sea survey missions in major ocean regions, including marine dynamic environment survey, geological environment survey, ecological environment survey, seabed resources exploration, comprehensive energy detection, comprehensive observation and experiment of military oceanography. These activities aimed at providing technical support for resolving critical issues related to marine resources, energy, national defense security, and mitigation of the impact of natural disasters at that time. At the same time, it was believed that the vessel would serve as an advanced platform for Chinese scientists to participate in major international marine research programs and further to enhance China's influence in the international marine research community.

The detailed design of the vessel commenced in 2011, with construction undertaken by Wenzhou Sino-Europe Shipyard Co., Ltd. from June 2012. The vessel was delivered in March 2014. Due to the modification of the original "Xiangyanghong 10" to the "Yuanwang 4", the name "Xiangyanghong 10" (see Fig. 10) was retained for the new vessel.

The vessel was designed with a total length of 93.0 m, a molded breadth of 17.4 m, a molded depth of 8.8 m, a designed draft of 5.5 m, a designed displacement of 4,400 tons, and a capacity of 65 persons (25 crew members and 40 researchers). It features a

Fig. 10 Research vessel new "Xiangyanghong 10"

maximum speed of 14.7 knots, a self-sustaining period of 60 days, and an endurance of 12,000 nautical miles.

The vessel's laboratories include data processing room, comprehensive geophysical laboratory, acoustic equipment compartment, gravimeter room, general chemistry laboratory, biology laboratory, moving multi-element control and observation room, geological and geotechnical laboratory, aft deck working control room, hydrological experiment monitoring area, pre-processing room, sample library, dedicated water treatment laboratory, weather workshop, and mobile laboratory (20-foot container), among others. The arrangement of the experimental facilities and laboratories, as well as their relationship with the working decks, complies with relevant standards and regulations while fully considering the workflow.

The vessel's main scientific research equipment consists of a CTD winch with a rope capacity of 10,000 m, an armored optical cable winch with a rope capacity of 10,000 m, a geological winch with a rope capacity of 10,000 m, a coaxial cable winch with a rope capacity of 10,000 m, a 30-ton A-frame on the stern, a 20-ton A-frame on the side, a side bracket mechanism for corers, and lifting fin. The main detection equipment includes CTD, ADCP, multi-beam submarine geomorphy detection system for full ocean depth, sub-bottom profiler, gravimeter, magnetometer, deep-sea towing system, and ultra-short baseline locating system.

Design features of the vessel:

1) The First Research Vessel Form with Gondola Appendage

As seabed exploration becomes increasingly demanding in terms of water depth and precision, ensuring an optimal acoustic environment at the bow of the vessel becomes crucial for acquiring accurate data. Over the past few years, newly built research vessels, both domestically and internationally, have faced significant compatibility issues between hull openings of bulb bows or channel-type bow thrusters and deep-sea multi-beam sounders. Some newly built research vessels abroad were sent back to the shipyard for rework due to the malfunctioning of their multi-beam sounders, with the bows being modified and the bulbs being removed. Similarly, certain domestic research vessels have also encountered issues with their multi-beam sounders, leading to the closure of hull openings for the bow thrusters.

To address these challenges, foreign research vessels took a two-fold approach in improving both the profile design at the bow and the installation methods of acoustic equipment. A primary focus was placed on the profile design of the bulb bow. In the design process, special consideration was given to the flow field at the bow to ensure that bubbles are generated behind the transducers of the acoustic equipment. However, failures were found both domestically and internationally. Secondly, different installation methods for acoustic equipment were employed. Different from the traditional embedded installation, two alternative methods, the use of domes or gondolas, were optional. In terms of the effectiveness of measuring, the Gondola installation method provides the best multi-beam effect. However, this installation method creates significant resistance, which affects the cost-effectiveness of the vessel. Additionally, it increases the draft of the vessel, requiring higher water depths at the dock. During navigation, special attention must be paid to issues such as entangling with fishing nets and stranding. The Gondola ship form has only recently emerged in newly built or refitted vessels abroad in the past few years, and there is no practical experience with it in China yet. The incorporation of the Gondola appendage necessitates comprehensive considerations of the ship's flotation condition, resistance performance, the appendage's main dimensions, as well as the arrangement, strength, vibration of acoustic equipment, among other factors. After repeated verification, digestion, and adoption of advanced foreign technologies, China successfully developed a research vessel form with Gondola appendage, which was adopted by shipowner.

This vessel is the first high-performance research vessel in China that was designed and constructed domestically, with complete independent intellectual property rights. It features a Gondola suspension platform for installing the acoustic transducer. Through nearly two years of actual ship verification, this ship form has demonstrated high measuring accuracy, excellent resistance, good seakeeping capability, and a high level of comfort.

2) Optimization of the Hull Main Dimensions

In terms of optimizing the main dimensions, a significant amount of work was done in studying, comparing, and analyzing domestic and foreign research vessels of similar displacements. Efforts were also made in verifying the vessel's performance indexes such as speed, seakeeping, stability, special wind resistance, and displacement. Taking into account factors such as the layout of scientific research equipment, configuration of the propulsion system, layout of the research laboratories and accommodation spaces, as well as considering the shipowner's investment scale and return on investment, the main dimensional parameters of the vessel were progressively optimized and ultimately finalized. Table 2 presents a comparison of information of several domestic and foreign research vessels with technical parameters similar to this vessel.

Through optimization of the ship form and main dimensions, combined with the specific functional layout of the new "Xiangyanghong 10", and considering the trend of decreasing length-to-breadth ratio in recent years for research vessels built abroad, the ratio related to vessel speed was set at 5.34, slightly lower than the average value (5.69) for similar vessels abroad. The ratio of molded breadth-to-designed draft, which relates to vessel stability, was set at 3.08, higher than the average value (2.944) for this type of vessel internationally, to ensure stability. Through actual ship verification, the vessel's performance indexes, including speed, seakeeping, stability, special wind resistance, scientific exploration capability, and accuracy, had all reached international advanced levels.

3) Optimization of Ship Resistance

The design team faced the lack of forefront technologies to resolve the significant interference caused by hull openings with bulb bows or channel-type bow thrusters in the proper operations of deep-sea multi-beam sounders, which existed both domestically and internationally. Therefore, careful consideration was given to whether this

Table 2 Comparison of data of several domestic and foreign research vessels with technical parameters similar to this vessel

Ship name	G.O.SARS	JAMESCOOK	"Kexue"	"Xiangyanghong 10"
Registry	Norway	UK	China	China
Delivery year	2003	2007	2012	2014
Main function	Multipurpose research	Multipurpose research	Multipurpose research	Multipurpose research
Total length/meter	77.5	89.2	99.8	93
Length between perpendiculars/meter	68.4	78.6	88.8	82.8
Designed molded breadth/meter	16.4	18.6	17.4	17.4
Molded depth/meter	9.1	9.5	8.9	8.8
Full load draft/meter	5.8	5.5	5.88	5.65
Full load displacement/ton	3800	5401	5027	4500
Personnel capacity/person	45	54	80	65

vessel required a bulb bow. Based on past experience, the drag reduction effect of a bulb bow is only significant when the Froude number (Fr) of the design speed exceeds a certain value. However, higher speeds do not necessarily result in better drag reduction from the bulb bow. Typically, the influence of a bulb bow on high-speed vessels ($F_r > 0.3$) is minimal. With an F_r of 0.27, indicating a medium-speed vessel, it was initially inferred that a bulb bow would have a noticeable drag reduction effect on this vessel. To further quantify this reduction, computational fluid dynamics (CFD) software was utilized to analyze the impact of the bulb bow on vessel resistance, comparing the resistance of two hull profiles involving one with a bulbous bow and one without. The results showed that the profile with a bulb bow exhibited significantly better performance. As the acoustic equipment of this vessel was installed in the Gondola form, it effectively avoided the generation of bubbles caused by the bulb bow. Additionally, the bulb bow can effectively reduce resistance. Therefore, it was determined to adopt a bulb bow profile design.

For the propulsion system, a domestic equipment manufacturer was contracted for the overall supply. However, since the manufacturer had not previously designed or produced propulsion systems within this power range, domestic propellers generally had lower efficiency, and there was an addition of the Gondola appendage, the resistance was increased significantly. Given these factors, it was necessary to place sufficient emphasis on optimizing vessel resistance to achieve international advanced levels of speed performance while using domestic propellers and incorporating the Gondola appendage.

After confirming the adoption of the bulb bow profile design, an additional 18 CFD optimized solutions were evaluated to determine the final profile design.

To facilitate comparison, the tonnage, speed, and propulsion power measured during speed tests were converted into a comprehensive unit index called the speed performance index. This index represents the speed that can be achieved by the tonnage of the vessel for each kilowatt of propulsion power. Table 3 presents a comparison of speed performance indexes of domestic and foreign similar vessels.

Although the speed index of this vessel is slightly lower compared to the "Kexue" vessel, it should be noted that the Gondola appendage used for multi-beam applications increases the resistance by approximately 25%. Additionally, the "Kexue" vessel utilizes ABB podded propeller, which is more efficient by over 10% compared to the domestically-made propeller used in this vessel. Taking these factors into consideration, it can be considered that the line resistance performance of this vessel is on par with the "Kexue" vessel.

4) Seakeeping Study

Acoustic detection equipment, such as multi-beam system, requires the ship platform to minimize motion in roll, pitch, and heave. Moreover, this vessel was also required to meet certain seakeeping requirements for many drifting and low-speed scientific research operations. The hull lines and general arrangement of this vessel were carefully considered to improve seakeeping performance. To verify its seakeeping performance, the vessel underwent seakeeping tests.

The tests confirmed that the vessel exhibited excellent seakeeping performance, satisfying the expected requirements. Additionally, to reduce the roll amplitude of the vessel during low-speed (0–10 knots) operations, the vessel was equipped with one controllable passive anti-rolling tank on each side and at the bottom of the auxiliary engine room. This ensures that the vessel maintains robust navigational and operational capabilities in severe conditions, such as high winds and waves.

The vessel's actual performance tests confirm that it can meet the requirements for ROV launch/recovery and helicopter hover operations in sea conditions of grade 4, as well as drifting survey operations in sea conditions of grade 5.

5) Special Wind Resistance

Although the new "Xiangyanghong 10" was only required to satisfy calculation criteria for "intact stability and damaged stability" according to regulations,

Table 3 Comparison of speed performance indexes of domestic and foreign similar vessels

Ship name	JAMESCOOK	"Kexue"	"Xiangyanghong 10"
Delivery year	2007	2012	2014
Whether equipped with Gondola appendage	No	No	Yes
Speed index (ton-knot/kW)	16.20	19.6	17.8

the design also included compliance with the industry standard *Special Wind Resistance Requirements for Oceanographic Survey Vessels* to enhance the vessel's wind resistance.

Since research vessels could face severe conditions, including extreme storms and waves towering over ten meters, it is crucial to ensure sufficient wind resistance, in order to avoid severe consequences.

Adhering to the principle of "people-oriented, safety first", wind resistance was fully considered in the design of the vessel's main dimensions and general arrangements. By meeting the requirements of the *Special Wind Resistance Requirements for Oceanographic Survey Vessels*, the vessel can safely operate under wind speeds (Force 12) of up to 51.5 m per second.

6) Ship-Wide Noise Control

Acoustic signals are the only physical information capable of long-range propagation in seawater. Research vessels are equipped with numerous acoustic detection devices, which require effective control of the vessel's own underwater radiation noise to ensure these devices' operational accuracy and improve detection range. Additionally, for vessels involved in marine biological survey and survey tasks, it is important to consider the protection of marine ecosystems and biodiversity. Therefore, research vessels must also incorporate calming design.

Within the context of calming design, the control of underwater noise necessitates compliance with two main internationally recognized standards at present, including ICES CRR 209 and DNV SILENT Class Notation, or similar requirements. These standards provide explicit guidelines for the measuring and limitation of underwater radiation noise for research vessels. Currently, research vessels that meet the requirements of ICES CRR 209 and DNV SILENT-R for underwater radiation noise limitations typically employ traditional propulsion systems, consisting of propulsion motors, shafts, and propellers. Low-noise fixed-pitch propellers are generally used.

The shipowner, considering investment scale and maneuverability, opted for a main propulsion system utilizing domestically-produced all-direction rudder propellers. With regard to the bow thruster, a channel-type bow thruster remained being adopted. Thanks to the use of a Gondola, the impact on the acoustic transducer of the thruster was minimized. In addition, various vibration and noise reduction technologies were employed, such as vibration reduction control for vibration sources, vibration reduction control for fans and pipelines, sound insulation control for air conditioning systems and compartments, and the laying of damping materials.

The generator set, pump set, air compressor unit, refrigeration unit, ventilation duct, exhaust pipe, and their mufflers are the main vibration sources and noise sources on the vessel.

The vibration control scheme refers to the vibration isolation of high-power vibration source equipment. On this vessel, different isolation approaches were used for different power equipment.

(1) Double-layer vibration isolation for generator: Diesel generator sets with large vibration adopt double-layer vibration isolation devices instead of the single-layer vibration isolation devices originally provided by the manufacturer. The

vibration level difference of the single-layer vibration isolation is generally 10–20 decibels. Double-layer vibration isolation can isolate the high-frequency vibration part that single-layer vibration isolation cannot isolate. The vibration level difference of the double-layer vibration isolation in the low-frequency range can generally reach 30–40 decibels, effectively reducing the transmission of mechanical equipment vibration and structural noise to the hull. It also protects the mechanical equipment when the hull experiences impacts to ensure safer and more reliable operation. Therefore, the vessel adopted the advanced double-layer vibration isolation technology. Other power equipment adopted single-layer vibration isolation devices and sound insulation devices. Auxiliary pipelines, piping, and cables connecting the elastic-mounted equipment adopted flexible connections with displacement compensation function.

(2) Vibration and noise reduction control for fans: The vessel was equipped with 16 types of fan equipment, which provide mechanical ventilation and other functions for the entire vessel, with a wide range of flow rates and power coverage. Fan equipment is a significant source of vibration and noise on the vessel. Depending on the power, ventilation quantity, compartment location of the fan installed, and expected noise levels in the respective compartments, differential vibration and noise reduction measures were taken for the fans. For the engine room ventilation fan, which is an axial flow fan with high power and large air volume, measures such as installing mufflers and vibration isolators were adopted for vibration and noise reduction control.

(3) Vibration and noise reduction for ship-wide pipelines: Various pipelines were distributed throughout the vessel. The fast flow of liquids and gases through these pipelines could cause pulsation effects due to fluid–structure interaction, transmitting vibrations from the pipelines to the compartments, decks, and other structures. To address this, measures such as the use of elastic pipe clamps, flexible hangers, and rubber flexible connectors were employed for pipeline vibration and noise reduction.

(4) Vibration and noise control for the air conditioning system: To ensure the proper functioning of acoustic instruments and equipment, measures were implemented for the entire air conditioning system to reduce vibration and noise. These measures included the use of low-noise air conditioning equipment, increased thickness of equipment sound absorption materials adopting sound absorption and vibration reduction design, and other means to minimize equipment noise. Low-noise marine centrifugal and axial flow fans were employed to reduce noise at the source. Water chilling units and air conditioners were placed on the lower deck as possible to maintain a distance from living compartments and laboratories and prevent direct noise transmission and avoid resonance between the equipment and the hull. The air conditioners in some compartments adopted air-cooled split cabinet units, with the main noise sources such as refrigeration compressors and condensers located outside the compartments. Flexible connections were adopted in the inlet and outlet pipelines of ventilation fans and water pumps. In addition, the fans, water pumps, and motors were installed on vibration reduction bases to minimize noise propagation. Centrifugal and axial

flow fans were equipped with soundproofing static pressure boxes to reduce fan noise. Soundproofing accessories were installed in some ventilation ducts as necessary. Moreover, high-quality, low-noise air supply end devices were equipped to keep the noise within the required range in the air-conditioned rooms.

(5) Compartment sound insulation control: Sound insulation control in compartments primarily focused on controlling the structural vibrations and air noise generated by the operating power equipment. This was achieved by applying sound absorption and sound insulating materials, as well as damping materials, to the interior walls of the compartments. In addition, soundproof doors and windows were also adopted, making the compartments enclosed spaces. Besides, attention was paid to sealing and preventing sound leakage to achieve effective noise control within the compartments.

    Sound absorption materials and insulating materials were selected according to arrangement requirements. Various materials such as alumina plates, mineral wool composite boards, and glass wool were used. Different sound absorption measures and materials were used based on the specific compartments and their locations.

(6) Arrangement of damping materials: Damping materials can reduce energy transmission, suppress low-frequency vibrations of wallboards, absorb and isolate air noise, which can improve the overall acoustic environment of a ship. By covering the insulation layer and the cement or steel plates on the layer to create a floating floor, a strong noise reduction effect can be achieved. Damping materials were specifically laid in critical compartments aboard the vessel.

After implementing the aforementioned measures, the vessel underwent an effect evaluation using advanced vibration and noise prediction software. The evaluation results indicated that the vessel's vibration and noise levels fully met the requirements for comfortable navigation and working conditions. Additionally, it was confirmed that the noise levels within the compartments exceeded both national and IMO standard requirements. Actual ship noise tests further validated the predicted results, earning recognition from the shipowner and the classification society.

This vessel is the first multipurpose research vessel built in China with co-investment from private capital and state funds, and has been officially included in the national oceanographic survey vessel fleet. During the construction process, the shipowner, designer, constructor, and inspector worked closely to complete the project. The entire construction period, from project initiation to delivery, was just 36 months, marking a new record for the construction of research vessel projects.

After more than a year of actual operation, the vessel's performance indexes were verified and proven to be effective.

Except for some scientific research equipment, key equipment such as propulsion systems and main generator sets were all domestically produced. The vessel also adopted the domestically designed and built Gondola, which could reduce the need for equipment such as lifting fins while ensuring the measuring accuracy. The total cost of the vessel is approximately RMB 200 million, which is significantly more economical compared to similar domestic vessels with a cost of RMB 500 million.

In March 2014, the new "Xiangyanghong 10" joined the national scientific research fleet. Since its delivery, the vessel has been engaged in scientific research operations at sea. Its operation sea areas have covered a large area of the Western Pacific and the Indian Ocean, making it an important member of China's ocean-going scientific research fleet. It has made significant contributions to China's marine research undertakings alongside its substantial economic and social benefits. As of March 2016, the vessel has spent 487 days on marine research, generating a value of RMB 107.05 million and contributing RMB 40 million in taxes and profits. The actual operational performance demonstrates that the vessel has excellent operating capabilities, earning praise from the shipowner and the crew.

In July 2017, the new "Xiangyanghong 10" research vessel completed its 43rd ocean voyage. The voyage spent for more than 230 days on sea, divided into five segments for scientific research purposes. The main task was to fulfill the "Southwest Indian Ocean Polymetallic Sulfide Resource Exploration Contract" and carry out surveys on mineralization anomalies and delineation of ore bodies/mineralized bodies with in the sulfide contract area. Several sets of equipment independently developed by China achieved phased success during the practical application, providing preliminary insights into the spatial structure features of typical sulfide ore bodies within the contract area.

Notably, in this voyage, the new "Xiangyanghong 10" carried the independently developed 4,500-m deep-sea resources autonomous survey system – "Qianlong 2" autonomous underwater vehicle (AUV). "Qianlong 2" conducted eight submersible operations during this expedition, totaling 170 h of operation time, covering a total distance of 456 km, with a maximum diving depth of 3,320 m. This fully validated the stability and reliability of "Qianlong 2" in complex seabed terrain along the mid-ocean ridges.

In addition, the mid-deep hole rock core drill and transient electromagnetic detection system were also successfully utilized during this scientific expedition. The combined use of these technologies signifies that China's exploration of polymetallic sulfide resources is experiencing high-tech upgrades.

The 49th ocean voyage of the new "Xiangyanghong 10" lasted 250 days in total and covered a distance of 29,821 nautical miles. Significant achievements were made in the exploration of sulfide resources in the Southwest Indian Ocean contract area, the investigation into the tectonic evolution and hydrothermal systems of the slow-spreading Carlsberg Ridge, and the investigation into microplastic pollution in the Indian Ocean.

This voyage, organized and implemented by the Second Institute of Oceanography under the Ministry of Natural Resources, was divided into five segments, mainly in the Indian Ocean. With the "Jiaolong's Sea Explorations" project as its core, the voyage featured multiple tasks, resource and environment exploration, highly precise investigation, and targeted international marine hotspots. It yielded the following outcomes:

(1) New progress was made in sulfide exploration in the Southwest Indian Ocean contract area. During this voyage, a total of three mineralized zones, two mineralized anomaly zones, and nine anomaly zones were discovered. This enriched the knowledge of the control mechanism of polymetallic sulfide distribution, metallogenic regularities, and resource evaluation in the contract area, and further revealed the prospect of sulfide deposits in ultrslow spreading ridges.
(2) New insights were gained into the tectonic evolution and hydrothermal systems of the slow-spreading Carlsberg Ridge. A new high-temperature vent cluster was discovered, expanding the distribution range of hydrothermal fluid activity. New evidence was obtained that this hydrothermal system is controlled by the tectonics and petrology of detachment faults. Three new hydrothermal systems were discovered, bringing the total number of hydrothermal systems found by China in the region to nine. Among them, the Baochuan hydrothermal zone is the furthest (20 km) confirmed hydrothermal system from the central rift, which provides the possibility of creating a new model for mineralization control in slow-spreading ridges. In the fourth segment of the voyage, the scientific research team successfully retrieved 14 seabed seismometers, which were deployed during the second segment of the voyage in March 2018. The deployment area is located in the "Yuhuang" mineralized zone of the Southwest Indian Ocean polymetallic sulfide exploration contract area, at water depths ranging from 1,400 m to 3,300 m. These seabed seismometers form a temporary seismic array above the mineralized zone, allowing the monitoring of small earthquakes occurring in the seabed hydrothermal zone and the recording of global large earthquakes. Similar to a doctor's stethoscope, they could listen to the Earth's internal vibration signals and provide insight into the Earth's condition. By analyzing the data from the seabed seismometers, scientists could interpret the temporal and spatial information of microearthquakes occurring in the "Yuhuang" mineralized zone, and further understand the submarine structure of this area, thus providing important foundational information for sulfide exploration in the contract area.

In March 2019, approved by the Ministry of Natural Resources, the new "Xiangyanghong 10" departed from Zhoushan to carry out the 54th ocean voyage of survey mission in the Pacific Ocean, focusing on resource and environmental investigations. This voyage was led by the Second Institute of Oceanography of the Ministry of Natural Resources, which served as a crucial voyage in fulfilling the deep-sea resource and environmental investigation tasks of the Ministry in the Pacific region for 2019. It was characterized by a large spatial coverage, extended duration, and diverse tasks. The main objectives of this mission included fulfilling the polymetallic structure exploration contract between the China Ocean Mineral Resources R&D Association, China Minmetals Corporation, and the International Seabed Authority, and conducting a specialized investigation on global change and the air-sea interaction phenomenon.

The voyage was divided into five segments, with a total sailing time of 255 days and a total distance of approximately 22,000 nautical miles. It was planned to make stops in Pohnpei, Micronesia, and Acapulco, Mexico for resupply and crew rotation. The successful implementation of this voyage has proved to be effective support for China's obligations in fulfilling exploration contracts and enhanced our scientific understanding of deep-sea regions.

2.11 Multipurpose Research Vessel "Xiangyanghong 06"

In March 2011, the SOA commissioned CSSC Chengxi Shipyard Co., Ltd. (CSSC Chengxi) to refit a small container ship "Haoli", which had been in service for nearly 20 years, with a single engine, single propeller, and single rudder, into a research vessel. CSSC Chengxi completed the first refit in 2012, naming the vessel "Xiangyanghong 06" (see Fig. 11). In December 2015, "Xiangyanghong 06" underwent a second refit at the Rizhao Gangda Shipbuilding Heavy Industry Co., Ltd., Shangdong, which was completed in 2016. In November 2016, the vessel was delivered to the North China Sea Branch of the SOA.

During the second refit, a series of challenging engineering items were completed while maintaining the vessel's main dimensions, focusing on the segmented removal and replacement of the hull after FR 26. The refitting items involved the vessel's main power system, engine room automation control system, network system, and whole-ship electric power system. The original configuration of a single engine, single propeller, and single rudder was transformed into a dual-engine, dual-propeller, and dual-rudder configuration. Additionally, equipment such as deep-water multi-beam system, dynamic positioning system, magnetometer, bow thruster, and anti-rolling fin were installed to enhance the vessel's overall performance. After the

Fig. 11 Multipurpose research vessel "Xiangyanghong 06"

refit, the vessel was able to withstand winds up to Force 12, allowing for global scientific research expeditions. This refit transformed "Xiangyanghong 06" into a high-end marine survey platform primarily focused on geophysical exploration, with versatile capabilities to cater to the needs of multidisciplinary research missions. Its functions include deploying oceanographic observation buoys, deploying subsurface buoys, monitoring marine environments, conducting comprehensive ocean surveys, and undertaking law enforcement patrols.

The vessel was designed with a length of 91 m, a molded breadth of 14.7 m, and a molded depth of 7.6 m, fitting with a bow thruster. It has a full load displacement of 4,900 tons, a maximum speed of 13.5 knots, an endurance of 15,000 nautical miles, a self-sustaining period of 60 days, and a capacity of 60 persons.

"Xiangyanghong 06" features various laboratories and data information centers, covering a total area of 267 square meters. The vessel is also equipped with advanced domestic instruments and equipment, including Doppler acoustic beam profiler, myriameter depth sounder, 8,000-m geological winch, A-frame, L-frame, and other equipment.

Another feature is the navigable equipment, which enables the vessel to collect various samples while being underway, eliminating the need to stop and take samples at fixed points. The vessel also boasts sub-bottom profiling capabilities. When it passes through an area, the equipment can analyze the geological structure of the corresponding seabed. In addition, advanced communication and navigation devices are equipped. They enable high-speed data communication, allowing for the transmission of experimental data and multi-channel communication with the land.

From April to July 2020, "Xiangyanghong 06" embarked on a 96-day voyage in the Pacific Ocean, sailing for the 2020 Pacific survey of the North China Sea Branch of the Ministry of Natural Resources, called "Global Change and Air-Sea Interaction" project.

This voyage marked the inaugural ocean survey voyage undertaken by the North China Sea Branch of the Ministry of Natural Resources in 2020, amidst the global spread of the COVID-19 pandemic. Prior to the voyage, potential challenges were fully considered and extensive preparations were made, including pre-voyage nucleic acid testing, stocking up on epidemic prevention supplies, ensuring sufficient fuel and food supplies, and changing the planned multi-stop voyage into an uninterrupted one. In this way, the annual key survey tasks were successfully implemented.

A crew of 52 members from the North China Sea Branch and the Third Institute of Oceanography both under the Ministry of Natural Resources ventured into the forefront of ocean survey, full of resolution for scientific anti-epidemic with no fear. Over a span of more than three months, they overcame multiple challenges in terms of the vessel's endurance, equipment capabilities, crew members' perseverance, water depth in the basin, the large number of sampling stations, and continuous high-load operations. In the end, they successfully fulfilled the survey missions in the Pacific Ocean.

During this voyage, "Xiangyanghong 06" conducted multidisciplinary surveys in the Pacific Ocean high seas area, involving marine geology, geophysics, hydrology, biology, and others. It collected geological samples from an average water depth of 5,700 m at over 110 stations and acquired comprehensive geophysical survey data covering nearly 10,000 km. These valuable first-hand data significantly contributed to the study of climate evolution, environmental characteristics, and geophysical features in the Pacific Basin.

On November 24, 2020, "Xiangyanghong 06" set sail from Zhoushan, Zhejiang, to carry out the Winter Survey Voyage of the Indian Ocean Joint Marine and Environmental Research Program organized by the Second Institute of Oceanography, Ministry of Natural Resources. The aim of this voyage was to study the changes in the marine ecological environment in the Indian Ocean and its response to monsoons, climate, and human activities. It focused on frontier topics such as the formation mechanism of hypoxia in the sub-surface water of the Bay of Bengal and its impact on the surrounding marine ecosystems. The voyage aimed to promote the construction of a maritime community with countries along the Maritime Silk Road including Sri Lanka, Myanmar, Thailand, and Bangladesh.

International cooperation was one of the highlights of this voyage. The Second Institute of Oceanography of the Ministry of Natural Resources invited researchers from Myanmar, Sri Lanka, Thailand, and other countries to participate in the scientific expedition. Prior to the departure, a 5-day seminar was held in Hangzhou. In recent years, the Institute has conducted multiple exchanges with marine institutions from Sri Lanka, Myanmar, Thailand, the Philippines, the Maldives, and other south and southeast Asian countries. It has signed several bilateral cooperation agreements with some of these institutions. Moreover, the Institute has hosted many international and regional conferences, laying a solid foundation for marine scientific and technological cooperation and cultural exchanges.

This scientific expedition was a collaborative effort involving interdisciplinary and multiple organizations. The State Key Laboratory of Satellite Ocean Environment Dynamics and the Key Laboratory of Marine Ecosystems and Biogeochemistry of the Institute were the main units responsible. Other participating organizations included the National Marine Environmental Forecasting Center, the National Ocean Technology Center, the North China Sea Marine Engineering Exploration and Research Institute, the Shenyang Institute of Automation of the Chinese Academy of Sciences, and Deepinfar Co., Ltd.

2.12 Multipurpose Marine Test Vessel "Dianke 1"

"Dianke 1" (see Fig. 12) is a multipurpose marine test vessel primarily responsible for the development of marine information technology, electronic science experiments, and supporting the construction of bases on islands and reefs in the South China Sea. It serves as a comprehensive platform for demonstrating new marine technology equipment, conducting test and training activities, and providing engineering support.

While "Dianke 1" falls under the category of scientific survey vessels, it differs from China's "Xiangyanghong" series and other research vessels in that it focuses on

Fig. 12 Multipurpose marine test vessel "Dianke 1"

the research and development tests of ocean exploration, search, and sensing products. For a long time, China had lacked such a specialized platform for ship test. As the leading enterprise in China's electronics industry, China Electronics Technology Group Corporation (CETC) produces a wide range of electronic products. However, in the past, CETC had relied on renting non-specialized ships for testing during the R&D of new products, resulting in various shortcomings such as poor sailing capabilities, slow data processing speeds, and long testing cycles, among others. The introduction of "Dianke 1" effectively improved this situation.

The vessel was designed with a length of 89.16 m, a molded breadth of 18 m, a molded depth of 5.5 m, a navigational speed of 13.5 knots, a displacement of 3,000 tons, an endurance of 5,000 nautical miles, a self-sustaining period of 40 days, and a capacity of 45 persons (including 15 crew members and 30 scientific and technical personnel).

The vessel was designed by Guangzhou Marine Engineering Corporation, commissioned by Institute of Marine Information Technology, CETC. It was constructed by Yangzhou Longhe Shipbuilding Co., Ltd and delivered in August 2016.

"Dianke 1" features a steel hull with a continuous deck and forecastle, as well as a forward-leaning bow column in the shape of a bulb bow. It is equipped with bilge keels on both sides, and a bow thruster at the bow. It belongs to the category of twin-engine, twin-propeller, twin-rudder, stern-engined ships, and the propulsion is achieved through two marine diesel engines driving fixed-pitch propellers via a gearbox. The vessel's forecastle deck is equipped with three layers of deckhouses.

To facilitate electronic testing tasks, a spacious test deck, as well as a large crane, an A-frame at the stern, and a helicopter landing platform are equipped. This allows for the testing of equipment, such as radar, satellite communication device, unmanned aerial vehicle, unmanned surface vehicle, and conducting hydroacoustic and meteorological detection work. Additionally, the vessel is fitted with multiple laboratories, including the general command room, high-frequency cabinet room, computer room, and conference data room, among other specialized research compartments. The following systems are configured for carrying out various tests.

1. Test Deck

"Dianke 1" primarily conducts test tasks on the compass deck and the main deck's open test area, which has a total area exceeding 1,000 square meters. As the test tasks mainly focus on electronic equipment for sea and air, there is a substantial need for open deck space to accommodate lifting equipment, helicopter platforms, unmanned equipment support platforms, electronic equipment bases, and test operation areas. In terms of ship form design, "Dianke 1" has a longer open stern deck. Compared to other research vessels with similar displacement, it has a lower molded depth and a larger breadth-to-draft ratio, so as to better meet the spatial requirements of the test deck.

2. General Electronic Test Interface

To facilitate various types of sea trials, "Dianke 1" is equipped with an open mission electronic support system, providing a universal electronic test interface for different test equipment installations.

In addition to the required communication and navigation devices in line with the regulations of the ship classification society, "Dianke 1" reserves sufficient platforms and space to provide a sound sea trial environment for various electronic mission systems. The general arrangement of the vessel effectively divides navigation equipment and test equipment through the design of two masts. The navigation mast, employing a truss structure, carries navigation radar antennas for X-band and S-band, C-band satellite communication, Beidou receiver, anemorumbometer, navigation lights, signal horn, lightning rod, and other facilities to meet navigation requirements. The integrated test mast incorporates multiple platforms for conducting offshore tests of new developed radar, satellite communication device, radio communication device, and other products. The mast is fitted with a wave-penetrating antenna cover to maintain a temperature- and humidity-controlled environment for electronic testing.

A sufficient safety distance is maintained between the two masts to prevent interference, thus ensuring electromagnetic compatibility. This guarantees that the integrated test mast's operations do not affect the signal horns, navigation lights, and communication during navigation, and that navigation signals do not impact the precision of electronic equipment tests.

Meanwhile, the compass deck of "Dianke 1" is pre-fitted with universal interfaces for disc and whip antennas, as well as standard general test cabins. By incorporating reserved space and data interfaces, the vessel allows for different test tasks executed for each voyage, such as installing control cabins for underwater robots or acoustic transmitters, as well as other modules.

3. Hydroacoustic Test System

"Dianke 1" is equipped with a robust hydroacoustic test system to facilitate testing of underwater sonar systems, cable arrays, magnetometers, and deep-diving submersibles of all types.

A new type of sonar boom LARS is installed on the side of the vessel. This system allows for the launch and recovery of small underwater sonar equipment during low-speed sailing or mooring. The system is composed of a fixed base mounted on the starboard side of the deck, a bearing assembly, an electric hoist, a boom, an outboard limit device, and other components. It can be launched and recovered by slowly rotating the bearing assembly fixed on the base. To ensure the reliability and safety of the launch and recovery of the system, an additional crescent-shaped limiting device is installed on the outside of the hull.

At the stern, there is an A-frame with a lifting capacity of not less than 10 tons. This A-frame can be used in conjunction with a towing winch to conduct towing experiments of large acoustic equipment for geological, hydrological, chemical, and biological operations.

4. Unmanned Equipment Test System

"Dianke 1" can serve as the mother ship for testing unmanned aerial vehicles (UAVs) and unmanned surface vessels (USVs) at sea. For UAV testing, the vessel provides excellent storage, takeoff and landing, supply, and maintenance conditions. Additionally, it is equipped with an automatic dependent surveillance–broadcast (ADS-B) aircraft tracking system that enables the reception of aircraft position, speed, and other parameters. An unmanned surface vessel test support system is also available onboard, comprising capabilities for offshore oil and water replenishment, launching and recovery of USVs, video monitoring and tracking, as well as data processing and analysis. These efforts broke the limitation of USV only being able to conduct experiments in coastal areas, allowing for USV testing in remote offshore areas, ensuring a more reliable assessment of their performance in executing missions.

5. Integrated Bridge System

"Dianke 1" incorporates an integrated bridge system that combines navigation, control, detection, perception, display, monitoring, management, and communication functionalities. This system ensures the efficient and safe execution of testing tasks. By integrating electronic signals from the engine room, navigation equipment, and testing equipment, an intelligent computer server-centric network is established. Furthermore, the vessel can be connected via satellite communication to other research vessels and onshore laboratories, facilitating real-time two-way transmission and processing of test data at sea.

During the testing process, the integrated bridge system provides real-time comparison of test data from electronic equipment and perception data from onboard electronic equipment. This process enables smooth proceeding without interference with multiple tasks and processes underway, guaranteeing the sharing of information throughout the testing process.

6. Helicopter System

To enhance the capabilities of commanding beyond visual range, rapid transportation of personnel and supplies, as well as the deployment of equipment and experimental specimens at sea, among other functionalities, "Dianke 1" is fitted with a helicopter support system. Following the Category III safeguard requirements outlined in the *General Requirements for Helicopter Facilities of Non-aviation Ships* (GJB543A—1998), the vessel features a landing platform for Z-9 helicopters, a fast mooring device, a visual landing aid system, a communication system, and relevant service compartments. The landing platform area is fitted with collapsible guardrails that can be folded down during helicopter takeoff and landing, and raised as protective barriers when not in use for preventing deck personnel from accidentally falling into the water.

The "Dianke 1" is a specialized support vessel for electronic product testing. It has achieved significant technological advancements in ship form development and ship functionality. It has addressed various deficiencies encountered in previous electronic equipment tests, including unfavorable working conditions, harsh environments, slow data processing speeds, and lengthy testing periods, among others. As the first specialized test vessel in China constructed with investments from industrial sectors, "Dianke 1" has significantly accelerated the development of domestically-produced marine electronic information equipment.

2.13 Multipurpose Oceanographic Research Vessel "Xiangyanghong 01"

Tracing the development history of China's research vessels, the "Xiangyanghong" series began its development in the late 1960s, with the "Xiangyanghong 01" boasting the first vessel in this series. In March 1965, the SOA issued a technical assignment for the design of the 614-I type marine hydrological and meteorological survey vessel to the MARIC. The preliminary design was completed in June, followed by a design review meeting for this vessel on June 28 in Beijing. The construction design of the vessel was completed by the MARIC in December of the same year. The construction of the vessel began in February 1967 at Jiangnan Shipyard, and the first-generation survey vessel, "Xiangyanghong 01" (see Fig. 13), came out on December 14, 1969.

The first-generation survey vessel, "Xiangyanghong 01", was designed with a total length of 65.22 m, a molded breadth of 10.2 m, a molded depth of 4.8 m, a displacement of 823.94 tons, and a maximum speed of 15 knots. It was equipped with two 8350Z main engines, each with a rated power of 772 kilowatts (1050 horsepower).

This marine survey vessel has served in the East China Branch of the SOA for over 20 years, making significant contributions to China's early marine hydrological and meteorological surveys, forecasts, and research.

Fig. 13 First-Generation survey vessel "Xiangyanghong 01"

The new-generation multipurpose oceanographic research vessel "Xiangyanghong 01" (see Fig. 14) was designed by the MARIC, built by the Wuchang Shipbuilding Industry Group Co., Ltd., and operated and managed by the First Institute of Oceanography of the SOA. The vessel was designed based on the "Kexue" vessel of the Institute of Oceanology, Chinese Academy of Sciences, with optimizations and improvements being made. Construction of the vessel commenced in June 2013. It was launched in September 2015, and officially delivered for use in 2016. It boasts the new generation of Chinese vessel that possesses non-restricted waters navigation capabilities and all-weather observation capabilities, meeting the requirements for multidisciplinary research in deep-sea marine science as a modern multipurpose oceanographic research vessel. Its technological level and research capabilities have reached international advanced levels.

The vessel was designed with a length of 99.8 m, a molded depth of 8.9 m, a molded breadth of 17.8 m, a draft of 5.6 m, a full load displacement of 4,980 tons, a capacity of 80 persons, a self-sustaining period of 60 days and an endurance of 15,000 nautical miles.

The research equipment onboard "Xiangyanghong 01" includes the water detection system, atmospheric detection system, seabed detection system, deep-sea detection system, and remote sensing information field validation system. It covers multiple disciplines such as geophysics, marine physics, ocean remote sensing, marine acoustics, marine atmosphere, marine observations, marine biology, marine chemistry, and others. The vessel has comprehensive ocean exploration capabilities, enabling detection from the atmosphere, sea surface, water as well as seabed, with a detection depth of up to 10 km. It can meet the global demands for marine environmental and resource scientific surveys.

The bow of "Xiangyanghong 01" is equipped with two bow thrusters, and the stern is fitted with twin-pod all-direction propellers. This configuration allows for flexible maneuverability and precise positioning, thanks to computer systems and

Fig. 14 New-generation multipurpose oceanographic research vessel "Xiangyanghong 01"

sensors that accurately determine the ship's position, course, and track. In wind conditions of Force 5 to 6, the ship's positioning accuracy ranges from 0 m to 3 m. With its high level of automation and informatization, the vessel is equipped with electronic charts, an autopilot system, and the capability for one-person operation. It also features unmanned engine rooms and fully automated monitoring and alarm systems. Two independent networks are configured for ship operations and scientific research, enabling seamless data sharing and exchange. The satellite broadband network can met all offshore communication requirements and enable ship-to-shore video conferences. A digital video surveillance system is also equipped for the vessel.

The control support system for scientific survey operations of "Xiangyanghong 01" includes four winches (10,000-m), along with a myriameter single-beam depth sounder and a full-depth multi-beam system. Additionally, various specialized survey equipment, including ADCP, USBL, fish finder, sub-bottom profiler, continuous and automatic surface multi-element measuring system, and multi-channel seismic system, are provided. With these cutting-edge facilities, "Xiangyanghong 01" is set to contribute significantly to China's tapping into deep-sea marine research. It will serve as a major force for China's multipurpose oceanographic research. "Xiangyanghong 01" has undertaken major marine survey tasks in various deep ocean areas, covering multipurpose oceanographic survey, ocean dynamic environment survey, geological environment survey, ecological environment survey, seabed resource exploration, and comprehensive energy exploration, among others. These efforts have provided crucial support in addressing pressing national issues related to marine resources, energy, and natural disaster impacts. Additionally, "Xiangyanghong 01" has played a positive role in promoting China's participation in major international marine research programs and enhancing the country's influence in the field of marine science and technology, contributing to the realization of China's vision to become a marine science and technology powerhouse.

While the "Kexue" vessel, the mother ship of "Xiangyanghong 01", received praise from the shipowner and industry experts, post-delivery feedback highlighted aspects that required further exploration or improvement. In response, "Xiangyanghong 01" embarked on an extensive original design process that focused on four key aspects, including upgrading to meet international rules and regulations, renewing shipborne equipment, customizing solutions based on user needs, and improving the overall design of the mother ship.

Design of this research vessel was heavily influenced by the ordering of shipborne scientific research equipment. Compared to its mother ship, "Xiangyanghong 01" has undergone a comprehensive upgrade of its onboard equipment, particularly the control support equipment for scientific research, of which nearly all equipment suppliers were replaced. With higher performance requirements, a deck deck control support system with improved capabilities for deployment, maintenance, and reliability was employed. Due to significant variations in performance parameters of the main crane and scientific research winches on "Xiangyanghong 01", the hull structure of the vessel underwent extensive reinforcement and modifications on the premise that the vessel's weight was controlled in a reasonable range. The lifting fin of the mother ship was gravity-driven, and improper maintenance could lead to difficulties in its operation and positioning due to accumulation of oysters. Therefore, based on the existing structural limitations, the design team of "Xiangyanghong 01" proposed a hydraulic push method to drive the lifting fin, which has proven to be reliable in practice. They also modified the installation position of the podded propeller, as well as upgraded and updated various facilities, such as the cover of the channel-type side thruster, the CTD control system, the stern A-frame, and the gravity corer flip mechanism.

In addition, based on the needs of the the Institute of Oceanography, "Xiangyanghong 01" had adjustments to some of its compartments and laboratories to better cater to the shipowner's requirements and the operational procedures.

Taking into account the practical needs of scientists and the environmental requirements of scientific research instruments, the design team developed targeted solutions to the issue of high temperatures in the air compressor compartment and gravimeter room, as reflected by the "Kexue" vessel. In addition, "Xiangyanghong 01" was equipped with high-efficiency filters and high-power air feeders to realize a purification level of 100,000 in the general clean laboratory. Through thorough research and analysis, high-quality cables with thinner sizes and better heat dissipation were installed for the entire vessel to ensure safe and reliable operation of the vessel.

From the perspective of regulatory updates, many specifications for special-purpose vessels were not yet implemented in China when the "Kexue" vessel was originally designed. However, in the reconstruction of "Xiangyanghong 01", the classification society implemented stricter regulations, posing numerous challenges to the design team. Nevertheless, the measures taken to address these challenges paved the way for the future design of similar vessels.

On October 19, 2016, "Xiangyanghong 01" embarked on its inaugural voyage from Qingdao, commencing a more than two-month national special scientific research project, called the "Global Change and Air-Sea Interaction" autumn voyage for comprehensive investigation into south water bodies in the East Indian Ocean. The expedition covered a distance of over 13,000 nautical miles and conducted extensive, cruising, and drifting buoy observations in the south waters of East Indian Ocean. Multidisciplinary field investigations were carried out during this expedition, involving marine physics, marine meteorology, marine chemistry, and water biology, and marine optics. The main objectives of the expedition were to provide a data foundation for understanding the characteristics of monsoon circulation fields in the Indian Ocean and the impact of air-sea interactions on East Asian climate change. Additionally, the expedition also sought to deepen knowledge of the structures and influences of equatorial and boundary current systems during the monsoon transition period, the processes of air-sea interaction during this period, the transition from the South Asian summer monsoon to the winter monsoon, the relationships between various marine chemical elements and the biologically active layer during autumn, and the response of plankton community dynamics to monsoon circulations.

During this maiden voyage, "Xiangyanghong 01" successfully applied new technologies, such as the dynamic positioning system, anti-rolling tank, and automatic CTD winch system, in multipurpose deep-sea research and surveys. With the support of the dynamic positioning system, even in adverse sea conditions for operations, the horizontal deviation between the sampling equipment's entry and exit points did not exceed one meter. This enabled real precise fixed point operations and the collection of high-precision data and fixed point samples.

Through the research conducted during this voyage, valuable first-hand data on multiple elements in the south waters of East Indian Ocean was obtained. For the first time, a systematic investigation was conducted to acquire multidisciplinary data on key areas of Indian Ocean dipole events during their peak periods, providing rich data for studying the evolution, development, and extinction mechanisms of these events and their impact on climate change.

During this expedition, two sets of "Bailong" buoy systems were successfully deployed, marking the first "one ship, two buoy systems" operation in the tropical Indian Ocean region of China. The "Bailong" buoy system is the first domestically developed deep-sea (7,000-m) climate observation system by the Institute of Oceanography, based on the assimilation of international advanced technologies and independent research efforts. The system serves as a support platform for deep-sea climate observation, capable of carrying multi-element sensors to achieve high-frequency sampling of surface meteorology, oceanic elements, and internal marine elements. Meanwhile, it utilizes Iridium satellite communication to transmit real-time observation data to the shore-based data center located in Qingdao. The "Bailong" buoy system is the only deep-sea buoy system in China participating in the tropical Indian Ocean buoy array, which currently comprises buoys from China, the US, and Japan. During this voyage, the "Bailong" buoys deployed achieved real-time uploading of deep-sea buoy data in China to the global telecommunications system, enabling global data sharing.

On August 28, 2017, "Xiangyanghong 01" set sail from Qingdao, embarking on China's first-ever multipurpose oceanographic research expedition around the globe. This expedition marked the first time China had integrated oceanic scientific expedition with polar scientific expedition, also featuring the first mission of "Xiangyanghong 01" to conduct both oceanic and polar scientific investigations.

In November 2017, "Xiangyanghong 01" collected extraordinary samples of basalt with peculiar shapes such as rope-like, branch-like, and stalactite-like formations from the seabed of the South Atlantic. These samples are valuable for studying the processes and evolution of magma activity under different structural backgrounds. They also hold significant scientific value in understanding the development of oceanic crust, the Earth's deep interior, seabed materials, and energy exchange, effectively filling gaps in various marine surveys.

During the Antarctic scientific expedition, "Xiangyanghong 01" conducted multipurpose surveys in marine physics and meteorology, marine geology, marine geophysics, seabed topography and mapping, marine chemistry and bioecology, and the marine environment in the South Atlantic Antarctic Ocean Sector. The expedition successfully exceeded the planned objectives.

"Xiangyanghong 01" achieved multiple "firsts" during this Antarctic scientific expedition.

For the first time, China expanded the Antarctic section survey from the traditional western longitude of 45 degrees to the western longitude of 37 degrees, covering an additional 400 km to explore the "traffic artery" where water exchanges between the Weddell Sea and the South Atlantic Ocean take place. In the middle of the Drake Passage, the vessel successfully completed comprehensive observations of marine environmental elements at two stations. This marked the first time that China obtained a full-depth observation profile of comprehensive marine elements in the critical region of the central South Atlantic westerlies.

"Xiangyanghong 01" successfully deployed two sets of deep-water subsurface buoy systems in the Powell Sea, enabling the observation of variations in flow velocity, temperature, salinity, and other physical and chemical parameters of water masses at different depths. This endeavor set a precedent for China's long-term monitoring of the marine environment in the South Atlantic Antarctic Ocean Sector using subsurface buoys. The vessel also conducted the first extensive full-coverage seabed topographic survey in the South Atlantic Antarctic Ocean Sector, covering approximately 8,880 km of surveyed lines and producing clear seabed topographic maps of approximately 15,000 square kilometers in six key areas.

Additionally, "Xiangyanghong 01" conducted high-resolution mapping and sampling of the seabed in the waters surrounding the South Orkney Islands for the first time. The vessel also conducted high-resolution multi-channel seismic exploration in the South Atlantic Antarctic Ocean Sector, obtaining a 1,100-km high-resolution seismic profile. This seismic profile is the highest-resolution data collected by China in Antarctica to date. Of utmost significance, the scientific expedition team discovered, for the first time in Antarctica, the coexistence of seabed hydrothermal fluids and cold seeps. Moreover, the team obtained direct geological and geophysical evidence strongly correlated with the formation of natural gas hydrates and seabed hydrothermal activities.

Throughout the Antarctic waters, the scientific expedition team collected seabed sediments from various geological units at fixed points by virtue of equipment such as seabed seismometer, sub-bottom profiler, and TV grab. The team completed 15 box coring operations and 8 core sampling operations, collecting products of hydrothermal fluid and cold seep activities.

On April 12, 2018, during the Chinese 46th Oceanic Expedition Voyage, the scientific expedition team of "Xiangyanghong 01" made a groundbreaking discovery of widespread rich rare earth deposits in the Southeast Pacific.

This global scientific expedition mainly consisted of the Chinese 46th Oceanic Expedition Voyage and the Chinese 34th Antarctic Expedition (coordinated with the icebreaker "Xuelong"). This expedition aimed to conduct multidisciplinary comprehensive investigations encompassing fields like environment, resources, and climate throughout the entire voyage. It spanned 260 days and covered approximately 35,000 nautical miles (about 65,000 km, or equivalent to circling the Earth 1.5 times). On May 18, 2018, the multipurpose oceanographic research vessel "Xiangyanghong 01" smoothly returned to Qingdao Port, Shandong, after accomplishing China's first-ever multipurpose marine expedition around the globe.

On August 10, 2019, the research vessel "Xiangyanghong 01", set sail from Qingdao, embarking on China's 10th Arctic expedition. On August 30, the vessel entered the Arctic Circle for the first time, carrying out planned research tasks in the Chukchi Sea, Beaufort Sea, and Canada Basin.

Since the commissioning of the "Xiangyanghong 01" and the "Xiangyanghong 03" (information separately introduced in Chapter 6) developed simultaneously with the former, they have participated in multiple international joint scientific expeditions, growing into a core force in international marine research. Their involvement has contributed to China's participation in international major oceanic studies, enhancing China's influence and discourse power in the field of international marine science and technology.

2.14 Deep-Sea Submersible Mother Ship "Tansuo 1"

In 2015, the Institute of Deep-Sea Science and Engineering of the Chinese Academy of Sciences (CAS) commissioned the China Ship Scientific Research Center to refit the multi-functional work vessel "Haiyang Shiyou 299" into a 4,500-m manned submersible mother ship as well as a universal platform for deep-sea scientific expeditions, in a bid to facilitate marine resource exploration, geochemical investigation, marine organism sampling, and other operations. CSSC Wenchong Shipyard completed the refit project on May 5, 2016, and delivered the vessel to the Chinese Academy of Sciences, naming it "Tansuo 1" (see Fig. 15).

The vessel was designed with a length of 94.45 m, a molded breadth of 17.9 m, a draft of 6.9 m, a displacement of 6,250 tons, a speed of 14 knots, a capacity of

Fig. 15 Deep-sea submersible mother ship "Tansuo 1"

60 persons (including 30 crew members and 30 researchers), and a DP-2 dynamic positioning capability. The vessel boasts an endurance of 8,000 nautical miles and a self-sustaining period of 60 days. Its main engine has a power of 8,820 kilowatts.

The refit project of "Tansuo 1" included lengthening the main hull, reconstructing the living quarters, renovating pipelines of lots of compartments, installing and debugging new equipment. All refit designs took into consideration eco-friendly and energy-saving factors, such as energy-efficient navigation, noise reduction, and vibration damping, among others. These technologies providing robust support for the vessel's sustained and healthy development.

The refit incorporated domestic independently developed systems, including submersible launch and recovery system, high-speed underwater acoustic digital communication, and acoustic positioning system. The refit also involved systematic improvements to the A-frame and the acoustics environment of the 4,500-m manned submersible. Thus, China not only acquired the capability for independent development of deep-sea submersibles, but also the independent capacity to develop their associated surface support system devices and carry out related ship design, construction, and equipment installation.

On June 22, 2016, "Tansuo 1" embarked on China's first multipurpose abyssal (10,000-m) scientific expedition activities in the Mariana Trench. On August 12, 2016, after accomplishing 84 scientific research tasks in the Mariana Trench, "Tansuo 1" successfully concluded its maiden voyage and returned to Sanya, Hainan. This marked a significant complete success for the first abyssal (10,000-m) scientific expedition in China's history of marine scientific and technological development.

During this expedition, a range of high-tech equipment self-developed domestically was utilized, including the autonomous remotely operated vehicle "Haidou" (10,000-m), abyss landers "Tianya" and "Haijiao", the in-situ testing system (10,000-m) "Yuanwei Shiyan", the 9,000-m deep-sea seabed seismometer, and the 7,000-m deep-sea glider. The research team executed 84 scientific expedition tasks in the Mariana Trench, obtaining a wealth of valuable samples and data from sections at different depths. During the maiden voyage, the deep-sea equipment onboard collected over 100 L of water samples from the 10,000-m depths. This marked an unprecedent achievement for China's deep-sea equipment obtaining 10,000-m depth waters, surpassing similar international capabilities. This expedition emerged as China's first endeavor at deep-diving scientific expedition at a depth of 10,000 m. The in-situ detection data of complete depth sequences, as well as water samples, sediment samples, and specimens of large organisms obtained from this expedition, have filled the long-standing gap in China's ability to acquire data and samples from ultra-deep depths, particularly at a depth beyond 10,000 m.

On August 24, 2018, the "Haidou" abyssal scientific expedition pioneer team for the multipurpose abyssal (10,000-m) scientific research (TS09 voyage) from the Institute of Deep-sea Science and Engineering, CAS, set sail from Sanya to the Mariana Trench to carry out a series of scientific tasks. The tasks included the "Research and Tackling Hard Issues for 'Haidou' Abyssal Cutting-edge Technological Challenges" of the strategic pilot science and technology special program of the CAS (Class B), "Preliminary Research on the Development of Deep-sea Intelligent Equipment" of the pre-research project of the strategic pilot science and technology special program of the CAS (Class A), as well as technological tasks of special key project "Deep-sea Key Technologies and Equipment" of national key research and development program. The expedition lasted for 54 days, and the team returned to Sanya on October 16. During this expedition, the team completed 504 km of depth-sounding operations at 87 stations in the Mariana Trench, successfully completing deep-sea equipment sea trials and scientific research tasks.

Many groundbreaking discoveries were made during this expedition, setting international records. For the first time internationally, three lionfish samples from a depth of 7,000 m and two opossum shrimp samples from a depth of 9,000 m were successfully captured and preserved at low temperatures. Additionally, a bassozetus fish species was discovered at a depth of 7,012 m, representing the deepest known depth at which this genus survives. Another notable achievement was the first-ever in-situ enrichment and fixed sampling of microorganisms across the vertically stratified waters at full depth, reaching a maximum depth of up to 10,890 m.

The manned submersible "Fenzouzhe" completed a total of 28 dives during the expedition, ranging in depth from 7,700 m to 10,900 m. Seven of these dives surpassed the 10,000-m mark, with several scientific investigations conducted in the deepest region of the Mariana Trench, the "Challenger Deep". Figure 16 shows the launch and recovery operations of the "Fendouzhe" submersible onboard "Tansuo 1".

Leveraging the "Fendouzhe" submersible's deep-diving capabilities and high precision, the expedition collected a valuable array of samples including abyssal water, sediments, rocks, and biological specimens. These samples provide invaluable data and resources for multidisciplinary research on diverse aspects of the deep sea abyss such as special environments, geology, and lives.

Aligned with the mandates of the National Key R&D Program by the Ministry of Science and Technology, including the "Special Project on Key Technologies and Equipment for the Deep Sea", as well as the Sanya Yazhou Bay Science and

Fig. 16 Launch and recovery operations of the "Fendouzhe" deep-diving submersible onboard "Tansuo 1"

Technology City Joint Project of Hainan Science and Technology Program, the 53-day TS-21–2 expedition aboard "Tansuo 1" conducted routine scientific investigations utilizing the "Fendouzhe" manned submersible in the Mariana Trench for the 2021 2nd expedition voyage. Throughout this expedition, "Fendouzhe" completed a total of 23 dives, with 6 of them surpassing the 10,000-m mark, enabling scientific investigations in the deepest region of the Mariana Trench, the "Challenger Deep". Sea trials were also conducted for deep-sea instruments and equipment such as the "Wukong" full-depth unmanned submersible, full-depth glass sphere, and acoustic releaser beyond water depths of 10,000 m. The vessel safely returned to Sanya on December 5, 2021.

During the expedition, the participating scientific teams collectively launched the *Mariana Consensus* initiative, aiming to establish a standardized platform system for deep-sea scientific investigations, and thus to ensure the long-term preservation and sharing of samples and data collected from the deep sea, and supporting international collaboration in deep-sea research. Additionally, the team also initiated the "Mariana Trench Ecological and Environmental Research Program", calling for the participation of researchers from domestic and international institutions. This program focuses on key scientific issues related to the Mariana Trench, such as the formation and evolution of the deep-sea Earth system, the origin of life and environmental adaptation, and biodiversity and climate change.

The expedition involved 60 members from Harbin Engineering University, Hainan Deep-sea Technology Laboratory, Hainan Deep-sea Science and Technology Service Co., Ltd., Hainan Tropical Ocean College, Huada Group, Shanghai Jiao Tong University, Sanya Yazhou Bay Deep-sea Science and Technology Research Institute, Zhejiang University, Hainan Zhejiang University Research Institute, the North China Sea Branch of the Ministry of Natural Resources, China Ship Scientific

Research Center, and the Institute of Deep-sea Science and Engineering of the Chinese Academy of Sciences. Among them, 17 scientists (including 5 female scientists) from 7 institutions participated in a dive in the "Fendouzhe" for the first time. Eight individuals made their first dive to a depth of 10,000 m. Notably, 67-year-old scientist Wang Jian became the first Chinese scientist to complete expeditions to both the North and South Poles, Mount Everest, and the global ocean at depths beyond 10,000 m (the deepest) for scientific investigations. These records make Wang Jiang the oldest man in this field. To date, "Fendouzhe" has completed 21 dives to a depth of 10,000 m, with 27 scientists from Chinese universities, institutes, and enterprises reaching the ocean's deepest point. It makes China lead the world in the number of deep-sea dives to 10,000 m and the number of divers involved.

In November 2020, China's independently developed full-depth submersible, the "Fendouzhe", successfully completed its 10,000-m sea trials and reached the seabed. From then on, the door to the depths (10,000 m) of the ocean was open to the Chinese people. With the deployment of "Fendouzhe" for routine scientific applications, China's journey to becoming a strong maritime nation is unfurling with a beautiful tapestry of exploring, understanding, investigating, and developing the deep sea.

"Fendouzhe" successfully completed its first dive for the 14th Five-Year Plan period, undertaking tasks such as target search, topographic exploration, collection of abyssal seabed samples, onboard equipment testing, and others. These operations resulted in a series of new breakthroughs. Several deep-sea technologies were put to the test at a depth of 10,000 m, further showcasing China's ability of submersibles for high-frequency continuous deep exploration operations.

Currently, China possesses three deep-sea manned submersibles, namely "Jiaolong", "Shenhai Yongshi" and the "Fendouzhe", as well as a series of unmanned submersibles including "Haidou", "Qianlong", "Haiyan", "Haiji", and "Hailong", among others. Together, they have laid the foundation for a complete spectrum of submersibles across full-depth waters, leading to a significant leap in deep-sea equipment technology.

During the development of the "Fendouzhe" submersible, a range of significant technological breakthroughs were achieved in areas such as pressure-resistant structure design and safety evaluation, titanium alloy material preparation and welding, buoyancy material development and processing, acoustic communication and positioning, intelligent control technology, lithium-ion batteries, seawater pumps, and manipulator operation.

The development of China's deep-sea scientific exploration has not only driven technological progress in new technologies and materials through the creation of major national equipment, but also provides support for submarine resource exploration, geological surveys, deep-sea organism surveys, scientific research, underwater engineering, salvage and rescue operations, deep-sea archaeology, and many other aspects through multiple scientific expeditions. In this way, it fills the gaps in research fields such as marine physics, microbiology, organic chemistry, and geology, among others.

By the end of 2021, "Fendouzhe" had completed a total of 51 dives, establishing an efficient five-day-three-dive operating mode. This improved the launch and recovery capabilities to adapt to more diverse sea conditions, fostering more submersible pilots. These advanced capabilities enable this submersible to be recognized by scientists worldwide, offering them an opportunity to dive to the deepest parts of the ocean. Through relevant scientific programs and international collaborations, it will further promote humanity's understanding of the abyssal realm.

2.15 Multipurpose Geological Survey Vessel "Haiyang Dizhi 10"

The "Haiyang Dizhi 10" (see Fig. 17) multipurpose geological survey vessel serves as the supporting equipment for Marine Geology Assurance Engineering, boasting the most powerful vessel of this kind that were independently designed and constructed in China. It features the most advanced overall performance and multipurpose survey capabilities in China, filling the void in China's geological survey vessel design field. Besides, it meets the demands of regional geological survey and coastal environmental geological multipurpose survey in the country's marine territory. In 2012, the China Ship Development and Design Center participated in the preliminary scheme certification for this project. In September 2014, the Center won the bid for the overall ship design. With the support and guidance from the Center's leadership and the assistance from other brother units, a significant amount of human and material resources were devoted to the design work. Despite challenges such as small ship tonnage, complex functionality, high technical difficulty, extremely outdated equipment data, and short construction period, the team fulfilled the delivery schedule and design quality. The design began in 2014, and construction commenced in October 2015 at Guangdong Shipping Engineering Co., Ltd., COSCO, followed by the vessel's launch in June 2017 and delivery in October 2017. The vessel is managed and utilized by the Guangzhou Marine Geological Survey, Guangdong.

The vessel was designed with a length of 76 m, a molded breadth of 15 m, a draft of 5.2 m, a displacement of 3,400 tons, a maximum speed of 15 knots, an endurance of 8,000 nautical miles, a self-sustaining period of 45 days, as well as an capability for non-restricted waters navigation. Upon delivery, "Haiyang Dizhi 10" was put into service alongside survey vessels "Haiyang Dizhi 8" and "Haiyang Dizhi 9". Together with the existing vessels "Haiyang 6", "Haiyang 4", "Tanbao", "Fendou 4", "Fendou 5", and "Ye Zhizheng", they formed a new layout of nine survey vessels covering shallow waters, deep waters, and the open ocean, featuring a complete marine geological survey vessel and technical equipment detection system for the China Geological Survey. This development significantly expands China's marine geological survey capabilities, bolstering the exploration of natural gas hydrates in key sea areas, and facilitating the industrialization of China's natural gas hydrates. Moreover, it has also accumulated valuable experience for the independent design and construction of natural gas hydrate drilling vessels (ocean drilling vessel) in

Fig. 17 Multipurpose geological survey vessel "Haiyang Dizhi 10"

China. "Haiyang Dizhi 10" is equipped with the first domestically developed lifting-type marine drilling system. Through design optimization and technological innovation, the drilling capability of the vessel can reach twice its original capacity. During the shipbuilding process, the latest achievements in the design and construction of similar international survey vessels were incorporated. By introducing, integrating, optimizing, and upgrading internationally advanced survey technologies, the vessel was equipped with 20 sets of internationally popular and high-end survey equipment, including AUV, integrated navigation and positioning system, marine gravity and magnetic measurement system, single-beam and multi-beam measurement system, sub-bottom profile measurement system, CPT, high-energy electric spark seismic source system, four-beam side-scan sonar measurement system, ultra-short baseline underwater acoustic positioning system, CTD, ADCP, and digital compass system. The vessel features high precision, multi-functionality, and strong comprehensive operational capabilities in terms of the overall planning and layout of national surveys, as well as compatibility and performance enhancement of survey equipment.

The vessel adopts the world's leading modular scientific research equipment layout, allowing it to simultaneously accommodate a total of 20 sets of survey equipment from three major categories including geology, geophysics, and marine hydrology. Currently, it is one of the most advanced survey vessels in China, ensuring fruitful outcomes in one survey.

The drilling system on board "Haiyang Dizhi 10" is a lifting-type drill with a gantry-style tower. Its advantages include being lightweight, faster lifting speeds by using hydraulic cylinders compared to traditional winch-lifting systems, and higher drilling accuracy through a wave compensation system that can compensate within a range of 1.5 m. Additionally, the open structure of the drilling derrick allows for the accommodation of other underwater equipment, enabling high-precision operations beyond drilling purposes. In addition to being a skilled drilling vessel, "Haiyang Dizhi 10" also supports routine marine scientific surveys with associated equipment. It is fitted with a range of scientific survey equipment, including single-beam and multi-beam measurement systems, sub-bottom profile measurement system, gravity measurement system, and marine magnetometry system, among others. With these systems, the vessel showcases strong comprehensive operational capabilities.

Compared to the recently launched "Haiyang Dizhi 8" and "Haiyang Dizhi 9", the biggest feature of the "Haiyang Dizhi 10" falls on its drilling capability, beyond capabilities for high-precision comprehensive investigations into marine geology, integrated geophysics, hydrology, physical oceanography, and other aspects. This is attributed to its equipped independently designed and constructed lifting-type marine drill which is capable of drilling up to 1,200 m. In the future, this vessel is to be capable of conducting drilling and sampling operations in waters up to 1,000 m deep and penetrate deeper rock formations (up to 400 m) in shallow water areas. The primary tasks of this vessel will revolve around marine drilling, regional marine geological surveys, and comprehensive environmental geological surveys of coastal zones.

The nine types of ships owned by the Guangzhou Marine Geological Survey have distinct survey capabilities, each focusing on different objectives to carry out marine geological surveys tailored to different research purposes. They collectively form a solid technical system for deep-sea exploration in China. In the future, they will provide great support for China's strategy towards being a strong maritime nation and the implementation of the Belt and Road initiative.

On February 26, 2019, the "Haiyang Dizhi 10" survey vessel returned to the specialized marine geological dock in Guangzhou after completing a joint marine geological scientific survey in the Indian Ocean with Pakistan. This voyage marked the vessel's first long-distance scientific survey expedition since its commissioning, also representing a crucial scientific event deepening friendly cooperation between China and Pakistan. Departing from Guangzhou on November 28, 2018, the expedition lasted for 91 days and covered a distance of 14,100 nautical miles. Scientists from China and Pakistan worked together in close cooperation, successfully completing the mission and achieving the expected scientific goals. During the voyage, researchers from various organizations in Pakistan visited the vessel for exchanges. The research vessel became a technological envoy for deep-sea geological cooperation and communication between the two nations, and also enhanced the friendship between their peoples.

During this voyage, "Haiyang Dizhi 10" underwent a rigorous test of long-duration operations. Its advanced technical equipment operated safely and efficiently, showcasing its innovative advantages as a multi-functional vessel with outstanding comprehensive benefits. The vessel has proven to be excellent in long-duration multi-purpose oceangoing surveys, as well as reliable in overall performance and its survey equipment capabilities.

2.16 *Multipurpose Marine Geological and Geophysical Survey Vessel "Haiyang Dizhi 9"*

"Haiyang Dizhi 9" (see Fig. 18) is a technology-advanced survey vessel in China with capabilities for specialized seismic surveys and comprehensive geological and geophysical surveys. It is specifically designed for short-travel 2D (3D) multi-channel seismic surveys using seismic cables, incorporating various functions including geophysical measurements, hydrological environmental measurements, and geological sampling techniques.

The construction of the "Haiyang Dizhi 9" was supported by the supporting equipment project of National Marine Geology Assurance Engineering. The vessel had its construction commenced at the Shanghai Shipyard on October 28, 2015, keel laying on December 26, 2016, and launching on February 28, 2017. On December 28, 2017, it was delivered to the Qingdao Institute of Marine Geology, China Geological Survey.

The vessel was designed with a length of 87.07 m, a molded breadth of 17 m, a molded depth of 7.8 m, and a gross tonnage of 5,178 tons. It can reach a maximum speed of over 15 knots, with a self-sustaining period of 60 days and an endurance of 10,000 nautical miles. The vessel is suitable for global oceanic navigations (class B ice strengthening). It is equipped with a wide range of equipment, including state-of-the-art twin-cable high-resolution short-travel seismic operation system, integrated navigation and positioning system, marine gravity and magnetic measurement system, single-beam and multi-beam measurement system, sub-bottom

Fig. 18 Multipurpose marine geological and geophysical survey vessel "Haiyang Dizhi 9"

profile measurement system, high-energy electric spark seismic source system, four-beam side-scan sonar measurement system, ultra-short baseline underwater acoustic positioning system, deep-water Doppler current measurement system, and digital compass system. In addition, it also has auxiliary survey equipment such as deep-water winches and A-frames. Equipped with these facilities, the vessel is endowed with a comprehensive detection capability for waters, from the surface of the sea to the sub-bottom and mid-bottom of the seabed. The overall planning for surveys, as well as the compatibility and efficiency of the survey equipment, have incorporated the latest achievements in these of similar international research vessels. Its associated technical standards are at the leading edge among Chinese research vessels, making it another vital asset for deep-sea exploration.

From the very beginning, "Haiyang Dizhi 9" has shouldered the responsibility of accelerating the construction of China into a strong maritime nation as proposed at the 19th CPC National Congress, as well as the goal of leading internationally in deep-sea exploration set by the Party Group of the Ministry of Natural Resources. It also bears the mission of building the China Geological Survey into a world-class research institution, as envisioned by the Party Committee of the China Geological Survey of the Ministry of Natural Resources. This vessel also carries the dreams of several generations of geological survey workers from the Qingdao Institute of Marine Geology.

Since its commissioning on December 28, 2017, the vessel has been deployed to the East China Sea, South China Sea, Yellow Sea, and the Western Pacific. It has completed various tasks including deep-sea geological surveys in the Western Pacific, seismic investigations in the South Yellow Sea, wide-line dual-cable seismic surveys in the East China Sea, shared voyages for sea trials in the South China Sea of the Pilot National Laboratory for Marine Science and Technology, and other missions. Over the course of nearly a year, the vessel had safely operated for 258 days, covering a total distance of more than 25,000 nautical miles. It completed various survey operations, including multi-beam surveys, sub-bottom profiling surveys, and marine gravity and magnetic surveys, totaling 13,290 km for each. The vessel also conducted 942 km of 2D multi-channel seismic surveys, as well as 2,759 km of wide-line dual-cable multi-channel seismic surveys. The vessel has achieved the following significant accomplishments in marine basic geological surveys, exploration of offshore oil and gas resources, and the application of deep-sea exploration technologies:

Firstly, it conducted marine basic geological surveys in the South Yellow Sea. In the surveys, the vessel revealed the deep-seated structures and geodynamic processes in the South Yellow Sea, providing solid and reliable data support for addressing critical fundamental geological issues such as the tectonic evolution, exploration of oil and gas resources, and geological hazards in the region.

Secondly, it successfully carried out multipurpose surveys in the Western Pacific. In the surveys, the vessel thoroughly investigated fundamental geological elements including submarine topography, geomorphic features, geological types, and sub-bottom structures. It gained new insights into resource-environment interactions and sedimentary processes, collected samples of manganese nodules, and discovered the characteristics and regularities of deep-sea rare earth resource enrichment. This provided valuable physical samples for the study of environmental evolution scientific issues such as East Asian monsoons, Qinghai-Xizang Plateau uplift, and global climate change.

Thirdly, it bolstered the engineering application of marine wide-line seismic exploration technology. The vessel acquired high-quality wide-line dual-cable seismic survey data in two segments of the East China Sea, establishing a comprehensive set of technical methods for data acquisition, processing, and interpretation of wide-line seismic data. This provided significant technical support for marine oil and gas resource exploration.

Fourthly, it established an open and shared platform for marine scientific and technological innovation. The vessel undertook the shared voyage mission for the Pilot National Laboratory for Marine Science and Technology (Qingdao), achieving data sharing and further enhancing the its efficiency between research vessels. It served the needs of marine research and study, while creating an open and mobile marine experimental platform.

During its first year of operation, "Haiyang Dizhi 9" successfully weathered strong winds and rough seas, overcoming the challenges posed by more than ten typhoons. The overall performance of the vessel and survey equipment was validated to be effective. It has made consistent progress in Party building, cultural development, science outreach, and ship management, among other aspects. By establishing a Party branch on the survey vessel and flying the five-star red flag in the deep sea, it has fostered a distinctive geological culture. Additionally, it established a mobile marine science popularization platform, practicing the core values of "responsibility, innovation, cooperation, dedication, and integrity" for geological workers in the new era through practical actions.

In 2020, the "Haiyang Dizhi 9" successfully completed the 58-day survey equipment sea trial and shared voyage mission in the South China Sea. The voyage was jointly conducted by 23 domestic research institutes, universities, and enterprises, with a total of 56 research and technical attenders. It accomplished 22 test tasks, including the testing of 10 newly independently developed equipment by the Qingdao Institute of Marine Geology, a national key research and development program, a shared mission for the Pilot National Laboratory for Marine Science and Technology (Qingdao), and several newly introduced survey equipment tests. This South China Sea sea trial and shared voyage tested and evaluated a series of domestically produced marine survey equipment. The innovative achievements include:

Firstly, for the first time, simultaneous underwater multi-target positioning was achieved by "Haiyang Dizhi 9".

Secondly, the search for a deep-sea lander deployed in a water depth of 2,500 m was successfully completed, assisting in the successful launch and recovery of the lander.

Thirdly, the investigation for vessel bottom transducers and the monitoring of the status of the sea trial equipment was completed by an independently-integrated remotely operated vehicle (ROV) for the first time.

Fourthly, the automatic unmanned aerial vehicle (UAV) flight and beyond visual range sea surface target inspection were carried out for the first time during the operation of "Haiyang Dizhi 9". This further verified the vessel's collaborative operating capabilities and improved the measurement and control capabilities of the vessel and its onboard equipment under operation conditions.

Fifthly, a comparative test was conducted for four types of both domestic and foreign gravimeters on board. The measuring accuracy and stability of domestic gravimeters were validated, providing important evidence for the construction and quality assessment of domestic oceanographic equipment.

The sea trial and shared voyage mission in the South China Sea tested the practical application effect of domestic oceanographic equipment and provided important support for accelerating the localization of oceanographic equipment and their early deployment into practical applications. Meanwhile, it demonstrated the strong management, coordination, and organizational implementation capabilities of "Haiyang Dizhi 9" as a national key scientific research vessel platform.

On April 7, 2021, as a deep-sea mobile TV grab with a weight of about 2.4 tons was lifted by a steel cable and slowly retrieved to the deck, all the sea trial tasks for 11 sets of equipment for the second segment of the 2021 South China Sea sea trial mission came to the end. The second-segment sea trial targeted the testing of 11 sets of equipment: 5 sets of independently developed equipment, including a 3,000-m lightweight acoustic deep towing detection system, a deep-sea and far-sea daily magnetic variation observation system, a seabed sediment multi-parameter probe and deployment system, a 6,000-m seabed ocean multi-element boundary layer monitoring system, and a deep-sea mobile platform; 3 sets of newly purchased equipment, including an ultra-short baseline system, a moving CTD (conductivity temperature depth) detection system, and a deep-sea mobile TV grab; and 3 sets of shipborne equipment, including a marine gravimeter, a magnetometer, and the rest. The trial aimed to verify performance, reliability, and stability of all equipment. Prior to this, a sea trial for equipment such as the ultra-large-capacity combined gun array multi-channel seismic system, the deep-sea visualized and controllable sediment core sampling system, and multi-beam calibration system was completed in the first segment.

2.17 *Multipurpose Resource Survey Vessel "Dayang"*

"Dayang" is the first domestically designed and built 5,000-ton ocean multipurpose resource survey vessel in China. It also serves as one of the core foundational equipment for the China Ocean Mineral Resources R&D Association's initiative of "conducting deep-sea exploration, developing deep-sea technologies, and establishing the deep-sea industry". It features the country's first new type of multipurpose survey vessel designed and built according to principles of greenization, informatization, modularization, portability, comfort, and internationalization. It primarily focuses on the exploration of various resources in the ocean, while supporting deep-sea research, and abyssal resource and environment surveys in all four major oceans. These functionalities make the vessel one of the key platforms for China's ocean resource and environment surveys as well as deep-sea scientific exploration.

The multipurpose resource survey vessel "Dayang" (see Fig. 19) was designed by the China Ship Development and Design Center and built by Huangpu Wenchong Shipbuilding Co., Ltd., followed by its delivery in March 2019.

The vessel was designed with a total length of 98.0 m, a molded breadth of 17.0 m, a molded depth of 8.8 m, a designed displacement of 4,500 tons, a capacity of 60 persons, a self-sustaining period of 60 days, an endurance of 14,000 nautical miles, a maximum speed of 16 knots, and a capability for non-restricted waters navigation.

"Dayang" is a modern ocean multipurpose resource survey vessel primarily focused on exploring various resources in the ocean while also considering the research demands of related disciplines in the deep sea. It integrates multiple disciplines, functions, and technological means, equipped with global navigational capabilities, and ranking among the world's highest in overall performance levels.

The vessel incorporates advanced technologies such as DC busbar and straight blade propeller thruster, as well as the "Variable-Speed Diesel Generator – BPC DC Busbar Electric Propulsion System and Dual Straight Blade Propeller Thruster Integrated Scheme". Compared to conventional propeller thrusters, this new type of propeller not only possesses propulsion functions but also serves as steering and anti-rolling devices. Equipped with a new generation of electric propulsion system based on variable-speed generators and DC main power stations, it features good maneuverability, low fuel consumption, low maintenance costs, low emissions, low harmonics, low noise, fast grid connection, and space savings. The "Dayang" vessel

Fig. 19 Multipurpose resource survey vessel "Dayang"

is equipped with two sets of 2,750-kilowatt straight blade propeller thrusters with anti-rolling and low underwater noise capabilities, located symmetrically at the stern. In addition, a 1,000-kilowatt telescopic all-direction propeller is installed at the bow. It also features a DC busbar all-electric propulsion system, and the required power for the propeller is provided by four sets of variable-speed diesel generator units, which ensure optimal energy-saving and environmentally-friendly working conditions. The vessel has excellent maneuverability and DP1 dynamic positioning capability, as well as good seakeeping performance and wind-wave resistance capability. It is able to support deployment of equipment such as ROVs and deep-sea towing devices, as well as hover operations of helicopters in sea conditions of grade 4, and berthing surveys in sea conditions of grade 5.

This vessel features over 70 sets of various advanced survey instruments and equipment. It has a laboratory area of over 400 square meters, accommodating multiple movable container-style laboratories. It is also fitted with a range of internationally advanced shipborne scientific and technological equipment for seabed surveys, water body surveys, some atmospheric surveys, deep-sea extreme environment surveys, remote sensing information field validations, deep-sea technology experiments, and ship informatization. This enables high-precision and long-term marine geological, marine dynamic, marine ecological, and sea-air environment comprehensive marine observations and detections, as well as truth-preserving sampling and on-site analysis capabilities. The bottom of the vessel is equipped with dual lifting fins and a sonar transducer space with dome. The deck houses advanced and convenient operation support systems, modular comprehensive container systems, as well as sea-air and remote sensing detection systems. This enhances the vessel's comprehensive offshore testing capabilities and improve the functionality of its information systems.

For the first time on a domestic survey vessel, "Dayang" adopts an integrated control chair design, ensuring a broad operational field of vision in limited space, which allows a panoramic view of the entire aft deck and the operational space to starboard). The chair has electric lifting and rotating functions to ensure operators have a good operational field of vision, thereby improving operational safety and efficiency. The integrated control chair executes integrated control over six winches, three cranes, and three propellers within a space of less than four square meters.

The hull of "Dayang" is strengthened to meet ice class B3 requirements, which can meet the requirements of survey operations in all four oceans. The vessel is fitted with advanced equipment for comprehensive ocean resource and environment surveys, supporting various research tasks including ocean seabed solid mineral resource exploration, ocean biology (genes) resource exploration, ocean environmental information resource exploration, mid-ocean ridges and deep seabeds research, ocean dynamic processes and air-sea interactions investigation, ocean ecosystems and biodiversity investigation.

2.18 Deep Trench Research Mother Ship "Zhang Jian"

The "Zhang Jian" trench research mother ship (see Fig. 20) is China's first ship designed specifically for the exploration of abyssal trenches. It is also a research vessel jointly funded and constructed by the Hadal Science and Technology Research Center of the Shanghai Ocean University, along with several private enterprises. Furthermore, it serves as the mother ship of China's manned deep-diving submersible (10,000-m), the "Caihongyu".

Shanghai Ocean University, together with several private enterprises, has developed three full-depth landers, one full-depth unmanned submersible, and one manned submersible capable of reaching depths of up to 10,000 m. These scientific research equipment, along with "Zhang Jian", work in synergy to establish the Chinese "Mobile Laboratory for Hadal Science and Technology" for investigations across the full depth of the ocean.

Zhang Jian was known as a top scholar, industrialist, and educator in modern Chinese history. Besides, he is also the the founder of Shanghai Ocean University. The naming of the Chinese 10,000-m manned submersible' research mother ship as "Zhang Jian" is intended to carry forward his practical development philosophy of "education as prerequisite, and industry as foundation", further to promote the progress of China into a strong maritime nation.

The vessel has a total length of 97.0 m, a molded breadth of 17.80 m, a designed draft of 5.65 m, and a designed displacement of 4,800 tons, a navigational speed of 12 knots, an endurance of 15,000 nautical miles, a capacity of 60 persons, and a self-sustaining period of 60 days.

Fig. 20 Deep trench research mother ship "Zhang Jian"

The vessel was designed by Shanghai Hansail Marine & Offshore Design Co., Ltd. and constructed by Zhejiang Tianshi Shipbuilding Co., Ltd. The design process commenced in September 2013, and the vessel was delivered in June 2016.

The vessel is mainly equipped with the following equipment:

The equipment includes with one 11,000-m manned submersible, one 11,000-m unmanned submersible, and three 11,000-m landers. The vessel also features dynamic positioning capability and a complete range of laboratories such as wet and dry laboratories, gravity and magnetic laboratory, and network processing center. Supporting equipment includes an A-frame at the stern (with a dynamic load of 30 tons), an A-frame on the side (with a dynamic load of 3 tons), a 6-ton telescopic folding arm crane, a 10-ton telescopic folding arm crane on the aft deck, a 6,000-m photoelectric composite cable winch, a 6,000-m geological winch, a 10,000-m CTD/hydrological winch, deep-sea winches, and other control support equipment. It is also equipped with advanced testing equipment such as the full-depth multi-beam system, sub-bottom profiler, and acoustic Doppler current profiler.

The vessel's completion has provided support for sea trials and operations related to the 11,000-m submersible series products (including three 10,000-m landers, one 10,000-m unmanned submersible, and one 10,000-m manned submersible). In addition to conducting deep-sea scientific survey and research, it will also carry out general deep-sea oceanographic research, encompassing disciplines such as deep and far sea physical oceanography, geophysics, marine chemistry, seabed geology, marine biology, as well as sea trials of marine surveying instruments and equipment, offshore oil and gas engineering facilities, and submarine electric power communication facilities. In comparison to existing domestic submersible mother ships, "Zhang Jian" also boasts multiple functionalities including underwater archaeology and filming, deep-sea rescue and salvage, installation and maintenance of marine engineering equipment, as well as deep-sea exploration and tourism.

Zhang Jian" the only research mother ship globally equipped with full-depth and full-area operation-type manned submersibles, unmanned submersibles, and landers. It boasts the shortest construction period worldwide and is considered the fastest-built medium-sized research mother ship both domestically and internationally in recent years.

The vessel also serves as a full-depth "Mobile Laboratory for Hadal Science and Technology" capable of providing a public platform for continuous and systematic deep-sea scientific research to ocean scientists from China and abroad. It can facilitate systematic scientific surveys of 26 abyssal trenches with depths of below 6,500 m worldwide, allowing for the acquisition of precious samples, the establishment of a DNA database for abyssal organisms and the conduction of a series of scientific research for them. Through these activities, the vessel aims to explore the origins of life and make contributions to humanity's exploration of the ocean.

On July 12, 2016, "Zhang Jian" embarked on its maiden voyage from Shanghai. The first segment of the journey took it to the South China Sea, where it conducted sea trials and scientific expeditions. Following this voyage, it traversed the equator, entering the South Pacific through the Northwest Pacific, which represents the southern route of the "Maritime Silk Road". It is on this very route that the illustrious

Chinese navigator Zheng He's fleet made a triumphant arrival in Papua New Guinea. In Papua New Guinea, "Zhang Jian" awaited the arrival of over 40 travel enthusiasts who were to arrive here by air. Together with researchers, they set out for the Solomon Sea and arrived in the waters near the New Britain Trench, which plunges to depths of over 8000 m. Here the researchers from Tongji University, Shanghai Ocean University, the Chinese Academy of Sciences, and Papua New Guinea conducted deep-sea joint scientific investigations, including seabed sonar exploration, abyssal environmental survey, seabed sediment sampling, and seawater sampling. During the testing of the "Caihongyu" 10,000-m lander's deep-sea operations capability, the research team even observed mysterious submarine macroorganisms and obtained a collection of valuable deep-sea samples.

During the journey, the travel enthusiasts enjoyed the opportunity to appreciate rare natural landscapes and observe the entire process of sampling survey conducted by the researchers. They could also board the sightseeing manned submersible "Underwater Aircraft" carried on the "Zhang Jian" vessel, which could dive to depths of over 100 m, allowing for the observation of underwater scenery in the South Pacific such as coral reefs and World War II shipwrecks.

In the future, the "Zhang Jian" vessel will expand its business domain to offer more underwater experiences, enabling more people to enjoy the pleasures brought by deep-sea technology to human life. It aims to share the skills of deep-sea technology with the general public, especially young people, giving them the opportunity to experience the ocean and understand its nature. Through this form of dissemination, it hopes to foster a greater love and support for marine research undertakings among young people and the public.

During its maiden voyage, "Zhang Jian" carried out activities such as mining environmental impact assessments, sea trials, scientific investigations, and experiential tourism. The vessel was funded by private capital, mainly generating revenue by providing deep-sea technology services to domestic and international clients. This maiden voyage was commissioned by two large mining companies in Papua New Guinea to conduct deep-sea environmental impact assessments, marking the first time that China's deep-sea technology services have been brought overseas and promoted to the world.

The maiden voyage effectively tested the "Zhang Jian" vessel's oceangoing sailing performance and the reliability of various scientific research equipment onboard for carrying out scientific research tasks. Throughout the 74-day voyage, the vessel experienced strong typhoons and even tsunami warnings, but there were no major system failures. It enhanced the crew and the scientific research team's ability to cope with complex environments.

Abyssal science experts from both domestic and abroad conducted various scientific investigation activities in the abyssal trench. This included the first descent of Hadal Science and Technology Research Center of the Shanghai Ocean University's self-developed 10,000-m lander to a depth of 6,700 m. This lander captured various macroscopic marine creature behaviors with underwater cameras at a depth of 5,000 m, providing important data support for future abyssal research. The abyss is free from light and under tremendous pressure. How can organisms tolerate such high pressure, especially macroorganisms like fish? How do they adapt and survive in such extreme environments? This is a question of particular scientific interest.

After sailing for over 9,000 nautical miles in 74 days, "Zhang Jian" returned to Shanghai Port on September 23, 2016. Unlike previous marine research projects funded by the Chinese government, the funding for the sea trials and scientific expeditions of the unmanned submersibles came from the revenue generated by the mining environmental impact assessments and experiential tourism activities.

2.19 Polar Research Icebreaker "Xuelong 2"

As polar scientific exploration and tasks continue to progress, it became clear to the Polar Office of the SOA that the existing research icebreaker, "Xuelong", is no longer sufficient to meet the demands of polar scientific expeditions. Consequently, there arose a pressing need for a new, advanced polar icebreaker to form a polar research fleet capable of tackling the rigorous requirements of polar scientific exploration. In 2008, the SOA submitted an application to the State Council for the construction of a new polar research icebreaker. In June 2009, the project was approved at a special meeting of the State Council and was included as a major project, titled "Xuelong Polar Expedition", in the "13th Five-Year Plan".

Given the limited practical experience of China's ship research and design institutions in developing polar icebreakers, it was decided that the shipbuilding would follow the principle of "being co-designed by domestic and foreign companies and independently constructed by China". It necessitated the most advanced domestic and foreign design companies in the field of icebreaking scientific research to conduct the joint design. It was also required that the new vessel be classed by the CCS and another internationally renowned classification society with expertise in icebreaking, making the vessel an icebreaker with two classification certificates.

In February 2010, the MARIC was appointed as the technical support unit for the project. Working in collaboration with the Polar Office of the SOA and the Polar Research Institute of China, the three units jointly organized the project proposal and feasibility study.

After a year of research and demonstration, the key specifications for the new polar research icebreaker gradually shaped. A design specification was prepared in both Chinese and English and underwent a comprehensive review by an expert panel that included international experts.

In August 2011, CCS and LR were determined to be the joint classification societies for the new vessel, providing services for preliminary consultation, design, and construction. In July 2012, after an international bidding process, Aker Arctic Technology Inc. of Finland won the bid for the joint design as the foreign partner, followed by a basic design contract being signed. Aker Arctic Technology Inc., renowned for its extensive experience in icebreaker design, provided unique insights on the vessel's general arrangement, icebreaking methods, and operations in the Arctic environment. In October 2012, the SOA signed a detailed design contract with the MARIC. Sending staff to participate the basic design, the latter played a leading role in the vessel's open-water performance, scientific expedition design, intelligent design, classification society review, and in acquiring dual classification certificates. Besides, the MARIC also undertook the basic design technical review work on behalf of the shipowner. As the shipowner, the Polar Research Institute of China contributed important requirements and improvement suggestions based on its unparalleled experience in polar navigation and scientific exploration throughout the entire design, construction, and supervision process.

On November 30, 2016, the user signed a construction contract with Jiangnan Shipyard (Group) Co., Ltd. Construction of the vessel commenced on December 20, 2016, and it was delivered on July 11, 2019. The vessel was christened "Xuelong 2" (see Fig. 21).

"Xuelong 2" has a total length of 122.5 m, a molded breadth of 22.32 m (26.42 m at its widest point), a molded depth of 11.80 m, a designed draft of 7.85 m, a maximum speed of 18.0 knots, a nominal endurance of 23,400 nautical miles, an icebreaking capability of 1.5 m of ice plus 0.2 m of snow (at a speed of 2–3 knots), as well as a gross tonnage of 12,769 tons. It is classified by CCS and LR, and flies the flag of China.

The vessel is designed to meet the following environmental conditions:

Air temperature: −30 to 45 °C.

Sea water temperature: −2 to 32 °C.

Design temperature of the air-conditioned compartment: 26 °C (when the outside temperature is 35 °C);

20 °C (when the outside temperature is −30 °C).

The lowest environmental temperature for the main equipment to ensure the normal operation of the vessel: –30 °C.

Fig. 21 Polar research icebreaker "Xuelong 2"

Electric propulsion adopted for the propulsion system: two sets of side thruster pods with a power of 7.5 megawatts each.

"Xuelong 2" adopts a bidirectional icebreaker type, with an icebreaking bow, class PC3 ice strengthening, mooring equipment located in the forward sheltered area of the forecastle deck, and a container hold in the bow. The larboard of the forecastle deck is designed as an outer passage that runs from bow to stern, which can be used for transferring cargo pallets on the vessel. The arrangement of the pilothouse provides good visibility in the bow and stern directions. The bow has a foremast (science mast) and two channel-type side thrusters. The engine room is located in the midship, close to the stern, with a helicopter deck and a helicopter hangar. The aft exposed deck is used for scientific research operations. An advanced and reliable all-direction electric propulsion system is configured for the vessel, equipped with diesel-electric propulsion machinery and two podded propellers. Other systems include the DP2 dynamic positioning system, integrated navigation and positioning system, and ship anti-rolling system. The vessel has a prominent ductkeel at the bow bottom and a multi-beam transducer installed in the ductkeel area, with anti-icing and sound transparency capabilities at the bottom.

The vessel is fitted with state-of-the-art equipment for marine environmental and geophysical surveys. It can undertake multipurpose survey and research tasks on polar oceans, sea ice, atmosphere, and other environmental basics, and has the capability to conduct integrated observation and sampling of marine environments related to climate change. It can also conduct investigations on oil and gas, as well as biological resources in polar ice areas, and has the basic ability to conduct "rough investigation and potential estimation". Additionally, "Xuelong 2" can also undertake partial material transportation tasks for polar research stations.

The vessel incorporates the following features:

(1) The icebreaker type design ensures navigability in ice-filled sea areas and meets the continuous icebreaking requirement of handling 1.5 m of ice plus 0.2 m of snow.
(2) Considering the actual ice conditions in the operating sea areas, a bidirectional icebreaking ship type is chosen for optimal performance and enduring open-water navigation economical efficiency.
(3) A bottom ductkeel structure is adopted to maintain a harmonious balance between icebreaking navigation and the installation and operation of the bottom acoustic equipment.
(4) An all-direction electric propulsion system is adopted to guarantee high maneuverability and relatively compact interior space. Since there are great differences in working conditions during operation, this system also allows for improvement in organic coordination, unification and optimization between the usage cost and the economical efficiency.
(5) The vessel is equipped with a DP2 dynamic positioning system for redundancy in normal sea conditions and reliable performance in harsh sea conditions.
(6) Large laboratory layout is adopted maximize flexible compatibility and resource sharing for survey and experimental activities.

(7) A spacious vertical operational workshop is arranged alongside a moonpool system to ensure practical survey operations in extremely cold polar conditions.
(8) The vessel ensures a generous deck area for survey operations to effectively facilitate the geophysics research and related functions.
(9) A helicopter system is configured for supporting a wider range of expeditions in the Arctic and Antarctic.
(10) A container bank and a station oil hold are set up to enhance Chinese polar research station's scientific exploration, logistical support, and safety.

The design team from the MARIC and Aker Arctic Technology Inc. integrated their respective design experiences and innovative concepts, resulting in several innovations in the vessel design, including:

(1) Innovation in ship type design: The vessel features a polar research icebreaker type with bidirectional icebreaking capability at the bow and stern. This allows it to navigate through 1.5-m thick ice at a continuous speed of 2–3 knots. In the event of encountering challenging ice ridges, the vessel's podded propellers can rotate 180 degrees, effectively reversing the stern as the bow and enabling the stern propeller to effectively cut through ice and "hollow out" ice ridges as tall as 10 m as rotating below the sea surface. The bow incorporates a patented design of anti-bubble sinking ductkeel, which prevents important acoustic equipment on the vessel's bottom from being affected by bubbles and crushed ice, and achieves a balance between navigation efficiency and acoustic performance. The vessel adopts a 6,600-V medium voltage AC podded all-direction electrical propulsion system, which consists of diesel-electric propulsion machinery and two icebreaking-type podded propellers, achieving a capacity of up to 22,000 kilowatts for the power station. The vessel is equipped with a DP2 system, an integrated navigation and positioning system, and a ship anti-rolling system. This ensures a high level of reliability and redundancy, enhancing the safety of the vessel during operations in extreme weather conditions and limited maneuvering spaces commonly encountered in polar regions.
(2) Innovation in scientific research system design: The vessel features internationally recognized general laboratories and modular design, complemented by a "watertight moonpool". In addition, the vessel is also equipped with acoustic equipment such as deep-water, medium-water, and shallow-water multi-beam systems, deep-sea sub-bottom profiler, bio-stock assessment echo integrator system, underwater omnidirectional sonar, ultra-short baseline system, and a depth sounder (10,000-m). These acoustic devices fulfill the requirements for detailed seabed measurements and fisheries resource exploration. Moreover, in response to the special environment of polar regions with multi-ice and low temperatures, the vessel updated multiple scientific research and experimental systems, such as surface water collection, to achieve innovative results.
(3) Innovation in comfort design: The vessel boasts exceptional wind resistance, capable of withstanding gusts of up to Force 12, thereby ensuring safety when navigating the infamously treacherous westerlies. An anti-rolling tank

is equipped to enhance onboard comfort in low-speed scientific research operations. Human-centered design of living facilities, including heating facilities in compartments, double-glazed windows for insulation, heated towel racks in bathrooms, electrical cable heating and anti-slip treatments on major outdoor staircases, as well as ample freshwater supply, ensures the comfortable living of research personnel. Moreover, communication capabilities of the vessel are significantly upgraded. During the early design phase of "Xuelong 2", the design team established a rigid requirement to have seamless internet connectivity using mobile phones while on board the vessel. In light of this, the vessel was equipped with a VSAT[6] satellite communication system, as well as a network amplifier to achieve Wi-Fi coverage throughout the vessel. Considering the need for drinking freshwater during long voyages, a drinking tank re-purification system was specifically developed.

(4) Innovation in intelligent design: "Xuelong 2" is the world's first polar research icebreaker to receive the prestigious Intelligent Ship designation. It combines intelligent navigation, intelligent energy efficiency, intelligent engine room, and intelligent hull systems, incorporating over 7,000 intelligent sensing points in the hull and equipment. The intelligent engine room is capable of monitoring the main and auxiliary operation systems through sensors in real time. It analyzes and assesses the operation status and working conditions of the mechanical equipment in the engine room. Based on this analysis, it formulates operational decisions and maintenance plans for the machinery. This ensures the reliability of operation and maintenance of the engine room equipment in polar environments. The intelligent hull system refers to the vessel being equipped with a hull monitoring system and an auxiliary decision-making system. This system allows for the entire life-cycle management of the vessel's hull, as well as monitoring and assessing the structural integrity of the hull. In this way, it ensures the safety of the vessel's hull structure. When the hull comes into contact with the ice surface, the vessel is capable of providing automatic forecasting. The intelligent hull system is also able to perform automatic "health check-ups" and collect data on the friction between the hull plates and the ice surface, providing valuable reference for future vessel design and construction. The intelligent engine room and intelligent hull system not only provide decision-making support for safe navigation but also significantly enhance the maintainability of the vessel, extending its service life. By equipping various sensors throughout the vessel, it achieves omnidirectional intelligent sensing, acquisition, exchange, and display of information. Additionally, intelligent experiments and intelligent wearables are also employed onboard.

(5) Innovation in extreme cold protection: Through thermal balance analysis and cold protection design, the vessel incorporates technologies such as water glycol non-freezing solution preheating, steam coil, heat tracing cable, and others. With these technologies, the vessel effectively addresses the challenges posed by extremely low temperatures in polar regions, thereby ensuring the normal

[6] Very small aperture terminal.

operation of all onboard systems and a safe and comfortable living environment for crew members.

Driven by innovative thinking, the design team successfully tackled the following design challenges:

(1) Bow and stern bidirectional icebreaker type design: This vessel is China's first polar research icebreaker to employ bow and stern bidirectional icebreaking technology. It not only possesses the capability for long-range open-water navigation but also enables navigation in ice in both directions. Bidirectional icebreaking capability significantly enhances the vessel's maneuverability in ice-covered areas, allowing for quick turns and changes of direction. This is particularly crucial in the complex and narrow water areas near the Antarctic coast, where it greatly improves the safety of navigation. The bow and stern bidirectional icebreaker type design maximizes the coordination between the bow lines for icebreaking and open-water navigation, providing a crucial guarantee for the icebreaking capability of the vessel. Furthermore, it provides the vessel with the flexibility to adapt to complex ice conditions in polar sea seas, thus solving the problem of "Xuelong" being unable to proceed in ice-accumulated areas near Zhongshan Station. This icebreaking capability allows for continuous operations in polar waters, where there is a mixture of first-year and multi-year sea ice. Navigation in both bow and stern directions is ensured with a capability of breaking through over 1.5-meter-thick horizontal ice plus 0.2-meter-thick snow at a continuous icebreaking speed of 2 to 3 knots. The bow heading navigation mode allows for a capability to ram and break through the ice. Meanwhile, the stern heading navigation mode allows for independent navigation without getting stuck in ice ridges consisting of 20 m of first-year ice (including a 4-m stack layer) plus a 0.2-m snow layer.
(2) Sinking ductkeel design: Considering that the vessel is designed to navigate in the polar regions, suspended installation for deep-sea exploration multi-beam devices is generally not suitable. This requires that the bow bottom should be designed reasonably to arrange the transducers, while also ensuring that the presence of the bow's bubble layer does not affect the functionality of the transducers and that the hull strength is guaranteed. Through full-scale model-based viscous CFD analysis, the vessel's design team examined the water flow around the ductkeel, studying the distribution of bubbles and vortices at the keel's bottom during navigation and assessing it through Q criterion and streamline method. Based on these calculations, the design of the ductkeel provides sufficient space for the installation of necessary measuring equipment. The vessel adopts a ductkeel configuration to accommodate the transducers of the acoustic equipment. This allows for optimization of the vessel's performance and economy in open-water navigation, while ensuring both icebreaking capability and the hull strength required for PC3 icebreakers. This design enhances the vessel's resistance to winds and waves, while also meeting the installation requirements for scientific research and exploration equipment.

(3) Ice zone environment "watertight moonpool" system design: To collect hydrological, ecological and other data and samples from dense ice-covered areas in polar regions, "Xuelong 2" is equipped with a 3.2 m × 3.2 m square well connecting the moonpool workspace deck to the vessel's bottom and the surrounding water, along with supporting deployment tools in the moonpool workspace at the midship. It enables the operators to deploy and retrieve operating equipment and devices directly into the water from the moonpool workspace deck. The moonpool system's design takes into account various factors, such as the characteristics of operations in ice zones, ship safety, and safety in scientific investigation surveys. Through thorough research and feasibility studies on the "wave-absorbing bulkhead" structure, track pulley LARS, watertight design for the moonpool bottom cover, top cover design, hydraulic power system, automated electrical linkage control, and monitoring system, the design team developed detailed design and implementation schemes. During the 36th Antarctic scientific survey, the "watertight moonpool" system fulfilled its design requirements. It became the first system to achieve the application of a "watertight moonpool" bottom cover internationally, enabling scientific surveys in densely ice-covered areas and severe weather conditions while improving ship safety and navigation efficiency.

(4) Flexible and adaptable laboratory system design: Laboratory construction is a complex systematic engineering, especially when it comes to the construction of laboratories on research vessels. In the field of modern laboratories, advanced scientific instruments and well-equipped and flexible laboratory infrastructure are essential for enhancing the level of modern scientific and technological development and promoting the growth of research outcomes. The vessel's laboratory scheme design not only considers functionality, applicability, and rationality but also includes the overall planning, reasonable layout, and plane design, as well as the establishment of infrastructure and basic conditions such as electricity, water supply, gas supply, ventilation, air purification, safety measures, and environmental protection, among others. Considering the icebreaker's characteristics, such as rocking, vibration, and tilt during icebreaking and the diversified tasks required, the laboratory design underwent careful deliberation in terms of material selection, design and installation processes, environmental detection, data processing, and intelligent laboratory management. The bulkheads of the general laboratory are made of composite ceramics to effectively resist seawater corrosion. Modular assembly allows for flexible arrangement of the laboratory to accommodate different voyages and research scientific research tasks. By employing modern network control and intelligent control technologies, the vessel can ensure laboratory environment monitoring, data sharing, visualization application, and intelligent control. Thus, the vessel is fitted with a modern, flexible, and intelligent ship laboratory system that meets polar scientific investigation requirements. This received high recognition from the expert panel.

(5) Vibration and noise analysis and control technology: In addition to meeting the vibration and compartment noise indexes required by relevant regulations,

scientific experiments on research vessels also require certain vibration and noise requirements for data accuracy. Vibration and noise control on research vessels is a systematic engineering. The traditional approach of considering vibration and noise control in detailed design or construction stages is no longer sufficient to meet the requirements for vibration and noise indexes on modern research vessels. Indeed, vibration and noise control must be incorporated into the entire design and construction process of research vessels.

The unique icebreaker type, high-power main engine, special thruster, and high-strength propeller of "Xuelong 2" pose significant challenges in controlling vibration and noise. The vessel adopts high-precision construction and various measures for vibration and noise control, allowing the vessel's vibration and noise levels to approach or meet the highest international standards for new generation research vessels.

After tilt tests, dock mooring tests, trial voyages, and scientific expeditions, "Xuelong 2" met all the requirements outlined in the technical specifications.

In October 2019, "Xuelong 2" embarked on its maiden voyage to Antarctica, joining "Xuelong" in a joint venture to carry out the 36th Antarctic Research Expedition of China. Upon arrival in Antarctica, "Xuelong 2" navigated smoothly through the thick ice, opening a 14-nautical-mile ice channel in just three days. Subsequently, "Xuelong" followed this channel to reach the unloading point at the central station. This maiden voyage fully utilized various platforms, including ships, sea ice, oceans, land, air, and research stations. Through multidisciplinary observations of land, ocean, atmosphere, ice shelf, and marine biology, this expedition implemented major national scientific research programs, completed infrastructure projects, improved environment in the station area. Other tasks during the voyage include routine support, material transportation, engineering construction, and international cooperation.

In November 2020, "Xuelong 2" set sail for the 37th Antarctic Research Expedition of China, which focused primarily on investigations related to global climate change. It conducted hydrometeorological, ecological environmental, and other scientific surveys. Meanwhile, it also conducted operational monitoring of new types of pollutants such as microplastic floating marine debris in the Antarctic Ocean.

The commissioning of "Xuelong 2" expanded the breadth, depth, and duration of China's polar oceanic surveys, and endowed China with the capability to lead large-scale polar marine science and technology projects, greatly enhancing the country's influence in polar affairs. Furthermore, it provided a critical platform for expanding China's polar strategic fields and advancing understanding, protection, and utilization of the polar regions.

From the approval by the State Council in June 2009 to the official delivery on July 11, 2019, the "Xuelong 2" project spanned ten years. The successful delivery of "Xuelong 2" significantly advanced China's polar scientific research undertakings. Through the work undertaken during the decade, Chinese technicians have absorbed, mastered, and improved upon the design and construction technologies of high-grade icebreakers. More importantly, China now possess the major intellectual property rights and key technologies associated with the design of the vessel. Throughout the research, design, and construction process of "Xuelong 2", 38 patent applications (including 6 invention patents) were filed, of which 10 were granted (including 1 invention patent). Additionally, 37 research papers were published, and 1 software copyright was obtained. This progress has filled the gap in China's endeavors in design and construction of polar research icebreakers, laying a solid foundation for the design and construction of future heavy icebreakers. It also filled the gap in polar icebreaker regulations and standards in China, playing a significant role in promoting the design, construction, and equipment matching for professional icebreakers in the country. Moreover, it injected strong momentum for the effective implementation of international polar regulations in China, laying a solid foundation for future advancements.

2.20 *Multipurpose Marine Geological and Geophysical Survey Vessel "Haiyang Dizhi 8"*

In order to enhance China's capability in investigating marine geology, oil and gas resources, natural gas hydrates, and other mineral resources, and to support the major strategic project of natural gas hydrate trial mining in the South China Sea, further to substantially and normally safeguard national maritime rights and interests, the China Geological Survey developed a scientific and technological strategy for the development of national land resources, featuring "first-class equipment" and "deep-earth exploration, deep-sea exploration, deep space-to-earth observation, and technological innovation for land development", in order to fulfill the historical mission of "exploring the depths of the Earth". To achieve this goal, in 2014, the Guangzhou Marine Geological Survey commissioned the MARIC to design a new geophysical vessel, with the construction commissioned to Shanghai Shipyard. The vessel was delivered in November 2017 and was named "Haiyang Dizhi 8" (see Fig. 22).

"Haiyang Dizhi 8" is a 5,000-ton vessel equipped with fully 3D seismic field data acquisition and processing capabilities, as well as the ability to conduct marine gravity, magnetic, and bathymetric surveys.

The vessel has a total length of 88.0 m, a molded breadth of 20.4 m, a molded depth of 8.0 m, a designed draft of 5.6 m, a displacement of 6,500 tons, a towing force of 50 tons at a speed of 5 knots, a capability of towing eight 8,000-m geophysical cables, and a capacity of 60 persons. With ice strengthening capability, the vessel allows for a speed of 16 knots, an endurance of 16,000 nautical miles, a self-sustaining period of 60 days. It incorporates an electric propulsion system, dual-duct adjustable pitch propellers, and a double-hull design.

The vessel is equipped with an internationally advanced high-resolution short-travel 3D seismic survey system, which allows for high-precision short-travel seismic cable 3D (and even 4D) seismic operations using up to six cables (expandable to eight cables). In addition, it is also equipped with internationally advanced integrated navigation, positioning, control, and post-processing systems. Besides this, the vessel also

Fig. 22 Multipurpose marine geological and geophysical survey vessel "Haiyang Dizhi 8"

features multi-channel seismic data collection system, digital air gun seismic source and control system, shipborne seismic data processing and interpretation system, marine gravity measurement system, marine magnetometry system, single-beam bathymetric survey system, as well as digital compass system and comprehensive information system. With these capabilities, the vessel can perform surveys of natural gas hydrates across all sea areas and conduct regional geological investigations. To meet the specific requirements of the national strategy for oil and gas resources in the South China Sea and natural gas hydrate trial mining, the vessel employs a systematic engineering approach and a people-oriented development philosophy. It incorporates the latest achievements in geophysical equipment technology development and hull research from both domestic and international sources. The vessel's design integrates multiple advanced technologies, ensuring a reasonable layout, clear logic in geophysical functionality, and the capability to operate with multiple cables even in challenging sea conditions. As a result, it has become a new type of vessel with high-resolution seismic sources, high-precision short-travel cable acquisition signals, and excellent geophysical performance, as well as outstanding performance in geophysical operations, living, and working environment quality.

The main features of this vessel include:

1) Advanced Hull Form Design

The marine geological survey requires sufficient lateral space for the deployment of multi-cable equipment. The breadth occupied by each cable is usually around 1.88–2.3 m per cable. Additionally, the 3D geophysical vessel has a relatively high center of gravity. To meet the requirements of equipment placement and stability, the vessel breadth of the 3D geophysical vessel is generally large. However, the vessel has a length-to-breadth ratio of only 3.92, with steep changes in the bow and stern sections, which limits the speed of the vessel. To address this issue, this project team developed a hull form with a large bulb bow, a small round bilge, a long entrance flow section, and a variable molded breadth between the upper and lower parts. As a result, the optimization of the bulb bow matched with the short and wide hull. The height of the bulb bow was appropriately raised, and its length was enlarged to increase the effective waterline length of the vessel. The shape of the bulb bow was moderately streamlined. Ultimately, the resistance was reduced by more than 10%, compared to vessels with similar length-to-breadth ratios.

2) Safe Towing Operations

The vessel adopts a new type of stub wing section duct and adjustable pitch propeller, along with a flap rudder for electric propulsion. The combination of the new stub wing section duct and the adjustable pitch silent five-blade propeller, along with the frequency-controlled electric propulsion system with the flap rudder, meeting the requirements for high towing force, low noise, good maneuverability, high speed, and low resistance during high-precision 3D towing operations of the geophysical vessel. The vessel adopts an asymmetric shaft bracket form with the ability to eliminate eddy currents in the wing's cross section, effectively removing the impact of eddy currents and bubbles on the propeller blades.

Underwater towing equipment for geophysical survey operations can be quite expensive, and the vessel needs to have a certain speed to ensure that the towed air gun array and cable remain floating on the sea surface. If the vessel loses power, the towed air gun array and cable can gradually sink and even become entangled, making it impossible to retrieve the cable, air gun, and other equipment. This can result in losses of tens of millions yuan. Therefore, the design team took into account power redundancy, ensuring that a single failure in the power system will not result in the vessel losing power and causing the aforementioned serious consequences. Besides, the vessel is equipped with four main diesel generator sets, each rated at 690 V, 50 Hz, and 2,610 kilowatts. They drive two sets of duct-adjustable pitch propellers through distribution panels, frequency converters, and other equipment. The distribution panels are designed with segmented sections, with each section on one side connected to two main diesel generator sets. The two main propulsion systems, along with three seismic source air compressors and distribution transformers, are mounted on both sides of the 690-V distribution panels. In the event of a single failure, this configuration allows for segmented power supply on both sides as per the demand, ensuring the safety of the power distribution and propulsion system.

3) Reasonable Space Arrangement

The vessel's design includes an open working area at the stern without bulkheads, which effectively solves the challenge of accommodating a large number of equipment and the need for a spacious area for operations. The deck design employs a reinforced staggered-deck structure to accommodate numerous equipment and meet their varying height requirements for operations, thus addressing complex structural design challenges. Considering the diverse types of geophysical equipment, their large quantities, limited space, close proximity, and various installation methods, the vessel was designed with specific reinforcement measures to ensure the safety and reliability of each equipment based on their respective characteristics.

4) Integration of Informatization and Automation

In addition to the traditional communication and navigation equipment, as well as platform equipment, the vessel is equipped with a large number of general and specialized survey equipment, making the vessel highly digitized. They provide numerous interfaces for various disciplines and enable the vessel to handle all types of data with complex interrelationships. Through scientific planning and proper implementation of the vessel's informatization project, the vessel achieves seamless information exchange and resource sharing between various onboard operation systems, between the vessel and shore-based facilities, between vessels, and between the vessel and underwater platforms. This provides professionals from different disciplines and fields with a suitable working environment. Moreover, an integrated management platform is available for data collection, processing, and monitoring throughout the vessel, significantly enhancing the automation and data management capabilities of marine scientific survey activities.

5) Expandable and Powerful Geophysical Survey Capabilities

This vessel is equipped with a storage space for six seismic cables and an installation space for two single-linkage cable winches. With simple modifications, the vessel can be expanded into an eight-cable geophysical survey vessel.

The integration of geophysical equipment system is the core and key of geophysical vessel design. The equipment system of the geophysical vessel mainly consists of the digital seismic cable LARS, the seismic source equipment and extender and LARS.

(1) Digital seismic cable LARS: The digital seismic cable comes in two diameters: 55 mm and 59.5 mm, with a typical length of 150 m per section. Special connectors are used to connect the sections. Due to the data cables and sensors inside the digital seismic cable, a larger bending radius is required. Considering the characteristics of the digital seismic cable, the digital seismic cable LARS includes cable deployment winch (including cable guider), cable storage winch, cable towing point, tail buoy hoist, extension rope winch, and other equipment. This vessel is the world's first short-travel digital seismic cable 3D geophysical vessel capable of expanding to eight cables, with a maximum capacity of towing eight 8,000-m seismic cables.
(2) Extender and its LARS: The extender LARS mainly consists of an extender and an extender LARS. The extender LARS assists in deploying the extender into the water from the vessel. On a 3D geophysical survey vessel, the extender is used to provide sufficient extension force. The equipment deployed underwater by digital seismic cables and the gun arrays through deployment equipment are connected to the extender via connecting ropes, allowing for a desired safe

distance between the equipment. The vessel is equipped with the BARO46 extender, which fully meets the capability of extending eight digital seismic cables.

(3) Seismic source equipment and its LARS: The seismic source equipment mainly refers to the gun array system, which is used to generate seismic waves in geophysical exploration operations. The seismic source deployment equipment is used to assist in deploying and towing the source equipment in the water. The gun array can be classified into rigid floating and soft floating types based on the form of its floating body. The vessel is equipped with a soft floating gun array system, which can accommodate a maximum of eight gun arrays. The stern is equipped with a curved deployment rail, a ramp, and auxiliary winches for the launch and recovery of the gun arrays. This arrangement occupies less vertical space on the vessel, reduces costs, effectively lowers the ship's center of gravity, and minimizes wind area.

6) People-Oriented Design Concept

Taking a people-oriented design concept into account, the vessel optimizes the details of compartment structure, air conditioning system design, and lighting design. It addresses the structural vibration and noise issues of the entire vessel from multiple dimensions, such as matching of structure and layout, predicting of vibration and noise levels, considering of lightweight structure and continuity of the geophysical vessel. Specific vibration isolation and radiation damping designs are applied to localized strong vibration source electromechanical equipment such as the geophysical air compressors to reduce air and underwater radiation noise. With these designs, the vessel effectively addresses the challenge of high vibration and noise levels during geophysical operations, ensuring a comfortable working and living environment. The comfort index meets the requirements of COMF (NOISE 3) and COMF (VIB 3), and the underwater radiation noise meets the DNV silent mark requirement, SILENT-S.

"Haiyang Dizhi" served as the world's first high-precision, six-cable (expandable to eight cables) 3D geophysical vessel that used short-travel signal cables at the time. Due to its short and wide hull, excellent stability, and flexible maneuverability, the vessel is particularly suitable for navigation or surveying and exploration in inter-island areas, complex waters, or adverse sea conditions.

Shortly after its delivery, the vessel carried out operations between the South China Sea islands, such as Wan'an Tan. Throughout the entire survey process, the vessel demonstrated excellent overall performance, reasonable system design, accurate geophysical survey system, advanced and reliable seismic system, and strong geophysical survey capabilities. It has been proven that all technical indexes of the vessel meet or outperform the technical specifications and relevant standards and regulations. This vessel has made a significant contribution to safeguarding China's maritime rights and interests in its waters. Since its commissioning, the vessel has been operating at full capacity and has achieved a series of significant results in high-precision seismic surveys, natural gas hydrate exploration, and other areas.

2.21 Lightweight Research Vessel "Tan Kah-Kee"

Xiamen University, a prestigious higher education institution with a long-standing history in marine-related majors in the southeastern coastal provinces of China, has a wealth of experienced marine research personnel. The university has close exchanges with many Southeast Asian countries in the field of oceanography and even has a branch campus in Malaysia. To leverage Xiamen University's capabilities in oceanographic research, enhance cooperation with counterparts in Southeast Asian countries, and promote the economic development of the "Maritime Silk Road", Xiamen University applied to the Ministry of Industry and Information Technology (MIIT) and the Ministry of Finance to construct a medium-sized, lightweight oceanographic research vessel. In November 2011, the Ministry of Industry and Information Technology and the Ministry of Finance approved the project proposal for the *Development of the Ship Type of Medium-sized Lightweight Marine Environmental Research Vessel.*

Commissioned by Xiamen University, the MARIC provided a feasibility report and the basic concept for the project. Based on this, Xiamen University signed a basic design contract with the US GLOSTEN in December 2012, followed by a detailed design contract with the MARIC. In January 2014, Xiamen University signed a construction contract with Guangzhou Shipyard International Company Limited. The vessel was delivered in February 2017 and named "Tan Kah-Kee" after Mr. Tan Kah-Kee (Chen Jiageng) (see Fig. 23), a renowned patriotic individual who made significant contributions to the development of the university.

The vessel has a length of 77.7 m, a molded breadth of 16.24 m, a molded depth of 8.00 m, a draft of 5.30 m at full load, a maximum speed of 14.0 knots, and a capacity of 54 persons (including 18 crew members and 36 researchers). With an endurance

Fig. 23 Lightweight research vessel "Tan Kah-Kee"

of 10,000 nautical miles at a speed of 11 knots, the vessel is propelled by generator set and propeller electric propulsion.

The vessel is equipped with several key laboratories and facilities, including the atmospheric laboratory, main laboratory, general laboratory, wet laboratory, clean laboratory, electronic laboratory, gravimeter room, sample refrigeration freezing room, data room, seawater sampling station, and acoustic equipment compartment.

The scientific research equipment on board includes the 10,000-m CTD winch with its deploying boom, 8,000-m coaxial cable winch, 15,000-m steel cable winch, 10,000-m optical cable winch, hydrographic winch, stern A-frame, long cylinder piston corer with its deploying devices, lifting fin.

The acoustic equipment on board includes the deep-sea multi-beam system, middle- and shallow-water multi-beam system, sub-bottom profiler, acoustic Doppler current profiler, fish finder, ultra-short baseline system, CTD (conductivity temperature depth) detector, hydrophone.

Design features of the vessel:

The design team, following the design goal of a medium-sized lightweight research vessel, strived to maximize the scientific research capabilities and operational space of a larger-tonnage scientific research vessel while keeping the overall tonnage relatively small. Through all-round lightweight and convenient design, the design team overcome challenges related to weight and center of gravity control, and also improved the rational utilization of space. This approach enabled the vessel to achieve a displacement of 3,000 tons, comparable to 4,000–5,000-ton research vessels in terms of total laboratory area, working deck area, survey equipment capability, and scientific research capacity.

There are breakthroughs in key technologies such as silent design and construction process optimization. As a result, the vessel features energy efficiency and environmental friendliness, convenience and comfort, modularization, low noise, high scientific research capacity, high operational efficiency, fine sampling and multidisciplinary in-situ survey capability.

1) Lightweight Design

Throughout the design and construction process, the total weight and center of gravity changes of the vessel were predicted and dynamically evaluated to achieve the goal of lightweight design.

(1) Optimized hull structure: The weight of a research vessel's structure typically accounts for over 50% of the total weight. Thus, controlling the weight of the structure is crucial for controlling the vessel's overall weight without load. The structural weight control focuses on weight reduction and lowering the center of gravity.

    Weight reduction design: Several lightweight design methods were used, including the use of lightweight superstructures (aluminum alloy materials for decks above the 02 deck), longitudinal frame structures for primary component grillages, and large open-hole lightweight rib plate designs for the raised stern area.

Comprehensive control of the structural center of gravity: Various conventional and unconventional methods were introduced to control the center of gravity. These include: lowering the center of gravity of the empty vessel by approximately 0.13 m through lightweight superstructure design, adopting unconventional structures for the bottom rib plates with large openings and light weights, increasing the number and area of bottom rib plate openings as possible, and increasing the thickness of the bottom outer plate in the 0.5 L (L is the vessel length) range of the midship section by 12 mm, which increased the weight by approximately 23 tons and lowered the center of gravity by approximately 0.1 m.

(2) Weight control through equipment selection: Lightweight design principles were implemented in equipment selection. Priority was given to selecting lighter equipment that met functional requirements, such as the main propulsion motors, main generator sets, bow and stern side thrusters, hot water boilers, and domestic sewage treatment systems, among others.
(3) Optimized system piping and outfitting pieces: Piping routes and outfitting pieces dimensions were optimized while meeting functional requirements. Lightweight materials were preferred. For example, plastic pipes were used for water supply and drainage in the superstructure. Iron outfitting pieces such as pipe brackets, checker plates, and grating were properly arranged, and small-sized structural steel was used as support brackets wherever possible while meeting the strength requirements.

2) Hierarchical Division of Functional Spaces

The optimization of spatial relationships, the hierarchical division of functional spaces, and the overall utilization of area were emphasized.

The vessel's layout incorporated concepts such as a maximized utilization of scientific research systems, centralized and versatile laboratory designs, as well as compact and partitioned living quarters. The layout addressed the challenge of accommodating large laboratories and extensive working deck areas within a smaller-tonnage vessel. A mainstream layout of left-arranged chimney and L-shaped laboratories on the starboard side was adopted. A large multipurpose laboratory was set up, boasting the largest laboratory area (net area of 126 square meters) among domestic research vessels at the time, and allowing for both wet and dry experimental conditions. The aft working deck features a large working area of 430 square meters, facilitating scientific research operations and allowing for future functional expansion. The sheltered aft working room provides a favorable indoor working environment for scientific research operations during the winter season.

Crew and senior research personnel are provided with single rooms, and general researchers with double rooms, thus elevating their living comfort.

3) Convenience Design of Scientific Research Spaces

(1) Convenience design for the working deck area: This vessel was the first multipurpose research vessel in China to fully implement the concept of removable and convenient design, with a focus on modular design. For the

first time, embedded anchor bolt grids (610 mm x 610 mm) was uniformly distributed throughout the entire working deck range. The working deck area is equipped with multiple fixed devices for 20-foot[7] container laboratories, providing sufficient temporary storage space for equipment that are used in special oceanographic surveys and scientific research tasks. Temporary tie-down points are strategically placed on the bulwark at suitable heights, enabling easy and safe operation of underwater equipment. The aft main deck is equipped with removable bulwarks, winches, and bitts that can be installed or removed as needed during different voyages, vastly enhancing the vessel's capacity to carry new types of detection and sampling equipment. Each deck is fitted with power sockets and quick-connect joints for rapid access to freshwater, seawater, and compressed air necessary for scientific research.

(2) Convenience design for the laboratory area: With the exception of the atmospheric laboratory and seawater sampling station, the rest laboratories were concentrated on the main deck, situated on the same layer as the open main working deck. This facilitated a seamless integration between the laboratory area and the deck workspace, not only reducing safety hazards during offshore operations, but also improving work efficiency and convenience.

The general laboratory, wet laboratory, scientific research workspace, and main laboratory (alongside its passage) were all connected with the aft deck, facilitating the entry of water and sediment samples into the respective laboratories. The threshold between the main laboratory and the aft deck was designed to be removable and watertight, significantly improving loading and unloading efficiency of instruments and materials. The laboratories on the entire vessel were primarily designed to meet the functional requirements of dry/wet and non-clean/clean operations. With a focus on versatility, the functional zones were clearly divided, making it convenient for the maintenance of cleanliness in the laboratory spaces. These designs catered to the requirements of different disciplines and facilitate operations. Research equipment storage rooms are arranged at the bow and stern, making it more convenient for the storage and access of experimental supplies.

Laboratory instruments and equipment or systems are arranged in proximity for easy access. Each laboratory is equipped with embedded anchor bolts (610 mm × 610 mm) on the floor for temporary fastening of experimental equipment. Experiment tables, laboratory wallboards, and ceiling panels are equipped with fixed guide grooves at 610-mm intervals. This design not only meets the 3D fixed requirements for flexible and convenient laboratory space but also allows for flexible division of the laboratory space. Electrical, gas, and water interfaces are designed on all experiment tables and ceilings. The laboratories are equipped with complete scientific research cable racks and cable through-bulkhead openings to meet the expanding needs of instrument cables. The experiment tables are all designed to be

[7] Foot, a unit of length. 1 foot = 30.48 cm.

detachable, allowing for more flexible use of space and easier installation of equipment and instruments.

(3) Convenience design for vertical passages: A passenger-and-freight elevator is arranged to connect the laboratory area to the main deck of the vessel. This not only facilitates personnel movement between the laboratories and the various residential decks but also allows for the vertical transportation of scientific research instruments and equipment.

4) Lifting Fin for Sampling Clean Seawater for Laboratories

A single lifting fin is installed at the front 2/5 position, incorporating various transducers and acoustic equipment, as well as a clean seawater sampling system. This design facilitates the vessel's successful sampling of clean seawater through lifting fin unprecedentedly in domestic, improving the convenience and safety of seawater sampling.

5) The target-oriented noise control technology that combines underwater radiation noise and compartment noise control was introduced for the first time. It incorporates key techniques for a comprehensive system design such as double-layer vibration isolation for generator set, silent drive for propulsion motor, and efficient silent large skewed propeller.

This design ensures both rapidity and low noise. It optimized both noise sources and transmission paths, and incorporated both absorption and isolation measures, ultimately achieving a high level of noise control for compartments.

(1) Underwater radiation noise control technology: The indexes for controlling underwater radiation noise in China was introduced for the first time. It follows the DNV SILENT-A+S (8 knots) standard for underwater radiation noise ship classification.

By decomposing the indexes into requirements for major equipment according to calibration standards, noise control measures and application areas were determined. The statistical energy method was employed to calculate reversely the energy distribution of various noise sources on the vessel by using the underwater radiation noise standard as input for the database-driven (based on a large number of ship trial results) empirical model. The transmission paths and contributions of each noise source were considered, and noise control indexes were allocated accordingly. Appropriate control measures were then adopted to achieve the maximum permissible noise levels and corresponding noise reduction measures for major noise sources.

(2) Application of self-designed silent large skewed propeller: Propeller is the primary source of underwater radiation noise. This vessel is the first in China to adopt the independently developed silent large skewed propeller designed by the MARIC. The propeller utilizes tip vortex cavitation control technology to

achieve a balance between propulsion efficiency and low noise characteristics. The successful application of the propeller has played a crucial role in controlling underwater radiation noise.

(3) Hybrid vibration absorption design for the main generator set: As a core component and one of the primary noise sources of a ship, the main generator set holds the key in the control over underwater radiation noise. This vessel's main generator set employed a hybrid vibration absorption design that includes double-layer vibration isolation for the diesel engine and single-layer vibration isolation for the generator. This design strikes an optimal balance between weight control and vibration reduction. Damping materials were also applied at critical locations on the base. Biaxial elastic joints, elastic brackets and elastic bulkhead-penetrated parts are used for the external piping of machines to minimize the transmission of diesel engine vibrations to the hull structure along the piping.

(4) Application of low-speed, low-noise variable frequency AC main propulsion motor: Customized low-speed, low-noise variable frequency AC motor was selected for the propulsion motor, along with single-layer vibration isolation and high elastic output damping design. The base design underwent rigorous calculations and included targeted local reinforcement and the application of damping materials.

To ensure effective control over underwater noise, a set of measures, including follow-up control, inspection, calculation, and correction, were implemented through out the whole process of design and construction. Results from ship trials indicated that the vessel's underwater radiation noise fully mets the requirements of DNV SILENT-A+S and even complied with the most stringent SILENT-R requirements in the high-frequency range. As a result, the ship was awarded the silent rating certificates SILENT-A+S (11 knots) and SILENT-R+F (8 knots) by DNV, thus demonstrating the highly successful control of underwater radiation noise of this vessel.

(5) Air noise control technology: Key noise control techniques were introduced for controlling compartment noise. Combining with the lightweight design concept, they play a crucial role in achieving optimal noise control at the minimum cost of lightweight.

During the initial design phase, equipment that might generate vibration noise on the vessel was categorized, and crucial vibration noise-generating equipment such as the main generator set, propulsion motor, air compressor, high-power pump, refrigeration unit, high-power fan, and hydraulic equipment were identified.

Through simulation and acoustic performance calculations, critical compartments and areas were analyzed, and targeted vibration and noise reduction measures were implemented for the relevant equipment and systems at even the link of manufacturing. Sound insulation techniques were applied to enhance the sound insulation performance of bulkheads and decks, thereby reducing the sound levels in critical compartments.

Targeted noise control measures were implemented for all aspects, from structural design, power system design, mechanical equipment selection, compartment layout to air-conditioning ventilation system design. Actual ship measurements confirmed that the noise levels on the vessel met the expected targets, with about 80% of crew compartments achieving a noise level of around 45 decibels, which surpassed the requirement of 49 decibels for the highest comfort level according to CCS COMF (NOISE 1). As a result, the vibration noise control effectiveness of the vessel received acclaim from the shipowner and DNV experts who participated in the sea trials.

6) First Mechanically Isolated Clean Power Supply System for Research Vessels in China

This system provides a reliable and high-quality AC and DC dual-power variable frequency clean power supply for precision laboratory instruments.

The system includes an innovative low-harmonic scientific research power supply, which utilizes a coaxial motor generator to physically isolate the vessel's everyday power grid from the scientific research power supply by means of the conversion of electrical mechanical energy. This significantly reduces the distortion rate of the open-circuit voltage of the research power supply, ensuring that the power supply system generates no harmonics and remains unaffected by harmonic interference. As a result, the laboratory's precision instruments can operate in a more stable and reliable manner. The voltage and frequency output of the power supply system can be adjusted during the design phase to cater to the specific needs of the vessel. To guarantee uninterrupted power supply to the scientific research equipment, a backup uninterruptible power supply was installed, and a static switch was used to realize seamless switching between the two power sources. With its high power quality and reliability, the power supply system can greatly ensure the stable operation of scientific research equipment. This high-quality laboratory clean power supply system, which represented an innovative design, was used for the first time on domestic research vessels.

In practice, the vessel has already undertaken 24 voyages, during which numerous sophisticated scientific research instruments were used, and there was no malfunction of these instruments due to power supply issues.

7) A New Model Trial Method for Validating Bubble Venting on Research Vessels

For research vessels, the bow shape significantly affects bubble generation. During pitch motion, the bow may experience slamming, leading to bubble formation. At certain speeds, these bubbles can be carried into the acoustic equipment installation area at the bottom of the bow. Due to the gas–liquid interface of bubbles and the noise generated during their collapse, they can interfere with acoustic equipment. Therefore, it is essential to prevent bubbles from flowing through the acoustic equipment installation area.

However, in ship designs, most profile designs that take into account the influence of bubbles are based on the designers' experience, with few quantitative or CFD-based optimization studies being conducted. One of the main reasons lies in the lack of a relatively definitive experimental method to accurately measure the impact of bubbles. During the design phase of the vessel, an innovative approach was proposed, utilizing dye to simulate bubble flow in pool tests, along with a bubble detection method. This technique allows for the determination of the flow paths of bubbles generated at different drafts of the bow and their trajectories along the hull bottom. To some extent, it addresses the challenges of simulating and tracking bubbles, thereby providing valuable insights for the placement of acoustic equipment, such as multi-beam systems, in the bow area.

8) New Technologies Introduced during the Construction Process

Through breakthroughs in construction techniques and processes, the vessel's design secured multiple technical patents. Key manufacturing and installation challenges were overcome, including the high number of securing points on the working deck and in laboratories, stringent precision requirements, welding deformation, and the double-layer vibration isolation of power equipment.

(1) Securing system installation quality control technique: The extensive and diverse nature of scientific research projects results in significant variations in the equipment and scientific instruments carried and used during different voyages. This places high demands on the openness and flexibility of the working deck and laboratory layout capabilities. Considering the convenience of deck securing, embedded anchor bolts (with a spacing of 610 mm x 610 mm) were installed on all

working decks and laboratories throughout the entire vessel, totaling 1,568, the largest among other domestic research vessels in terms of its number and distribution range. During the assembly of the anchor bolts, it was required that the deviation of the center distance must be controlled within $\pm$ 1.5 mm. Such a high precision requirement along with huge installation workload indeed posed significant challenges to the shipyard's construction process. To ensure installation quality, extensive testing and research were conducted, leading to the development of a feasible solution: "installing solid stainless steel cylinders and drilling and tapping holes on the vessel." Specialized installation procedures were established, and custom tooling was designed to minimize the effects of hull and welding deformation, ensuring the precision of bolt hole positioning.

(2) Optimized water-lubricated bearing integral stern tube installation process: The vessel's stern is equipped with a propeller propulsion system, an integrated stern tube, and water-lubricated bearings. The shaft end features a shaft sleeve, while an A-frame is installed near the propeller. Inside the hull, there are left and right shaft tunnels. The stern shaft system, including its line shape, appendages, and internal compartment structures, is highly complex. Additionally, the limited space posed significant challenges during construction. To address this, the shipyard developed a well-conceived installation plan and construction workflow, while also providing the necessary tooling to ensure precise installation of the shafting system.

(3) Optimized removable mooring winch installation process: In order to cater to the flexibility requirements of scientific research equipment and operation, as well as maximize the utilization of scientific research space, the vessel was designed with a substantial number of removable devices, such as removable mooring winches, scientific research winches, bitts, and lighting poles. This approach is relatively uncommon in domestic research vessels, as the reliability of their installation directly impacts the reliability and safety of the vessel's operation. The detachable winch, in particular, bears significant loads, making safety and reliability critical. To address this, the shipyard developed multiple installation plans for comparative analysis. Through close collaboration with equipment manufacturers and the shipowner, the final solution involved enlarging the winch base mounting plate, increasing the number of assembly bolts, and using red lead canvas as an intermediate shim to enhance the contact area

and friction between the winch base and the deck surface without compromising deck precision. Additionally, a dedicated common transition base was designed. This approach effectively resolved the installation challenges of the detachable winch. The well-considered installation plan not only meets the safety and reliability requirements but also satisfies the shipowner's strength specifications.

The successful development of the "Tan Kah-Kee" research vessel represents a significant breakthrough in several key technologies for the design and construction of medium-sized lightweight and low-noise research vessels. It provides valuable reference for the development of similar ship types. Some vital research outcomes, such as muting technique, lightweight design, and the removability concept, were successfully applied to the design and construction of subsequent research vessels, such as the "Shenhai 1", "Dayang", "Shiyan 6", and "Dongfanghong 3".

From its commissioning until May 30, 2020, the "Tan Kah-Kee" research vessel has successfully completed 24 scientific research missions, covering a total of 617 days and approximately 64,320 nautical miles. It has undertaken numerous scientific research tasks, providing reliable ship-time support to nearly one thousand researchers and students from marine research institutions, colleges and universities from countries such as China, the US, Germany, Japan, Canada, and Malaysia. Notably, the vessel has effectively carried out several key missions, such as the shared voyage of the National Natural Science Foundation of China, the South China Sea observation voyage, and the national special marine environmental monitoring voyage.

Supported by Xiamen University Malaysia, "Tan Kah-Kee" has served as a platform, hailed as "Maritime XMU", for promoting global marine change research, marine science education, and scientific and technological exchanges across the ASEAN region. It is set to contribute to the advancement of marine research and education at Xiamen University and even globally. More importantly, it will fulfill its honorable mission of promoting Chinese culture, advancing the construction of a strong maritime nation, and supporting the Belt and Road Initiative. On August 17 and 18, 2019, "Tan Kah-Kee" held an open day event in Port Klang, Malaysia, which attracted nearly one thousand local Chinese residents and other visitors. Following the conclusion of the event, over 30 teachers and students from institutions such as the China-ASEAN College of Marine Sciences, the University of Malaysia, University Malaysia Sabah, University Putra Malaysia, and the University of Delaware joined the vessel to explore the mysteries of the ocean. The vessel enables the teachers, students, and experts from Xiamen University and its Malaysian campus to conduct scientific research in deep oceans, facilitating the cultivation of high-quality international talents for China. It also contributes substantially to the promotion of cooperation and exchanges in marine research and ocean education between China and the ASEAN region, acting as a bridge to deepen the friendship between China and Malaysia.

2.22 *Pelagic Fishery Resource Survey Vessel "Songhang"*

Pelagic fishery is a strategically significant resource industry closely related to human life and economic development. Conducting pelagic fishery resource surveys is a necessary measure to achieve sustainable development in pelagic fishery, promote China's transition from a major country to a strong country in pelagic fishery, and safeguard more of the country's marine rights and interests.

Pelagic fishery resource survey vessels ought to strive to collect systematic and comprehensive data to provide sufficient basis for the development of fishing industry and scientific research.

In May 2013, the former Ministry of Agriculture and the Shanghai Municipal Government jointly applied to the National Development and Reform Commission[8] for the construction of a pelagic fishery resource survey vessel. Upon approval, the duo co-funded the construction of this vessel, which would be used to conduct oceangoing surveys and cultivate professionals in pelagic fishery resource surveys for Shanghai Ocean University. The vessel was designed by the China Ship Development and Design Center and built by CSIC Tianjin Xingang Ship Co., Ltd. Being constructed from October 2015, the vessel was launched in March 2017 and was delivered in September 2017.

Prior to this vessel, Shanghai Ocean University had a training vessel named "Songhang" which was unfortunately destroyed by the Japanese army during the "1·28 Songhu Anti-Japanese War". To honor history and preserve maritime culture, while embodying the scientific and striving spirit of "a century-old institution dedicated to the ocean, serving the industry, and contributing to the nation", Shanghai Ocean University named the China's first pelagic fishery resource survey vessel "Songhang" as well (see Fig. 24).

The vessel has a total length of 87 m, a molded breadth of 14 m, a molded depth of 7.6 m, a draft of 4.8 m, a cruising speed of 12 knots, an endurance of 10,000 nautical miles, a capacity of 59 persons, and a self-sustaining period of 60 days.

Designed for non-restricted waters navigation, the "Songhang" pelagic fishery resource survey vessel primarily operates in the North Pacific, South Pacific, Southwest Atlantic, and other sea areas. It undertakes various tasks such as investigating and researching important pelagic fishery resources like tuna, squid, bamboo fish, and Antarctic krill. It also engages in national pelagic fishery resource and new fishing ground development. The vessel also undertakes studies on the mid-water, bottom, and variable water-depth trawling for distant fishing, as well as longline fishing for tuna and light squid fishing, and explores new operation modes, and supports marine environment observations and remote sensing data receptions.

The vessel plays an active role in establishing a long-term mechanism for nurturing senior applied talents in China's pelagic fishery sector. It serves as a base and key platform for developing versatile talents in marine-related fields, thereby promoting

[8] National Development and Reform Commission.

Fig. 24 Pelagic fishery resource survey vessel "Songhang"

the development of China's marine fishery science, and providing an essential scientific research platform for the construction of internationally influential high-level universities with distinct characteristics.

The vessel is equipped with two major scientific research systems covering fishery resource and marine hydrology survey systems, as well as three major fishing operation systems including mid-water and bottom trawling, tuna longline fishing, and light squid fishing. It also has five laboratories for marine biology, hydrobiochemistry, monitoring, general experiments, and network information. This allows the vessel to conduct sampling of pelagic deep-sea fish and plankton, continuous detection of ocean environmental parameters through fixed-point or cruise surveys, continuous observation of sea surface conventional meteorology, and exploration of seabed topography. Finally, the vessel can systematically record, analyze, and integrate the survey data.

The vessel is powered by a propulsion system consisting of Wärtsilä diesel engine and ABB propulsion motor. It is fitted with Germany-made straight blade propeller, which is introduced in China for the first time. With the help of accurate and sensitive pitch control, it integrates propulsion, maneuverability, and anti-rolling functions, providing a strong guarantee for precise positioning and vessel attitude control.

Prior to the advent of "Songhang", there were hardly any precedents in China for the use of straight blade propeller. There was nearly no experience available in terms of propulsion performance, shipment plan, and control operation for this type of propulsion system. The design team conducted simulation calculations and model tests to study the impact of various installation parameters of the straight blade propeller on overall performance. They analyzed the influence of factors such as the propeller disk installation position, the longitudinal and transverse inclination angles of the propeller disk, the width of the stern board, and the immersion depth of the stern board on the performance of the straight blade propeller. As a result, they identified an optimal balance point that minimizes hull resistance while maximizing propulsion efficiency. The trial voyage results showed that in addition to meeting the predetermined propulsion performance targets, the straight blade propeller was excellent in reducing vibration, noise, and roll motion.

The vessel is equipped with seven major scientific research systems, including marine hydroacoustic sounding system, winch system, spectral detection system, marine organism and fish survey system, marine atmospheric environment monitoring and meteorological remote sensing system, marine hydrology sampling and testing system, and auxiliary equipment and deck survey operation support system.

In accordance with the requirements for research vessels, corresponding vibration and noise reduction measures were implemented, yielding excellent results. Sea trial measurements indicated that the vibration levels in crew cabins, engine rooms, centralized control rooms, and living areas all met design specifications and were significantly below standard limits. The average value of the maximum measured vibration across the entire vessel was 0.4 mm/s, demonstrating a remarkable reduction in ship vibration.

From November 8 to December 15, 2017, the "Songhang" pelagic fishery resource survey vessel completed its maiden voyage. The route covered depths ranging from 100 m to 4,000 m in order to accommodate the diverse depth requirements of different equipment. During this voyage, the vessel tested and verified the seven major scientific research equipment systems and three major fishing systems onboard. Additionally, it utilized scientific research equipment to investigate the hydrometeorological and environmental characteristics of the surveyed area. Moreover, acoustic equipment and fishing gear sampling were also employed in this voyage to study changes in fish species and resource density in the middle and surface water of the surveyed area.

The primary objective "Songhang" is to conduct public welfare investigations of pelagic fishery resources. This represents a significant milestone in China's history of fishery science and technology. Aside from belonging to Shanghai Ocean University, the research data from this vessel are shared by fisheries, marine, environment, and many other research institutes. From the perspective of international pelagic fishery, it demonstrates China's responsible attitude to the world. The Chinese government, in an open and cooperative manner, is intensifying the research on the utilization of fishery resources and the support for the conservation of these resources. This will undoubtedly enhance China's influence in international conferences and negotiations, as well as elevate its capacity and status to participate in the formulation of international conventions and relevant regulations, and the implementation of international conventions on the conservation of fishery resources.

Since its delivery in 2017, the vessel has successfully completed multiple scientific research missions, covering a distance of over 60,000 nautical miles.

2.23 *Submersible Support Mother Ship "Shenhai 1"*

In order to advance China's deep-diving submersible delivery technology, provide crucial high-tech equipment for oceanic international seabed resource surveys and scientific research, and develop common technologies for deep-sea exploration and underwater operations, the SOA initiated a major scientific expedition project for manned submersible deep-sea research in 2002.

The manned deep diving technology system relies on the organic combination of the pilot, manned submersible, and submersible support mother ship to achieve scientific exploration of the deep sea. The support capabilities and functions of the submersible support mother ship include carrying, maintaining, and servicing the manned submersible; launching and recovering the manned submersible; providing water surface command, communication support, and water surveillance during the operations of the manned submersible; conducting investigations of the operating environment while supporting oceanic scientific auxiliary surveys; and facilitating rapid analysis and experimental research of deep-sea samples.

In 2007, the manned submersible mother ship, converted from "Xiangyanghong 09" multipurpose oceangoing survey vessel, was found to be largely unfitted to submersible operations after more than 10 years of sea trials. For instance, "Xiangyanghong 09" had a relatively high freeboard, which reached 4.3 m when fully loaded. This caused the suspended submersible to swing violently when the ship experienced adverse sea conditions. Consequently, for safety reasons, the launch and recovery of the submersible had to be carried out in favorable sea conditions, compromising the efficiency of scientific research operations. Furthermore, when the submersible descended, the mother ship was expected to remain directly above the diving position. However, "Xiangyanghong 09" lacked a dynamic positioning system, causing the ship to be affected by wind and ocean currents when the submersible submerged for 12 h, resulting in drifts of over ten nautical miles. This took a heavy toll on scientific research operations. Therefore, a new support mother ship was needed for advanced manned submersibles. The China Ocean Association (China Ocean Mineral Resources R&D Association) decided to build a new manned submersible support mother ship with improved support capabilities. The MARIC, along with other relevant units, were entrusted with the preliminary demonstration work.

In 2010, experts from the MARIC participated in technical activities organized by the China Ocean Association for the establishment of oceanic installations. They were involved in the preliminary design and project biding, under the auspices of the China Ocean Association. In November 2016, they were commissioned by the China Ocean Association to undertake the design of the ship. In November 2016, the China Ocean Association signed a construction contract with Wuchang Shipbuilding Industry Group Co., Ltd. Being constructed from September 2017, the ship was delivered in November 2019, with a name "Shenhai 1" (see Fig. 25).

This vessel has a total length of 90.2 m, a molded breadth of 16.8 m, a molded depth of 8.3 m, a designed draft of 5.5 m, a cruising speed of 12 knots, a maximum

Fig. 25 Submersible support mother ship "Shenhai 1"

speed of 16 knots, a capacity of 60 persons, a self-sustaining period of 60 days, and an endurance of 12,000 nautical miles.

The vessel features a long forecastle, with the lifeboat deck and lower areas made of steel, and the material above the lifeboat deck made of aluminum alloy. It is equipped with DP-1 dynamic positioning capability, twin-pod electric propulsion unit, B3 class ice strengthening capability, and a retractable full-circle propeller at the bow.

Throughout the design process, the consensus reached during the technical feasibility study was consistently upheld: The new vessel must embody the design principles of internationalization, comfort, convenience, green technology, and informatization. Corresponding technical measures were implemented to ensure these principles were effectively realized.

1) Optimized Main Dimension

   To ensure the safety of submersible launch/recovery operations, the influence of the ship's main dimensions on general arrangement, seaworthiness, speed, economy, and displacement limitations were comprehensively considered.

2) Anti-bubble Leakage and Low-resistance Axe Bow

   The hull form design of the vessel must meet the requirement of quickly transporting the manned submersible to the experimental area within a short time frame. Additionally, it must ensure the proper functioning of the transducers installed on the hull bottom, minimizing interference from bow wave-induced bubbles on the transducers' performance.

   Although a conventional bulb bow could reduce wave-making resistance, the high-speed movement of the bulb bow relative to the sea surface flow during frequent entry and exit from the water could result in large amounts of bubbles entering the ship's bottom, potentially affecting the detection quality and range of the transducers. Therefore, a unique axe bow shape was designed for the ship, which reduced both the generation and leakage of bubbles while maintaining speed. This design has been granted a patent for invention. Resistance and self-propulsion experiments conducted at the Netherlands MARIN trial tanks have shown that this axe bow design outperforms the conventional bulb bow design in terms of both speed and reduction of bubble generation at the bow.

3) Support Capabilities and Functional Compatibility

The strength and compatibility of the support capabilities and functions of the submersible are crucial factors for the success of the design of the manned submersible support mother ship. Therefore, special consideration must be given to the key aspects such as the carrying and maintenance of the manned submersible, launch and recovery, water surface command, communication support and surveillance, deep-sea sample rapid analysis and experimental research, and investigation of the operating environment and oceanic scientific auxiliary surveys.

(1) Manned submersible carrying and maintenance system: This system primarily relies on the transportation track, track vehicle, and maintenance system.

① Transportation track and its associated track vehicle system: The track runs along the longitudinal direction of the ship, passing through the working area beneath the A-frame at the stern of the main deck, extending to a 24-m transportation track and its associated track vehicle within the shielded working storage area. This enables the manned submersible to be transported along the track on the mother ship, positioned and secured on the working deck or inside the shielded storage, and lifted as needed in the storage.

② This system includes the dedicated maintenance platform and overhead traveling crane inside the shielded working storage area. Together with the shielded working storage area, these facilities facilitate surface maintenance and upkeep of the manned submersible, as well as the recovery of deep-sea samples, disassembling or replacing equipment, and the safe entrance and exit of submersible crew members within the designated conditions.

③ This system also includes the dedicated maintenance supporting equipment, tools, and spare parts for the manned deep-diving submersible system, such as batteries and their charging and discharging systems, kentledges and their loading and unloading systems, high-pressure air cylinders and their inflation systems, buoyancy materials and their installation tools, among others, which are stored in the shielded working storage area and spare parts storage.

(2) Launch/recovery system: The launch/recovery system refers to the system used to transfer the manned submersible between the sea surface and the

transportation track and track vehicle on the mother ship. This system consists of three subsystems:

① Manned submersible-specific A-frame system: This system is custom-designed for the submersible, featuring a lifting device with constant tension compensation capability. Its lifting load capacity (30 tons) and the capability for deploying/retrieving the submersible in various operational sea conditions surpass those of the A-frame system on "Xiangyanghong 09", further enhancing the safety of offshore operations.

② Auxiliary towing winch and its associated towing cables: These are used to tow the submersible to the stern of the mother ship during recovery, and then to the location under the A-frame outboard part. They also assist in positioning the submersible during launch/recovery, providing stability and controlling horizontal displacement and momentary movements during lifting to avoid collisions with the A-frame.

③ Offshore high-speed work boat: This vessel carries the surface operators who hang the lifting cable of the A-frame onto the top lifting point of the submersible or detach the lifting cable from the lifting point on the support mother ship. The surface operation personnel also need to secure the auxiliary cable on the auxiliary winch to the submersible's horizontal towing point.

Water-surface hanging or detaching and cable securing operations are considered highly risk with respect to personnel safety. Therefore, the agility of the work boat and the proficiency of the operators are required to be at a high level.

(3) Surface command, communication support, and surface surveillance: The support mother ship for manned submersibles should meet the requirements for surface command, communication support, and surveillance. Therefore, it needs to have a high level of informatization, wide professional coverage, multiple interfaces, and should involve underwater positioning, tracking, and data analysis of the manned submersible. Additionally, it should establish a stable and efficient informatization platform to support the integration and interpretation of various data.

   ① The communication system with the manned submersible consists of an ultra-short baseline system, a multifunctional underwater acoustic communication apparatus system, an underwater telephone, and an acoustic array system. The ultra-short baseline system determines the position of underwater acoustic transponders and the manned submersible relative to the mother ship, facilitating the positioning and tracking of the underwater equipment. The multifunctional underwater acoustic communication apparatus system, underwater telephone, and acoustic array system enable real-time communication between the manned submersible and the surface support mother ship, as well as the bidirectional transmission of instructions from the mother ship and various data, voice, image, and other information from the submersible.

② By establishing a secure, efficient, and stable informatization platform, the ship enables rapid analysis of comprehensive data and operational support during the manned submersible's complex seabed detection, observation, sample collection, and other operations. Meanwhile, it also enhances the overall level of information management on the ship. This informatization platform facilitates information exchange and resource sharing between the ship and shore-based facilities, between other ships, and between the ship and underwater platforms. It creates an integrated data collection, processing, monitoring, and management platform for the entire vessel, enhancing automation and data management capabilities in marine research. It also establishes a business system management and command dispatch platform for the entire vessel, effectively ensuring the ship's collaborative work capabilities and business processing abilities. Specifically, during submersible diving operations, the ship significantly improves diving efficiency, ensuring the accuracy, timeliness, and safety of information.

(4) Deep-sea sample rapid analysis and experimental research: The ship's onboard laboratory system supports the rapid analysis and storage of deep-sea samples and facilitates offshore experimental research in marine geology, biology, chemistry, physics, and atmospheric science, among other disciplines. The laboratories include the geology laboratory, biochemistry laboratory, gravimeter room, submersible operation center, data processing center, water sampling room, low-temperature laboratory, sample preservation warehouse, laboratory warehouse, moving seawater laboratory, atmospheric chemistry laboratory, and seven container laboratories.

(5) Manned submersible operating environment survey and oceanic scientific auxiliary survey system: The support mother ship for manned submersibles must provide full-process support for submersible operations, including investigations of the operating environment. The operating environment survey equipment system configured for this task includes not only the equipment directly used to support the launch/recovery of manned submersibles, such as the A-frame at the stern, but also general deck equipment system and auxiliary specialized survey equipment system. The configuration of this equipment also takes into account auxiliary surveys for oceanographic research, thus improving the efficiency of equipment utilization.

① The ship's general deck equipment system includes a stern A-frame (30t), a midship main crane (25t × 12.5 m), a starboard auxiliary crane (4t × 16 m), a larboard auxiliary crane (1t × 7 m), a starboard A-frame (15t), and a bow material crane (1t × 7 m).

② The auxiliary specialized survey equipment system includes a 6,000-m ROV (remotely operated vehicle) winch and its launch/recovery system, installed at the starboard of the shielded working storage, which is specifically used for coordinated operations with the manned submersible. In addition, the vessel is equipped with a 10,000-m geological winch, a CTD winch, a hydrology winch, and gravimeter, and ADCP. The dense securing system on the stern deck facilitates the installation of various mobile sampling devices, such as TV grabs, visual multi-tube samplers, and visual box corers. It can simultaneously carry several key underwater equipment, such as great-depth autonomous remotely operated underwater vehicles, 6,000-m ROVs, and AUVs.

The aforementioned equipment effectively supports operating environment assurance surveys and oceanic scientific auxiliary surveys in hydrology, geology, biology, and others disciplines. Thanks to the inclusion of these equipment, the ship could serve as the mere mother ship in China capable of simultaneously accommodating all types of scientific research submersibles.

4) Environmentally Friendly Design of the Mother Ship

   The ship's design has achieved desirable results in terms of power station, propulsion configuration, as well as conformity to the Clean standards and the latest Tier-III emission standards for generators.

   (1) Power station, propulsion configuration, and economical speed: The ship is equipped with two sets of 2,610-kilowatt and two sets of 760-kilowatt diesel generator sets. With linear and system optimization design, the ship can meet the entire power load and achieve an economical speed of 12.6 knots with only one 2,610-kilowatt diesel generator set running, significantly reducing overall fuel consumption.
   (2) Conformity to Tier-III emission requirements: For the first time in China's research vessel designs, the ship was fitted with a selective catalytic reduction (SCR) device. This ensures that all the ship's diesel generator sets meet the latest Tier-III standards set by the International Maritime Organization (IMO) for nitrogen oxide emissions. It fulfills the future demands for international navigation and demonstrates the ship's orientation towards environmental friendliness and sustainability.
   (3) Conformity to additional CCS class markings: In addition to meeting the IMO's Tier-III emission requirements, the ship also meets the design and construction requirements for additional CCS class markings, including Ice Class B3, PSPC (B), Underwater Noise 2, COMF (NOISE 3), COMF (VIB 3), GPR, AUT-0, DP-1, OMBO, and Clean, among others.

These designs elevate the ship's compliance with environmental regulations to an advanced level, surpassing similar ships in its category.

5) Underwater Radiation Noise Control

During deep diving operations, acoustic communication is the only means of communication between the submersible and the surface mother ship. The ship is equipped with acoustic equipment, such as multi-beam and ultra-short baseline systems, necessary for multipurpose surveys. These systems impose high requirements for controlling underwater radiation noise.

Throughout the entire design process, various measures were implemented to control underwater radiation noise for this ship.

(1) Preliminary assessment of the initial design: During the preliminary design phase, the design team conducted simulation calculations to pre-assess the vibration and noise levels of the mechanical equipment and structures of the entire ship. Based on the simulation results, the design scheme for local structures was adjusted.
(2) Control of major electromechanical equipment: When selecting major electromechanical equipment such as diesel generator sets, quantifiable indexes were established for selecting high-quality products with lower vibration and noise levels.

    Data on electromagnetic noise from the podded propeller was requested. Furthermore, propeller testing was conducted in tank to ensure that the underwater radiation noise in line with control requirements.
(3) Vibration isolation design for the bases: Double-layer vibration isolation or floating raft installation methods were employed for the bases of major vibration noise sources such as generator sets, fans, compressors, and pumping units to minimize noise transmission to the hull.
(4) Optimization of whole structure and equipment base design.

    According to the final ship trial, the ship met the additional marking requirements for underwater noise (Underwater Noise 2) set by CCS.

6) Vibration and Noise Reduction Design

Through comprehensive vibration and noise reduction designs, including overall area division, design of simulating calculations, limited technical specifications for electromechanical equipment, double-layer vibration isolation of units, and structural optimization, the ship received the additional markings of COMF (NOISE 3) and COMF (VIB 3) from CCS.

On November 10, 2020, "Shenhai 1", carrying the "Fendouzhe" full-depth manned submersible, ventured to the Mariana Trench in the Western Pacific. The submersible successfully descended to a depth of 10,909 m, only 125 m shallow from the deepest point as measured at 11,034 m by far, setting a new record for China's manned deep dive history. This achievement marked China's ability to conduct deep-sea scientific research and resource exploration in most of the world's ocean areas, and demonstrated that China's manned submersible technology has generally entered the forefront internationally.

In March 2021, the "Shenhai 1" submersible support mother ship carried an unmanned submersible for the first time to participate in the Western Pacific Deep-sea Scientific Expedition Voyage organized by the National Deep Sea Center. During this expedition, the Harbin Engineering University's research team successfully completed the third phase of deep-sea trials for their full-depth untethered AUV (autonomous underwater vehicle) key technology research project. The "Wukong" full-depth AUV conducted trial dives to depths of 5,000 m and 7,000 m in the high seas of the Western Pacific. During the trial dives, the submersible reached a maximum depth up to 7,709 m, setting a new record for China's unmanned submersibles in terms of depth.

2.24 Small-Waterline Twin-Hulled Research Vessel "Shen Kuo"

In recent years, as China's national research capabilities and funding in the field of oceanography expand, private enterprises and capitals in the country also actively response to it. They collaborate with relevant national research institutions, fully leveraging their own edges to contribute to the advancement of comprehensive strength in China's marine scientific and technological innovation. Following the successful construction of the "Xiangyanghong 10", "Zhang Jian", "Changheze", and "Shanghe" by Tehe Ocean Technology Group, another research vessel, the small-waterline twin-hulled vessel named "Shen Kuo", was constructed under the investment of Shanghai Rainbow Fish Science and Technology Service Co., Ltd., a subsidiary of the Tehe Ocean Technology Group. The completion and operation of the "Shen Kuo" research vessel once again proves the vast potential and bright prospects of collaboration between private enterprises and national research institutions in the oceanography field. The "Shen Kuo" research vessel will become one of the main vessels for national oceanographic survey and monitoring, playing a vital role in understanding the ocean and smart ocean development (see Fig. 26).

Designed by the China Ship Scientific Research Center and built by Zhejiang Tianshi Shipbuilding Co., Ltd., the "Shen Kuo" research vessel incorporates several new military-civilian integrated technologies. It is the first vessel in China to adopt the DC networking electric propulsion system and high-power low-speed permanent magnet motor developed by Shanxi Fenxi Heavy Industry Co., Ltd. With excellent seakeeping capability, low vibration and noise levels, and a large deck area, "Shen Kuo" is well-suited for oceanographic surveys and ocean engineering operations.

The small-waterline twin-hulled research vessel features a submerged torpedo-shaped lower hull, an above-water platform, and pillars that connect the upper and lower hulls crossing the water surface. Its design offers the advantage of a smaller waterline area, resulting in less wave disturbance and superior seakeeping performance in waves. The vessel exhibits minimal swaying in windy and wavy conditions,

Fig. 26 Small-waterline twin-hulled research vessel "Shen Kuo"

boasting a large righting moment. With the adjustable fin, it ensures exceptional wave resistance stability.

For the first time, the vessel employed a domestic high-power DC bus electric propulsion technology and high-power permanent magnet motor. It provides excellent seakeeping quality, outstanding acoustic performance, high speed, low energy consumption, and long endurance. Particularly, it is well-suited for conducting underwater acoustic research and sea trials of marine acoustic equipment. The "Shen Kuo" small-waterline twin-hulled research vessel can safely navigate in sea conditions up to grade 10 and perform regular operations in sea conditions of grade 4.

This vessel was designed with a length of 63 m, molded breadth of 23 m, molded depth of 9.4 m, displacement of 2,180 tons, and a capacity of 60 persons, the vessel boasts a speed of 12 knots, an endurance of 5,000 nautical miles, and a self-sustaining period of 25 days.

The vessel's construction undertaken by Zhejiang Tianshi Shipbuilding Co., Ltd. started in May 2017, and it was launched in April 2018 and delivered in March 2019. It represents a deep and far-sea scientific survey vessel that integrates multiple functionalities and technological means, satisfying the requirements of multidisciplinary and interdisciplinary research in marine science.

Following its delivery, "Shen Kuo" embarked on its maiden voyage, Mariana Trench sea trial, as part of the "Rainbow Fish Challenge to the Abyss" project, led by the joint efforts of Shanghai Ocean University and Westlake University. The scientific expedition team boarded the "Shen Kuo" research vessel from Shanghai Luchao Port and traveled to the Mariana Islands areas to conduct a series of deep-sea equipment tests and scientific sampling in the Challenger Deep, the deepest part of the global ocean. It served as a significant voyage for the "Rainbow Fish

Challenge to the Abyss" project. The voyage involved 60 expedition team members and crew members, which was expected to last 45 days. During this voyage, the sea trial and expedition team was to conduct corresponding tests including sea trials of "Caihongyu (Rainbow Fish)" 10,000-m manned submersible USBL system, sea trials of two second-generation "Caihongyu" 10,000-m landers, sea trials of a 4,500-m large-depth buoy, and others. Additionally, the team was to collect samples from the Mariana Sea, including seawater, sediments, macroorganisms, microorganisms, and underwater imagery, for further research by scientists from Shanghai Ocean University and Westlake University.

The Mariana Trench sea trial and scientific research voyage represented an important milestone in the "Rainbow Fish Challenge to the Abyss" project. Each achievement in the sea trial and scientific research was of great significance to the development of China's marine science and technology. With a shared goal of serving the national strategy and regional development, Shanghai Ocean University and Westlake University have joined hands to strengthen cooperation and exchange mechanisms for scientific research and talent cultivation, and create a new model for collaborative development through leveraging their respective strengths and sharing high-quality talent resources. In addition, they are actively exploring new mechanisms for joint construction of a research center and have officially signed a strategic cooperation agreement. The strategic cooperation agreement marks a proactive exploration and reform innovation in implementing the spirit of the 19th National Party Congress, serving the national strategy of becoming a strong maritime nation, realizing Shanghai's goal of building a globally influential scientific innovation center, promoting high-quality integrated development of the Yangtze River Delta region, and exploring regional cooperation in education.

During this expedition with researchers from Shanghai Ocean University and Wastlake University aboard the vessel "Shen Kuo", the team conducted scientific equipment testing and investigation in the open ocean, and achieved fruitful results, which represented a preliminary verification of the vessel's overall performance. By naming the vessel "Shen Kuo", after the renowned ancient Chinese scientist, the intention is to perpetuate the spirit of scientific exploration embodied by Shen Kuo himself.

2.25 Deep-Sea High-End Talent Incubator "Dongfanghong 3"

In the 1960s and 1990s, the Ministry of Education approved Ocean University of China's construction of two vessels respectively, namely, the 2,500-ton "Dongfanghong" and the 3,500-ton "Dongfanghong 2". These two multipurpose marine training and survey vessels have provided more than 10,000 trains for multidisciplinary marine students in China, totaling over 3,000 days of operations at sea and covering nearly 300,000 nautical miles in China's surrounding seas and the Western

Pacific. These accomplishments have made significant contributions to the cultivation of talent in marine sciences in China.

The original "Dongfanghong" was decommissioned on January 21, 1996, after serving for exactly 30 years. It was a pioneer in China's marine industry and witnessed the birth of the first generation of marine science and technology workers in the country.

Since its commissioning in 1996, "Dongfanghong 2" has been in service for more than 20 years so far, with an average duration of nearly 300 days at sea each year. On par with the most utilized research vessels internationally, the vessel has become a key enabler for major maritime research projects and national defense projects of China.

To better safeguard national interests and boost the marine economy, and to play a greater role in marine research and multipurpose surveys, national defense construction, marine resource development, marine engineering, international cooperative surveys, as well as maritime teaching practices and the training of marine professionals, the Ocean University of China decided to build a new deep and far-sea multipurpose research vessel, named "Dongfanghong 3".

The construction goals for "Dongfanghong 3" (see Fig. 27) are as follows:

(1) To serve the establishment of deep-sea oceanic research teams and cultivate high-end talents: gathering and cultivating talents in basic marine sciences, enriching and optimizing deep-sea oceanic teaching modes and training systems, and nurturing various types of innovative marine science and technology professionals, as well as advanced, composite management talents.

Fig. 27 Deep and far-sea multipurpose research vessel "Dongfanghong 3"

(2) To support fundamental research in deep-sea oceanic marine science: researching marine multi-scale oceanic dynamic processes and air-sea interactions, studying biogeochemical processes and the environmental evolution of marine ecosystems, systematically researching the effects of marine environments and seabed resources, and investigating marine organisms, genetic resources, and biodiversity.
(3) To support deep-sea oceanic high-tech R&D and their application: simultaneous stereoscopic and regionalized observation technology, key technologies for resource exploration, development, and transportation, standardized sea trial technology for high-end instruments and equipment, integrated satellite, aviation, ship, and shore-based detection, as well as experimental data processing and data network application technologies.

The project was officially approved in September 2013. In November 2014, the MARIC won the bid to be the vessel's designer. In November 2015, Jiangnan Shipyard was selected as the constructor. Construction of the vessel began in October 2016, with a trial voyage taking place in December 2018. The vessel completed its underwater radiation noise trial in April 2019 and obtained the DNV SILENT-R certificate, the highest level for underwater radiation noise, followed by its delivery in May 2019.

The vessel has a length of 103.8 m, a molded breadth of 18.00 m, a molded depth of 8.70 m, a scantling draft of 5.50 m, a maximum speed of 15.55 knots, and a capacity of 110 persons (including 28 crew members and 82 researchers). It features an endurance of 15,000 nautical miles at 12 knots, a self-sustaining period of 60 days, and a designed displacement of 5,500 tons.

The underwater radiation noise requirement is set to be compliant with DNV SILENT-R (at 11 knots).

Installation method for the multi-beam system: The bottom surface of the multibeam transducer is flush with the baseline of the hull and is in embedded installation.

Stability and unsinkability: The vessel's resistance to wind stability meets the requirements of gusts up to Force 12 (wind speed of 51.5 m/second). The damaged stability complies with the latest SPS (2008) regulations.

Dynamic positioning capability: The vessel is capable of withstanding sea conditions of grade 4 (significant wave height not exceeding 1.8 m), a current of 1.5 knots, and achieving stationkeeping in winds of Forces 5–7. The positioning accuracy is between 0–3 m. When a favorable heading angle is chosen, the vessel can reliably maintain a heading within $\pm$ 10 degrees and achieve a track keeping accuracy standard deviation of less than 3 m in high-speed automatic track keeping mode under the mentioned conditions.

Seakeeping capability: The vessel is designed to withstand sea conditions of grade 4 (5–7 force winds), meeting the requirements for ROV deployment and helicopter hovering. It can also operate effectively in sea conditions of grade 5 (7–8 force winds) for drifting survey operations. The vessel is capable of safe navigation in sea conditions of grade 7 (9–10 force winds).

Based on the shipbuilding goals of Ocean University of China, the "Dongfanghong 3" research vessel is designed to acquire reliable and accurate detection data, and achieve a world-leading level in terms of its functionality as a platform of sci-tech innovation and construction as a talent practice base. The vessel is also intended to ensure the economic viability of vessel operations and the safety and reliability of scientific research operations. Additionally, it is designed to have expandable capabilities for both vessel and scientific research activities. The overall design concept of "Dongfanghong 3" is based on the following principles: a. "One Leading": In line with national strategic needs, the "Dongfanghong 3" research vessel is designed by drawing on advanced design concepts from both domestic and international newly-built and under-construction research vessels. It integrates Chinese characteristics to create a leading shipbuilding scheme both now and in the future that allows for stereoscopic and regionally synchronized rapid detection and observation for deep and far seas, as well as integrates field comprehensive investigation and high-end talent cultivation. b. "Two Integrations": The ship-based information network is integrated with the land-based information network, and scientific research is integrated with the innovation platform. c. "Three Major Functions": The vessel is equipped with the capability to conduct multipurpose scientific research and resource surveys in deep and far sea areas, ranging from upper atmosphere observations to investigations that extend to the seabed penetrating the air-sea interface. It can also form an observation array in collaboration with unmanned aerial vehicles, research vessel fleets, and other deployed observation instruments. Additionally, the vessel has the capability to conduct sea trials and supports high-tech R&D and innovation for deep and far sea state-of-the-art marine instruments and equipment, alongside the capabilities of conducting multidisciplinary offshore training in an effective manner and fostering deep and far sea innovative research personnel.

The "Dongfanghong 3" research vessel is equipped with eight main systems: water detection, seabed detection, atmospheric detection, acoustic research, remote sensing information observation and validation, chemical experimental analysis, biological experimental analysis, and sediment analysis. The vessel is primarily used for conducting comprehensive marine environmental surveys, teaching and training, and collecting marine environmental data for national defense needs. It serves as a sea-floating comprehensive laboratory that can support real-time and synchronized observations in hydrology, chemistry, biology, geology, geophysics, atmospheric science, and related interdisciplinary fields. In addition, it also supports field scientific experimental research and remote transmission of information and data, featuring advanced and complete scientific research facilities.

The equipment configured for the vessel:

(1) Main laboratories: atmospheric environment laboratories (left/right), atmospheric detection and remote sensing laboratory, air-sea flux laboratory, general dry/wet laboratories, in-situ direct seawater analysis laboratory, temperature-controlled laboratory, chemical clean laboratory, general clean laboratory, CTD laboratory, geophysical laboratory, instrument centralized control room,

computer room, data processing laboratory, salinometer room, electrical maintenance room, gravimeter room, chemical storage room, sample refrigeration room, sample freezing room, CTD operation room, CTD water sampling room, winch operation control room, acoustic instrument compartment, and shielded working deck, among others.

(2) Lifting equipment: a telescopic main crane with a lifting capability of 15 tons × 10 m, a telescopic folding-arm auxiliary crane at the bow with a lifting capacity of 2 tons × 14 m, a telescopic folding-arm auxiliary crane at the stern with a lifting capacity of 8 tons × 10 m, and two double-track overhead cranes on the shielded deck with a lifting capacity of 5 tons each.

(3) Scientific research LARSs: a CTD deployment device with a safe working load of 6 tons, an A-frame on the side with a safe dynamic load of 12.5 tons, an A-frame at the stern with a safe dynamic load of 15 tons, one set of long cylinder piston corer deployment device.

(4) Various winches: a 12,000-m CTD winch, a 10,000-m CTD winch, a 12,000-m fiber cable winch, a 10,000-m steel cable winch, a 10,000-m optical cable winch, and a 3,000-m biological winch.

(5) Scientific research lifting fins: two sets, capable of extending 2.8 m outside the vessel's bottom.

(6) Acoustic equipment: full-depth multi-beam system (0.5 degrees × 1 degrees), medium and shallow water multi-beam system, deep-water sub-bottom profiler, scientific fish finder, and single-beam depth sounder.

(7) Other scientific research instruments: 10,000-m CTD profile detection system (36 bottles), in-situ continuous and automatic surface multi-element measurement system, two multi-channel seismic air compressors, various sediment samplers, long cylinder piston corer, TV grab, surface ocean current and wave measurement system, moving carbon dioxide measurement system, shipborne scientific research meteorological station, marine gravimeter, and marine magnetometer.

As the latest generation of large multipurpose research and training vessels in China, this vessel features several innovative designs:

(1) The vessel adopts an integrated hull design concept, which ensures a harmonious balance between speed, economy, silence, and scientific research and training capabilities. Its speed index, calculated based on the propulsion power per ton of displacement, is at the forefront internationally. With an economical speed fuel consumption of only 11.4 tons per day, it boasts one of the most fuel-efficient vessels of its kind. The vessel's compartments have a high level of noise control, with 92% of the living compartments meeting the highest comfort requirements specified by the CCS. Additionally, it boasts the largest capacity (110 people) among global multipurpose research vessels.

(2) For the first time, the vessel is equipped with a rainfall-type air conditioning air supply and return system in public areas, significantly enhancing living comfort. It is also the first time that a whole-ship centralized air-conditioning network monitoring system is adopted for a domestically built research vessel,

with the air-conditioning allowing for whole-ship networking and centralized monitoring from the engine control room. Besides, the vessel marks the first to employ a cover in same profile design with the detachable side thruster on the complex curved shell plate.

(3) The vessel innovatively utilizes a 3D model structure weight tracking and control method and integrated it with whole-ship finite element structural strength analysis, to meet the requirements of structural strength assessment and weight optimization. An underwater radiation noise target-oriented design and control method was innovatively developed, breaking through key technologies such as double-layer vibration isolation for generator sets, silent drive for propulsion motors, and silent propulsion with efficient propellers. Thanks to these technologies, the vessel obtained the first DNV SILENT-R certificate for underwater radiation noise in China, achieving world-leading levels of underwater noise control.

(4) It is the first research vessel to fully utilize a 3D experience platform for digital design and construction, with a high level of integration of CAD,[9] CAE,[10] CAM,[11] and other technologies. It constructs a 100% 3D model and realistically simulates the actual living and working scenarios on board the vessel.

The vessel employs a modular and unitized laboratory design concept to maximize the utilization of scientific research space. It achieves a world-leading laboratory area index in terms of laboratory area per unit displacement. For the first time, it achieves the application of a clean laboratory on board a ship with a cleanliness level of Class 1,000.

The vessel achieved breakthroughs in the design, installation, and hull deformation control technology for ultra-large multibeam transducers. As a result, the world's deepest and highest-resolution full-depth multi-beam sounding system was successfully installed on a domestic research vessel for the first time.

(5) The vessel is the first in China to systematically apply electromagnetic compatibility technology on a research vessel. In addition, a ship electromagnetic environment monitoring system with the core goal of realizing whole-ship electromagnetic compatibility has been developed and applied.

(6) The vessel is the first in China to use a "tripod-shaped" CTD side door with both upward and downward opening methods. It has enhanced the workshop environment during prolonged offshore water sampling operations.

As a new generation large multipurpose research and training vessel, "Dongfanghong 3" boasts state-of-the-art qualities. Various tests conducted during the trial voyage confirmed the vessel's successful design and construction. It not only met the set targets for speed and maneuverability, but also surpassed international advanced levels in areas such as empty weight control, underwater radiation noise, overall layout comprehensive considerations, wind stability and loading capacity, power

[9] Computer aided design.

[10] Computer aided engineering.

[11] Computer aided manufacturing.

station load and fuel consumption, power system reliability, and living environment comfort. This is mainly reflected in the following aspects:

(1) Effective control of empty weight, center of gravity, and flotation condition: Similar vessels abroad emphasize the need for effective control of empty weight and center of gravity. This not only helps with optimizing the distribution of flotation condition for a greater effective loading, but also allows for greater capacity to modify the vessel, thus extending its life cycle. From design to construction, weight control had been a guiding principle for this vessel. The final inclination test confirmed that the weight and center of gravity were very close to the estimated design state, without the need for temporary weights fixed measures. This laid a solid foundation for the delivery and operation of the vessel. During the trial voyage, the vessel demonstrated great flexibility in adjusting ballast and flotation condition. All liquid compartments, including consumable compartments and ballast tanks, fulfilled their intended functions, showcasing the accuracy of early estimations as well as the successful weight and center of gravity control (for steel plate tolerance, cable tray, damping dressing, piping, iron outfitting pieces, compartment interior decoration) achieved during the construction process. The vessel's scientific research payload exceeds 220 tons, reaching an advanced level for similar international research vessels.
(2) Leading internationally in terms of speed and economic indexes: Determined by the trial voyage, the vessel reaches a speed of 12.18 knots with a single generator running (the design target is 12 knots), while the maximum speed reaches 15.55 knots with both generators running (exceeding the design value of 15 knots). Comparing to the internationally recognized speed index, which measures the speed achieved by the displacement of the vessel per kilowatt of propulsion power, the vessel's speed is already at a leading level in the world (see Table 4).

    According to test results, the diesel generator power station load during major operating conditions falls between 75% and 87%, which represents the optimal operating range for the power station and the best fuel consumption range for the diesel engines. Based on calculations, if the vessel operates at an economic speed of 12 knots with a single generator, the daily fuel consumption would be approximately 11.4 tons, making it one of the most fuel-efficient ship types among similar electric propulsion vessels of the same tonnage level.
(3) Realization of good course stability, maneuverability, and seakeeping capability: The trial results showed that, at its maximum speed of 15 knots and with a maximum rudder angle of 35 degrees, the vessel had a maximum heel angle of approximately 8.5 degrees (below the typical requirement of 12 degrees). The diameter of its turning circle was around 1.49 times its length, significantly lower than the required 3 times the length (3×103.8 m). The vessel's turning performance surpassed that of single-propeller propelled ships and even some all-directional propelled ships. Sea-stopping trials also demonstrated the vessel's flexibility, as realized a stop within 579 s at a high speed of 15 knots,

Table 4 Comparison of speed performance indexes with similar ships

Ship name	Delivery year	Designed displacement/ton	Length-to-breadth ratio *(L/B)*	Propeller/unit	Single propulsion power/kW	Maximum speed/ knot	Speed index (ton-knot/kW)
UK JAMESCOOK	2007	5400	4.739	2	2500	15.0	16.20
UK DISCOVERY	2013	5368	4.933	2	2200	15.0	18.30
Australia INVESTIGATOR	2014	4575	4.805	2	2600	15.0	13.20
"Kexue"	2012	4661	5.405	2	1900	15.88	19.48
"Dongfanghong 3"	2019	5536	5.495	2	2100	15.55	20.50

with an inertial sliding distance of approximately 1,681 m. In case of an emergency stop, it can halt within 117 s, sliding a distance roughly 5.31 times its length.

The vessel's excellent course stability is attributed to its dual propulsion motors, dual propellers, and dual flap rudders. At a speed of around 15 knots, the vessel can change its heading by 3–9 degrees in 3 min. Even if the heading remains unchanged, it only requires 2–6 steering maneuvers per minute, with a maximum steering angle of merely 3–5 degrees.

The vessel is equipped with roll period instruments and tilt displays, revealing a roll period of 9–12 s and minimal roll acceleration. These features ensure crew comfort and fulfill seakeeping requirements.

(4) Well-equipped scientific research functionalities to ensure smooth operations: The vessel currently holds the title of the world's quietest and highest-capacity comprehensive research and training vessel. It excels in economic efficiency, vibrational and noise control, electromagnetic compatibility, and other performance indexes. The deck working area and laboratory workspace both exceed 600 square meters, making it the most well-equipped vessel for scientific exploration in terms of deck working area and laboratory space utilization.

The research vessel "Dongfanghong 3" has achieved or even surpassed the advanced level of similar research vessels in terms of winch, control support, and the configuration of major acoustic equipment both domestically and internationally. Moreover, the laboratory area per unit displacement of the vessel is also at the forefront internationally. It is the first research vessel in China and the second globally to be equipped with a 0.5-degree × 1-degree ultra-high precision multi-beam transducer, thereby achieving a significant leap in underwater detection capabilities.

(5) Obtaining the stringent SILENT-R certification for underwater radiation noise from DNV: The vessel is the first vessel of this kind in China and the fourth globally (including two military vessels) to be designed, built, and tested in compliance with DNV's most rigorous noise classification standard, SILENT-R. By means of targeted underwater radiation noise design, innovative solutions along with measures such as double-layer vibration isolation for generator sets and a quiet propulsion system with low-noise propellers driven by silent propulsion motors were introduced. Through independent research and collaboration with renowned domestic and foreign companies, mastery of underwater radiation noise control technology was achieved after extensive studies via modeling, analysis, inspection, and testing. Eventually, an SILENT-R certificate regarding underwater radiation noise was obtained for the vessel after reasonable planning, consistent design optimization, rigorous process control, and arduous efforts. When the vessel passes over the water, its noise level is equivalent to that of a meeting room. The low underwater noise level ensures the optimal functionality of various acoustic equipment on board.

The success of underwater noise control for the "Dongfanghong 3" research vessel meets the shipowner's requirements and provides invaluable experience for future underwater noise design on research vessels. It will lead the way in computational assessment and control technology research aimed at reducing underwater radiation noise in ship design. It also paves the way for research into testing methods for underwater radiation noise on completed ships, and further strengthens China's influence in the formulation of new standards and regulations for underwater radiation noise. These efforts will provide ample technical support to meet these new requirements for more future ships in terms of underwater noise.

(6) Significant vibration and noise reduction, overall improvement in living environment comfort: The whole-ship vibration and noise reduction had been emphasized since the beginning of the vessel design work. Through meticulous control from design and construction processes, it achieved remarkable results as confirmed by its trial voyage. Vibration and noise measurements conducted with both main engines running at full speed (15 knots) showed significantly lower levels in the engine room, central control room, and machinery repair room compared to international standard values. Particularly in the frequently manned engine control room, the measured noise level was only 65.6 decibels, well below the requirement of 70 decibels for the highest comfort level specified by the classification society. The average measured noise level in the pilothouse during full-speed operations was 55.4 decibels, also meeting the highest comfort level standard. The trial voyage also demonstrated that noise reduction was particularly effective in compartments surrounding the chimney and engine room fans, apart from open decks.

With a capacity of 110 persons, the vessel was designed with a total of 61 living compartments. Under cruise conditions of 12 knots, all the compartments met the level 2 comfort requirements specified by CCS. Among them, 56 met the highest comfort requirements specified by CCS (not exceeding 49 decibels), accounting for 92% of all living compartments. Additionally, 36 living compartments met the highest comfort requirements for senior crew members (not exceeding 45 decibels), accounting for 59% of all living compartments. At the maximum speed of 15 knots, all living compartments met the level 2 comfort requirements specified by CCS. Remarkably, 50 living compartments met the comfort requirements for the crew (not exceeding 49 decibels), accounting for 82% of all living compartments, and 39 met the highest comfort requirements for senior crew members (not exceeding 45 decibels), accounting for 64% of all living compartments. In DP[12] operations, all conditions met the requirements of MSC.337(91). Forty-six living compartments met the highest comfort requirements specified by CCS (not exceeding 49 decibels), accounting for 75% of all these compartments, and 56 met the level 2 comfort requirements specified by CCS (not exceeding 52 decibels), accounting for 92% of all living compartments. Notably, the measured noise level in compartments adjacent to the chimney was only 51 decibels. Moreover, the compartment with the lowest noise level on the entire vessel measured only 38.9 decibels.

[12] Ship dynamic positioning.

The vibration measurements also manifested the vessel's reasonable structural design, well-executed welding process, and effective double-layer vibration isolation measures, as no instances of excessive vibration had been observed. Additionally, during the trial voyage period, the vessel's air conditioning and ventilation systems operated well, with low noise levels. The indoor temperature could be adjusted flexibly with obvious cooling effect, greatly enhancing the vessel's comfort level.

Since its trial run in June 2019, the "Dongfanghong 3" research vessel has successfully completed multiple voyages and significant scientific research missions. Leveraging its superior vessel performance, excellent onboard environmental conditions, and advanced scientific research equipment, the ship has acquired a wealth of precise scientific data and achieved numerous groundbreaking results that were previously unattainable on domestic research vessels. This success highlights the construction capabilities of China's new generation of multipurpose deep-sea scientific research and training vessels.

2.26 Multipurpose Marine Fishery Survey Vessels "Lanhai 101" and "Lanhai 201"

In the 1980s, China had a total of 58 fishery resource survey vessels, mainly small and medium-sized ships, including 50 marine fishery survey vessels and 8 freshwater fishery survey vessels. This established a preliminary, scientifically arranged and reasonably structured system of fishery resource survey vessels, covering the major fisheries area along the coast. However, due to insufficient funds and other reasons, the survey vessels of fishery research institutes and colleges gradually aged and retired, leaving only a few remained, along with the mere large fishery resource survey vessel "Beidou".

Entering the twenty-first century, China's fisheries transitioned into a new phase of scientific management, making the investigation of fishery resources, especially offshore resources, more urgent than ever. However, domestic fisheries research institutions faced the dilemma of lacking dedicated vessels. To obtain data and conduct scientific surveys of fishery resources, research teams were then forced to rely on portable scientific instruments and borrow or share vessels. This greatly limited their ability to control survey stations, data accuracy, and sample storage and transportation. Marine fishery constitutes an important part of the national marine strategy, with fishery resources and ecological environment surveys being the foundation and prerequisite for fishery development. To address this situation, in 2010, the National Development and Reform Commission and the Ministry of Agriculture and Rural Affairs, in line with the "strong maritime nation" strategy proposed at the 18th CPC National Congress and the task of "adhering to the coordinated development of activities on land and at sea, and accelerating the construction towards a strong maritime nation" emphasized at the 19th CPC National Congress, constructed the 1,500-ton

"Nanfeng" marine fishery resource survey vessel for the South China Sea Fisheries Research Institute, Chinese Academy of Fishery Sciences (CAFS). Although a project for building an additional four 300-ton fishery resource survey vessels was approved in 2013, they still fell short of meeting the demand. Therefore, in 2014, the CAFS applied for the construction of two advanced multipurpose marine fishery survey vessels, which received approval subsequently.

These two vessels are the most financially invested, largest in tonnage, and most advanced in terms of facilities among all marine fishery scientific survey vessels commissioned by the Ministry of Agriculture and Rural Affairs. They are considered as the "national treasures" of China's marine research, an integral part of the establishment of China's marine fishery research vessel system, and one of the "symbolic projects of modern agriculture". They are dubbed the "aircraft carriers in fishing industry" and bear the significant mission of national offshore and pelagic fishery resource investigations and research. These vessels serve as crucial offshore scientific research platforms for the comprehensive implementation of China's strategy to become a strong maritime nation. Their completion marks a further step from offshore to deep-sea oceans in China's marine fishery scientific research.

These two vessels were designed by the China Ship Scientific Research Center and built by Hudong-Zhonghua Shipbuilding (Group) Co., Ltd. under the commission of the CAFS.

They were designed from September 2015 to September 2017 by the Center, and were delivered on June 28, 2019 by the Shipbuilding Group. Being named "Lanhai 101 (see Fig. 28)" and "Lanhai 201", these two vessels serve as a pair of twins, with one ship type in two vessels. They were received by the Yellow Sea Fisheries Research Institute and East China Sea Fisheries Research Institute respectively.

These two vessels were designed with a length of 84.5 m, a molded breadth of 15 m, a full load displacement of 3,200 tons, an endurance of 10,000 nautical miles, a maximum speed of 15 knots, a self-sustaining period of 60 days, and a capacity of 60 persons.

Both vessels are equipped with a range of specialized laboratories, including the fishery resources laboratory, bioenvironmental laboratory, physicochemical environment laboratory, acoustic evaluation and remote sensing laboratory, fishery biology laboratory, and comprehensive laboratory. The main research equipment on these vessels include scientific fish finder system, wireless trawl monitoring system, CTD, shipborne Doppler current profiler, satellite remote sensing and receiving system, side-scan sonar system, sub-bottom profiling system, bottom trawl, Agassiz trawl, plankton trawl, geological sampling equipment, omnidirectional color scanning sonar, fishery resource acoustic assessment system, and multiple internet devices, among others, with a total of over 20 sets of scientific research instruments and fishing equipment. These vessels can meet the requirements of navigation in non-restricted waters, as well as the South and North polar seas (except for ice zones). They have the ability to simultaneously detect, analyze, and process comprehensive data of marine fishery resources, hydrology, physics, chemistry, acoustics, and remote sensing information. Additionally, they are capable of collecting investigation data, sampling and analyzing samples on site, and conducting scientific research tasks such as fishery

Fig. 28 Marine fishery resource survey vessel "Lanhai 101"

resource survey and assessment, fishery ecological environment monitoring and evaluation, fishery satellite remote sensing application, and research and experiments on fishing gear and fishing methods. Through a mechanism of sharing resources and usage, they provide a common investigation platform and talent development platform for China's marine fishery scientific research, international fishery technology cooperation and exchange, and training of marine fishery research team talents.

In addition to being reliable, durable, fuel-efficient, lightweight, and easy to operate, multipurpose marine fishery survey vessels also need to have strong adaptability to rolling motion and the potential to withstand overload. During the design and construction process of these vessels, advanced technologies from both domestic and foreign sources were adopted and key technical challenges were overcome. They not only meet the safety requirements for offshore investigation operations but also improve the energy-saving and environment-friendly performance of the vessels.

1) Innovation in Ship Form

   These two vessels feature a unique vertical bulb bow ship form that provides significant fuel savings, with an energy efficiency of approximately 21% during cruising speeds and 15% during high-speed travel compared to conventional vessels. During trial voyages, the vessels achieved a maximum speed of 15.00 knots with the hulls freshly painted at the designed draft, wind conditions below Force 2, and engines running at 100% (full capacity). In wind conditions of Force 4 or below, and sea conditions of grade 4 or below, with engines operating at 100%, the maximum speed reached 14.85 knots.

2) Environmental Friendliness

The two vessels' main engines and generator sets are equipped with Selective Catalytic Reduction (SCR) exhaust systems. The fuel systems are also equipped with low sulfur fuel tanks and supporting equipment, allowing them to burn low sulfur fuel. This enables the vessels to meet the strictest emission limits set by the International Maritime Organization (IMO) for nitrogen oxides and sulfur oxides under its Tier III regulations. The stern tube lubricating oil adopts environmentally friendly biofuel, preventing any pollution from oil leaks. Furthermore, the vessels are fitted with ballast water treatment systems that comply with IMO's D2 standards and meet the special requirements for ballast water management of some non-signatory countries of the Ballast Water Management Convention.

3) Cost Effectiveness and Practicality

Both vessels are equipped with a single engine, a single ducted propeller, and a single flap rudder, providing high towing capacity (maximum bollard pull of 50 tons) and excellent maneuverability (turning diameter of 1.6 times the ship's length at high speed and full rudder). They feature DP-1 dynamic positioning capability (under sea state 3, with a current of 1 knot and wind force 4, the vessel can automatically maintain position and heading, with a positioning accuracy within 1.0 m and a heading angle within $\pm$ 3 degrees), ensuring flexible and precise positioning. Additionally, the vessels are designed with a square keel, bilge keels, and a controllable passive anti-rolling tank, offering excellent seakeeping performance and cost-effectiveness, fully meeting the requirements for fishery survey and exploration operations.

4) Vibration and Noise Reduction

Numerous vibration and noise reduction efforts were made to improve the living and working environment for the crew and enhance their comfort. These include double-layer vibration isolation for the main and mooring & emergency generator sets, floating raft vibration isolation for the pump sets, soundproof enclosures for the engine room fans, vibration and noise reduction treatments for compartments, and importation for low-noise fans. Adopting these measures, the vessels meet both the comfort level requirements of the CCS and the latest requirements of the *Code on Noise Levels on Board Ships* (MSC.337(91)), which specifies that the noise levels in crew living quarters should be controlled to below 55 decibels.

5) Lifting Fin Fishery Exploration Marine Survey System

The retractable fin system is equipped with scientific research instruments such as the EK80 (38 kHz, 70 kHz, 120 kHz, 200 kHz) and ADCP (75 kHz, 600 kHz). The system is designed to withstand the maximum hydrodynamic torque generated when the vessel operates at its maximum speed of 14.5 knots and a maximum rudder angle of 35 degrees. The fin system can be deployed or retracted at a maximum vessel speed of 4 knots. To meet the requirements of scientific research operations, the fin can extend 0.8 m or 2.5 m below the hull (two positions, adaptable to different instruments), avoiding interference from surface bubbles and the hull boundary layer on acoustic scientific equipment.

This enhances measurement accuracy and ensures the quality of scientific survey data.

Each of the two vessels is equipped with two 25-ton mid-depth net winches, one 24-ton mid-depth net drum, and an automatic trawl control system. This control system, via visual operation, utilizes the display system of the wireless trawl monitoring system to adjust the postures of the net board and trawl by controlling the winding speed of the net winch. Additional features include four squid cranes and a longline system. The vessels' automation level of the exploratory fishing operations and the efficiency of the survey operations are high, which can effectively reduce the workload of the operators. Moreover, the vessels can conduct four types of exploratory fishing operations, including bottom trawling, controllable trawling, longline fishing, and squid fishing.

Control systems include an electric variable frequency CTD winch, a 5,000-m hydrological winch, a 1,000-m plankton trawl, a CTD dedicated deployment device, and an A-frame (5t × 3.5 m). They allow for centralized control of the winches and automated operation under pre-set working conditions, greatly reducing the workload of scientific researchers and improving the efficiency of research surveys.

"Lanhai 101" primarily undertakes routine, special, and emergency surveys and monitoring of marine fishery resources and the fishery environment, as well as multipurpose survey and research of the ocean. Together with "Lanhai 201", it will serve as the backbone of China's marine fishery scientific surveys over the next decade. They will complement existing and planned marine fishery research vessels in the three major sea areas, forming a comprehensive national marine fishery resource survey fleet. This will significantly enhance China's marine fishery scientific survey capabilities, elevate the level of marine fishery science and technology research, and provide robust support for the scientific management of fisheries in China and worldwide.

The successful construction and smooth delivery of the two fishery research vessels mark a milestone in the development of modern fishery facilities and equipment in China. This achievement represents another significant step in implementing the strategy to accelerate the building of a strong maritime nation. The two fishery resource survey vessels boast exceptional navigation performance, cutting-edge technology, safety and reliability, quiet and comfortable condition, environmental sustainability, and complete functionality. They will greatly elevate China's level and capacity in fishery research, offering crucial scientific and technological support in the ongoing improvement of marine fishery management systems, exploration of resource conservation approaches, expansion of fisheries abroad, and the pursuit of sustainable development in marine fishery. They will contribute to the modernization and transformation of fisheries, securing a decisive victory in building a moderately prosperous society in all respects, and realizing the Chinese Dream of national rejuvenation.

Since its service in July 2019, "Lanhai 101" has successfully undertaken multiple survey missions, including the Yellow and Bohai Seas Shared Survey Voyage of the National Natural Science Foundation of China, Survey of Fishery Resources in

Waters under Provisional Measures of China and South Korea, Survey of Fishery Resources and Environment in Xisha, South China Sea, and Survey of Fishery Resources in Yellow Sea, among others. In a total of eight expedition voyages, the vessel has spent 102 days at sea, covering a distance of 18,600 nautical miles. Thanks to these voyages, the vessel has collected extensive observation data on fishery resources and ecological environment, effectively supporting research on marine ecosystem dynamics and fishery resource management, and playing a significant role in safeguarding China's national marine fishery rights and interests. In the future, it will form an enhanced system of marine fishery resource survey vessels alongside the "Beidou" and "Nanfeng" vessels of the CAFS. Together, they'll play a significant part in promoting the sustainable utilization of marine fishery resources and environment and propelling China's transition from a big fishery country to a fishery power.

On July 25, 2019, "Lanhai 101" embarked on its inaugural voyage to carry out the Shared Survey Voyage to Yellow and Bohai Seas of the National Natural Science Foundation of China (NORC2019-01). During this voyage, the vessel provided marine multidisciplinary comprehensive research conditions in the Yellow and Bohai Seas for 64 national natural science fund projects with onboard requirements. Each leg of the voyage lasted 15 days, aiming to understand environmental changes in the Bohai and Yellow Seas. The mission focused on field observations of marine hydrology, meteorology, chemistry, ecology, and geology, as well as in-situ marine biological cultivation experiments. Samples of atmosphere, seawater, sediment, and organisms, along with relevant environmental parameters, are collected to provide scientific insights into the processes and mechanisms of environmental changes in the Yellow and Bohai Seas under the influence of global change and human activities.

To complete the Scientific Expedition Experimental Research Shared Voyage in Yellow and Bohai Seas of National Natural Science Foundation of China (NORC2020-01), "Lanhai 101" set sail from Qingdao on May 30, 2020, and returned on June 11, covering a total of 13 days and 2,800 nautical miles. It completed scientific research investigations at 66 stations, collecting a total of 21,298 on-site observation data points on hydrology, meteorology, chemistry, biology, and other fields. These data have provided scientific evidence for studying the ecological environment and biological resources' changes in the Yellow and Bohai Seas under the background of global climate change and human activities. They also offered scientific support for effectively addressing the challenges in governing the Bohai Sea, preventing and controlling brown and green tides in the Yellow and Bohai Seas. This interdisciplinary research opened up new paths for biodiversity conservation and the restoration of critical habitats for endangered wildlife in the Yellow and Bohai Seas.

Targeting the international frontier in marine research, the content of this expedition closely aligned with the strategic requirements of national ecological civilization construction and fully embodied interdisciplinary efforts. To ensure the quantity and quality of data obtained during the voyage, the project team solicited input from principal investigators of each onboard project multiple times before the voyage began. A dedicated online video conference was held to gather their requirements, suggestions, and expert advice. Guided by the principle of addressing scientific questions, the voyage project team actively supported exploratory research on the origin of brown tides in the Yellow and Bohai Seas beyond the planned survey transects. The related sampling and subsequent analyses will provide critical data to address this challenging issue.

On January 7, 2021, the "Lanhai 101" marine fishery resource survey vessel was sent to the south-central sea areas of the Yellow Sea to carry out the Winter Overwintering Ground Survey Mission in the Yellow Sea. This expedition marked the first voyage of this vessel in 2021.

According to the survey requirements, a total of 50 multipurpose survey stations were established during this 20-day expedition, covering various aspects such as fishery resources, physical and chemical environment, biological environment, and marine acoustics. During the voyage, "Lanhai 101" encountered a blizzard. Demonstrating the spirit of "team collaboration, responsibility, rigor, and dedication" upheld by the Yellow Sea Fisheries Research Institute, the scientific team continued to collect high-quality first-hand survey data. The results of the survey and assessment provided strong technical support for the conservation and management of fishery resources in the Yellow Sea.

2.27 Multipurpose Geophysical Research Vessel "Shiyan 6"

With the introduction of the Belt and Road Initiative and major strategic plans such as "adhering to the coordinated development of activities on land and at sea, and accelerating the construction towards a strong maritime nation" and "managing the ocean", higher expectations have been placed on the development of the marine industry in the new era. As the most crucial platform for exploring and researching the ocean, research vessels have become a significant assurance for implementing the strategy of becoming a strong maritime nation.

Strong sea power requires a priority on its development. To meet the evolving demands of China's marine science development, the Chinese Academy of Sciences (CAS) has established a research vessel fleet. This initiative aims to further integrate resources, coordinate various marine observation projects, improve the utilization efficiency of research vessels and their equipment, enhance interdisciplinary collaboration in marine sciences, and promote the sharing of survey data, thereby facilitating the production of significant marine scientific and technological achievements. The CAS Research Vessel Fleet currently consists of seven active vessels, including multipurpose research vessels and specialized research vessels. Among them, the 3,243-ton "Shiyan 3" multipurpose research vessel and the 1,153-ton "Shiyan-2" scientific survey vessel have been in service for 37 years and are beyond their intended service life. They lack the space for renovation and upgrade, making their safety a concern. Therefore, it is urgent to update and replace these vessels. Once they are decommissioned, the CAS will face the challenge of having no 3,000-ton multipurpose research vessel and geophysical research vessel for its Research Vessel Fleet. This will result in an inadequate tonnage sequence, incomplete scientific research capabilities, and a significant weakening of the marine survey capabilities and scientific and technological competitiveness of the CAS fleet. It will also hinder the ability to meet the increasing demands for oceanographic survey and scientific research tasks, thereby constraining marine scientific and technological innovation capabilities of the CAS.

Guided by the strategic mission of advancing China's marine science and technology development, serving national major needs, and fulfilling the responsibilities of "Three Orientations" and "Four Pioneering Initiatives", the South China Sea Institute of Oceanology of the CAS, proposed the construction of a new multipurpose geophysical research vessel, "Shiyan 6". This vessel would feature advanced ship performance, powerful detection capabilities, and efficient integrated management and operation. The purpose of constructing the "Shiyan 6" vessel was to integrate and replace the functions of "Shiyan 2" and "Shiyan 3", achieving a renewal and upgrade of the ship type.

The "Shiyan 6" vessel (see Fig. 29) boasts the largest tonnage ship type owned by the South China Sea Institute of Oceanology, CAS. It is the first 3,000-ton new-type multipurpose research vessel in China, primarily dedicated to geophysical surveys while also accommodating multidisciplinary scientific research needs. With these capabilities, the vessel is considered to be one of the advanced and representative ship types at this level of research vessels, both domestically and internationally.

"Shiyan 6" has the capability for global navigation and all-weather observations. It possesses not only prominent geophysical survey capabilities but also capabilities of fulfilling multidisciplinary comprehensive research needs. It is capable of conducting scientific surveys in offshore shallow water areas and South China Sea island reef regions, as well as detecting and sampling under extreme environmental conditions.

Fig. 29 Multipurpose geophysical research vessel "Shiyan 6"

After being put into operation, the vessel will greatly enhance China's capabilities in marine research, facilitating China's research in marine science and extreme environment in abyssal oceanic regions by providing advanced offshore mobile laboratories and technical experimental platforms for exploration equipment. In this way, it will significantly drive the development of China's deep and far sea marine science undertaking. More importantly, it will also contribute to a deeper integration between the military and civilian sectors in maritime field.

The vessel also plays a vital role in ensuring marine environmental safety, safeguarding national sovereignty and maritime rights, and exploring new opportunities for the development of the blue economy, serving as a platform for bolstering major strategies of the country in the new era.

The vessel was designed by the MARIC and constructed by CSSC Huangpu Wenchong Shipbuilding Co., Ltd. Construction of the vessel commenced in November 2018, and it was successfully delivered on December 18, 2020.

The vessel was designed with a length of 90.60 m, a molded breadth of 17.00 m, a molded depth of 8.00 m, and a draft of 5.5 m at full load, a capacity for 60 persons (including 27 crew members and 33 scientific researchers), an economic/maximum speed of the vessel of 12.0/17.2 knots, an endurance of 12,000 nautical miles at a speed of 12 knots, and a self-sustaining period of 60 days. The vessel is classed by the CCS.

The vessel is equipped with a wide range of systems and instruments, including the 10,000-m CTD profile detection system (24 bottles), in-situ continuous and automatic surface multi-element measurement system, two multi-channel seismic air compressors, long cylinder piston corer, TV grab, surface current and wave measurement system, navigational carbon dioxide measurement system, shipborne scientific meteorological station, marine gravimeter, magnetometer, full-depth multi-beam system, shallow-water multi-beam system, deep-sea sub-bottom profiler, scientific fish finder, single-beam depth sounder, and lifting fin.

Other equipment includes a CTD deployment device and an 8,000-m CTD winch, as well as a 10-ton side A-frame and a 20-ton stern A-frame, a deployment device for the long cylinder piston corer, a 13,000-m steel cable winch, a 13,000-m optical cable winch, a 6,000-m hydrobiological winch, among others.

Main laboratories include the atmospheric laboratory, geophysical laboratory, general dry laboratory, general wet laboratory, biochemical laboratory, clean laboratory, in-situ direct seawater analysis laboratory, temperature-controlled laboratory, truth-preserving sampling laboratory, dedicated water treatment laboratory, sample refrigeration room, sample freezing room, CTD water sampling room, network center, comprehensive instrument room, scientific research electrical maintenance room, and acoustic equipment room, covering an total area of over 330 square meters. In addition, the vessel also features an additional 470 square meters of auxiliary scientific investigation space. The open working deck area exceeds 600 square meters, meeting the standards of a 4,000-ton research vessel.

The vessel was primarily designed for geophysical exploration and seismic survey operations, while also meeting the needs of interdisciplinary research in hydrology, biology, marine chemistry, and other fields. The level of scientific research equipment on board and the configuration of laboratory space reach or approach the level of international advanced research vessels.

This vessel filled the gap in the domestic medium and small-sized multipurpose geophysical research vessels and achieved a wide range of scientific investigation capabilities with a smaller tonnage. By adopting high-efficiency and energy-saving generator sets, optimizing the layout of living quarters, rationally distributing electrical loads, streamlining scientific survey workflows, and implementing vibration and noise reduction measures, the vessel achieves features such as small tonnage, flexible operation, comprehensive scientific survey capabilities, and low fuel consumption, demonstrating advanced technical and economic performance.

In terms of stability and wind resistance, strict and systematic control of the weight and center of gravity of the entire vessel ensures intact stability in compliance with international conventions and further in line with the standard requirements for withstanding Force 11 winds (wind speed of 46.5 m/s). This significantly enhances the vessel's ability to cope with harsh sea conditions.

In terms of seakeeping capability and maneuverability, the vessel features integrated ship form design technology for controlling air bubble interference. By combining with high web bilge keel and specially designed seakeeping ship type, the vessel enables ROV operations in sea conditions of grade 4, drifting surveys in grade 5, and safe navigation in grade 7. In addition, the vessel are equipped with a high-power all-direction podded electric propulsion system and side thrusters, providing excellent power positioning capabilities.

The vessel's dynamic positioning system allows for precise positioning in sea conditions of 1.5-knot ocean current and Force 6 winds. The positioning accuracy is between 0–3 m, and the vessel's heading can be reliably controlled within ± 10 degrees.

In terms of environmental noise reduction, it marks the first application of ABB new-generation D-type Azipod podded propulsion system and the new generation ACS880 low-noise frequency converter on a research vessel. Moreover, the advanced measures such as double-layer vibration isolation in generator sets greatly reduce underwater radiation noise of this vessel, making it meet the requirements of DNV's SILENT-A+S standards at 11 knots.

In terms of meeting the requirements for domestic production and control of key scientific research equipment, "Shiyan 6" took an important step by actively responding to and successfully employing domestically produced large-capacity seismic air compressors on board for the first time.

Thanks to the adoption of efficient and energy-saving generator sets, advanced frequency-conversion electric propulsion technology, optimized layout of living compartments and laboratories, reasonable distribution of various electrical loads for different scientific investigation projects, optimization of scientific research equipment configuration and work processes, and the use of lightweight materials, "Shiyan 6" possesses the advantages of moderate tonnage, flexible operation, powerful scientific investigation capabilities, and low operating costs.

Design features of the vessel:

1) Selection of Main Dimensions

The selection of a ship's length, molded breadth, and molded depth needs considerations on speed, stability, economy, seakeeping capability, and convenience for scientific research operations.

Since the vessel was mainly designed to navigate in the South China Sea, which was more complex in sea conditions compared to other sea areas of China, it was important to ensure a good seakeeping performance against severe sea conditions. Therefore, the total length was set at approximately 90 m in avoidance of a too-short ship length.

In addition, an oceangoing research vessel requires a well-considered wind stability, which places a higher requirement on intact stability, Chinese ship standards for withstanding Force 11 gusts, based on the code requirement (Force 10 gusts). This requires the waterline area of the ship's hull not to be too small, and the center of gravity of the empty ship not to be too high. Therefore, the vessel was designed with a width of 17 m and a molded depth of 8 m.

Taking into account factors such as harbor depth, propeller emergence, scientific research operation freeboard, and seakeeping requirement for minimizing rolling motion in rough weather, the draft of the vessel was designed to be 5.2 m, with a full load draft of about 5.5 m, a scantling draft of 5.7 m, as well as a reserved modification allowance of 0.2 m.

In terms of deck height, considering the high utilization rate of laboratories as working spaces and the prominent features of ventilation ducts and numerous cables on the research vessel, the deck height of this vessel was appropriately increased compared to typical research vessel types. This ensured a spacious working space for research personnel and sufficient space for cable and equipment installations for the shipyard.

2) Profile Design

The profile design of the research vessel is a result of incorporating valuable design experience from both domestic and foreign research vessels. The bow features a bulb bow, which strikes a balance between hydrodynamic performance and the requirements for minimizing underwater radiation noise. The midship has a slightly rounded cross-sectional profile with a gradual transition at the waterline towards the bow and stern, avoiding excessively sharp and thin characteristics. This design effectively enhances seakeeping capability. A raised platform is incorporated into the stern design to accommodate the installation of pod, increasing the immersion depth and ensuring a consistent draft of the propeller during the 360-degree rotation of the pod. This mitigated the risk of a "toe out" installation caused by the inclination of the vessel, thereby preventing the phenomenon of propeller racing when turning to the side due to its proximity to the water surface.

The research vessel requires ample space for accommodating large specialized survey equipment, necessitating a wide deck area and hold capacity. However, an excessively wide hull design would result in significant sailing resistance. To minimize sailing resistance, the vessel's design reduces the hull width as much as possible while ensuring sufficient working deck area, maintaining an optimal streamlined shape. This significantly lowers sailing resistance and effectively reduces both the amount of construction materials and the propulsion power needed.

In the hull bottom area, the profile design around the multi-beam system installation section incorporates the "integrated design technology to control bubble interference" developed by the MARIC. Through analyzing the bubble leakage path at the waterline at the bow and taking into account factors, such as the height of the bulb bow and the position of the bow thruster, the design optimized the bottom shape of flat lines to ensure that all multi-beam transducers are completely integrated within the main hull. This not only optimizes hull resistance but also ensures the undisturbed operation of the multi-beam transducers in specified sea conditions.

The design utilizes a high-web triangular bilge keel to improve its resistance to rolling at various speeds. The increased area of the keel provides better course stability and enhanced anti-rolling performance.

3) General Arrangement Optimization

The vessel took into consideration the requirements of both the ship system and the scientific research system.

(1) Optimization of scientific research facilities arrangement: Based on the principle of "sufficiency, practicality, and usability", the flow of personnel, samples, and information within the scientific research process was considered when planning the laboratory layout. Laboratories with similar functions were strategically positioned close to each other, allowing for more efficient operations. The arrangement adopted a mainstream layout of left placement of chimney and "L-shaped" layout of compartments on the starboard side, as well as a combination of dry/wet general laboratories with special laboratories. Additionally, a spacious low freeboard working space was reserved on the stern working deck. This not only facilitates scientific research operations and future expansion but also provides convenience for the installation of movable container laboratories. Moreover, the presence of a large stern shielded working deck space ensures an optimal working environment for scientific research activities.

For geophysical survey operations, a mobile installation approach was adopted. During non-geophysical survey voyages, relevant equipment can be stored onshore to maximize the utilization of deck working area and payload capacity.

The forecastle deck is equipped with an external walkway for facilitating real-time observation by scientific personnel during marine research operations while also providing additional outdoor space for the crew.

(2) Optimization of living environment: The vessel ensures all personnel to live above the main deck, where there is ample access to natural light and ventilation, while being shielded from the impact of water surface waves. Personnel are accommodated in concentrated areas, with crew members and scientific researchers residing on separate deck floors to facilitate management and minimize potential interference.

Each compartment is equipped with single beds, and independent sanitary facilities are available. Additionally, each deck within the living quarters is furnished with a pantry, a laundry room, a storage space, and a public restroom, significantly improving living conditions for personnel on board.

4) Propulsion System Optimization

Considering the high requirements for station-keeping and positioning capabilities during scientific research operations, the vessel is equipped with a podded propulsion system that offers superior maneuverability and strong power positioning capabilities. The latest D-type pod developed by ABB is equipped, which demonstrates the following significant advantages compared to traditional C-type pods (see Fig. 30 and Fig. 31):

(1) It adopts a hybrid cooling method that combines water and air cooling methods, enabling maximum torque. This allows for lower propeller speeds, resulting in greater thrust with a smaller body.
(2) The propulsion motor adopts an asynchronous induction motor, which offers high power at a lower cost.
(3) The length of the support arm in the middle section of the pod can be adjusted, allowing for maximum immersion depth of the pod's propeller by increasing the length of the support arm. This reduces the frequency of the propeller emergence.
(4) The main body of the pod has an increased side projection area and is equipped with a short fin at the stern, effectively increasing the rudder area and improving rudder efficiency.
(5) The pod's main body is made of a new type of cast steel material, which boasts high strength and light weight characteristics.
(6) It adopts a Fieldbus form, which is simpler compared to traditional hard-wired connections.

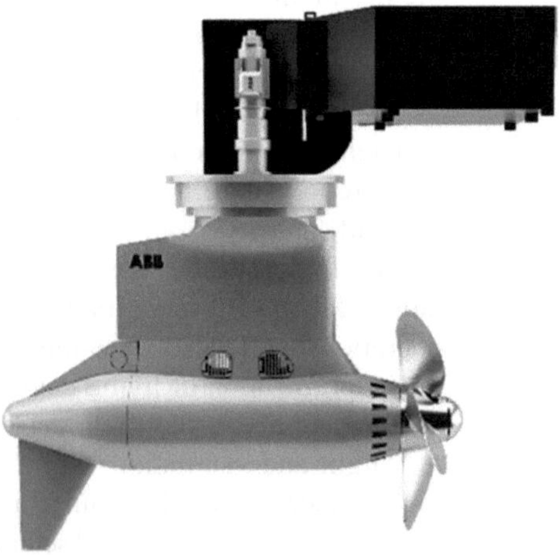

Fig. 30 D-type pod

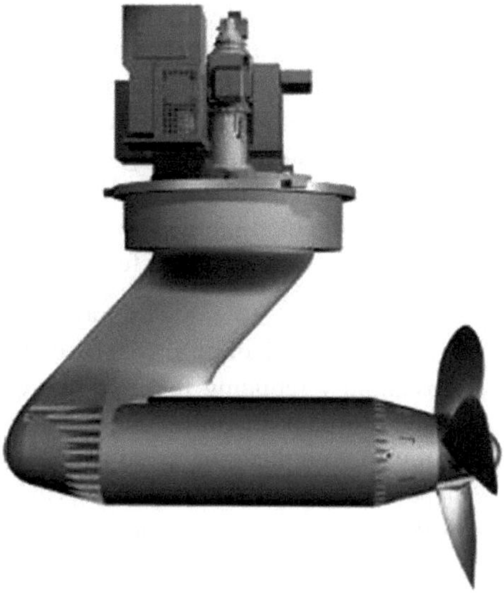

Fig. 31 Traditional C-type pod

(7) The external shape of the pod is optimized, enhancing hydrodynamic performance and achieving fuel savings of up to 10% when fuel consumption is at its maximum.
(8) The status of the pod can be monitored online, offering higher reliability and predictability compared to traditional manual monitoring methods.
(9) It has fewer major wear parts, allowing for sealing and bearing replacement without the need to dismantle the pod from the hull. Maintenance of the shaft system can be completed within 48 h in the dock, reducing aftersales service costs by around 30%.

5) Operational Efficiency

As a multipurpose research vessel with non-restricted waters navigation capabilities, the vessel supports diverse scientific research tasks. These tasks include requirements for stationary and precise positioning, low-speed seismic operations, moving measurement, transit at cruising speeds, and maximum speeds for special situations. Additionally, the vessel frequently employs a dynamic positioning system and deploys portable acoustic equipment and large-scale towed bodies. Therefore, the propulsion system must be adaptable to varying operational requirements and have high operational efficiency. Electric propulsion is chosen due to its advantages of flexible crew configuration and layout, excellent safety performance, brilliant maneuverability, good motility, minimal vibration, and low noise. The vessel's tonnage, hull form selection, and the integration of specialized and comprehensive functions all highlight its economic efficiency as a research vessel. This design allows for significant fuel savings, reduced operating costs, increased vessel utilization rates, and comprehensive utilization of favorable resources, ensuring optimal operational time and comprehensive maintenance.

The main power station consists of multiple energy-efficient and environmentally friendly diesel engines. The number of running generators automatically adjusts according to the electricity load, ensuring that the sets always operate in the most efficient way. This achieves optimal fuel economy in various navigation and operational conditions, ultimately improving the vessel's overall efficiency.

In addition to optimizing the hull shape, reducing resistance, and rationalizing the configuration of equipment in mooring, drifting, and cruising states, some other measures were taken to reduce fuel consumption and save energy, including:

a. Air conditioning zoning was optimized. According to the requirements of operational personnel, efforts were made to concentrate personnel on the same deck, substantially reducing the area that requires air conditioning when the vessel berthed in dock or port. Additionally, the central air conditioning system was equipped with medium-speed variable frequency control for optimal energy efficiency. b. The vessel incorporated LAN interfaces in all living compartments and working spaces of major compartments, enabling digital management platforms for file and data transfer, and developing towards a paperless work method. c. The quality of shipborne equipment and processes was improved to reduce maintenance workload during operation. New technologies were used for various piping fittings and accessories, so as to minimize wastage and pollution caused by leaks and spills. d. A seawater desalination system was configured to reduce the consumption of freshwater for fixed onboard purposes. Other than drinking water, seawater desalinated water was used for various purposes such as domestic use, engine room cooling, laboratory use, and deck washing whenever possible.

6) Environmental Protection

   (1) The nitrogen oxide (NOx) emissions of the vessel are currently in compliance with the latest and most stringent Tier III requirements set by the International Maritime Organization (IMO). Through post-processing method and selective catalytic reduction (SCR) technology, the vessel's NOx emissions are controlled to a level of 3.4 g per kilowatt-hour, enabling the vessel to navigate in the North American and U.S. Caribbean Sea NOx emission control areas.
   (2) The vessel has also been granted the Clean notation by CCS (China Classification Society). In addition to meeting the requirements of the International Convention for the Prevention of Pollution from Ships, the vessel complies with CCS regulations regarding pollution prevention structures, equipment, and operational procedures. To obtain the Clean notation, efforts were made to enhance the vessel's environmental rating and reduces its negative impact on the environment.

7) Harmonic Control of Empty Vessel Weight, Center of Gravity and Electric Propulsion System

For small-tonnage medium-sized research vessels, the control of the empty vessel weight and center of gravity is crucial. It not only affects the stability against wind but also directly influences the load capacity and the ability to carry out scientific research, playing a vital part in the proper functioning of the scientific research capabilities. Due to the high requirements for crew and scientific research capabilities, it is often challenging to achieve an optimal waterline and underwater projection ratio for medium-sized research vessels. As a result, the superstructure tends to be relatively large. To address this, measures were implemented to optimize the structural design, equipment layout, and procurement. This includes minimizing the number of bulkheads within the superstructure while ensuring sufficient supporting strength and vibration requirements, placing heavier equipment in lower positions, and using lightweight furniture, aluminum honeycomb panels, perforated external stair steps, and magnesium–aluminum alloy cable trays.

The vessel is equipped with an advanced all-electric propulsion system and specialized acoustic survey equipment. The key technological challenge for AC motors in ship propulsion lies in the speed control of AC motors, for which variable frequency control is a suitable method. However, the frequency converter using a large number of high-power switch components have disadvantages such as low power factor and difficulties in eliminating high-order harmonics. Therefore, harmonic calculation and analysis were conducted to minimize the impact of harmonic interference on various systems aboard the vessel.

It is apparent that harmonic interference and short-circuit current are two interacting parameters. The entire electric propulsion system, including generator, variable frequency transformer, frequency converter, distribution board, filter, and high-power equipment motor (such as the rudder and propeller integrated all-direction electric propulsion system, bow thruster), was supplied by the equipment manufacturer as a complete set. After conducting harmonic calculation and analysis, the equipment manufacturer adjusted the parameters of the equipment from an overall perspective, ensuring that the system meets the harmonic requirements while improving its overall cost-effectiveness.

Power cables with significant harmonic components were laid separately from other cables, especially signal cables for sensitive equipment such as radio communication, acoustic equipment, and network equipment. This minimized the impact of harmonic components on weak electrical signals.

8) Vibration and Noise Reduction

(1) In terms of compartment layout, efforts were made to concentrate compartments that generate noise, such as the emergency generator room and air conditioning room. Compartments that produce noise were separated from living compartments or positioned far away, and isolation measures were implemented. A certain distance was maintained between living compartments and experimental compartments to prevent direct noise transmission. In structural design, continuity was ensured as much as possible, with structural components aligned vertically. To minimize interference between living compartments and internal passageways, as well as interference between living compartments, soundproofing materials were applied

to the steel walls between passageways and living compartments as well as inter-living compartments.

(2) High-damping materials were installed in compartments with loud noise, main engine bases, auxiliary engine bases, and compartments where noise avoidance was needed, such as the emergency generator room, air conditioning room, engine control room, comprehensive instrument room, and acoustic equipment room, to reduce vibration and minimize noise generation or prevent its propagation.

(3) Double-layer vibration isolation was employed for main diesel generator sets and berth diesel generator sets, while elastic shock absorbers were used for emergency diesel generator sets. Floating floors were installed for the engine control room to reduce the impact of main engine vibrations. Elastic mounting was adopted for installing the bases of main equipment, such as air compressors and fans. High-power pump sets, such as seawater cooling pumps, freshwater cooling pumps, and air conditioning cooling water pumps, were connected using flexible hoses and elastic mounting.

Quality silencers were fitted in the exhaust piping of diesel generator sets. In addition, mid-pressure low-noise centrifugal ventilators were equipped with variable frequency speed control for ensuring low air supply noise, smooth operation, and minimal vibration, while meeting the requirements of air conditioning conditions and system resistance. Necessary noise reduction accessories were installed on ventilation ducts, and high-quality imported air supply terminals were selected to ensure that the noise levels in the air conditioning compartments were within the required range.

The vessel received recognition from various domestic marine research institutions for its moderate tonnage, advanced performance, spacious working deck area, and comfortable living spaces.

2.28 Multipurpose Marine Research Training Vessel "Zhongshan Daxue"

On June 26, 2021, the 6,000-ton multipurpose marine research training vessel, developed and designed by the MARIC for Zhongshan University, and built by Jiangnan Shipyard Group, was successfully delivered.

"Zhongshan Daxue" (see Fig. 32) is a multipurpose marine research training vessel approved as a major construction project by the Ministry of Education in June 2016. It served the national strategy and the "Maritime Silk Road" initiative, as well as the needs of national economic and social development. With a total length of 114.3 m and a designed displacement of 6,848 tons, it is currently the largest domestically built multipurpose oceanographic research vessel with the strongest comprehensive scientific research capabilities and the highest number of innovative design features.

Fig. 32 Multipurpose marine research training vessel "Zhongshan Daxue"

The vessel realized the "1234" strategy: a. It embodies the idea of "one main body", which is in line with national strategic deployments and focuses on fundamental and cutting-edge marine science questions, serving the development of national and local marine industries and cultivating innovative talents. b. It establishes "two major platforms": building an internationally leading offshore mobile detection and monitoring platform, and a world-class platform for marine science and marine engineering talent development. c. It possesses "three-in-one" capabilities: integrating multiple disciplines, functions, and technical means into one; incorporating sea, land, and air as a whole; and integrating information collection, transmission, and analysis processing. d. It forms "four distinct features": automation in deck operations, real-time information analysis, efficient scientific research operations, and demonstration of hull functions.

The vessel was designed with a length of 114.30 m, a molded breadth of 21.4 m, a molded depth of 9.25 m, a designed draft of 6.00 m, a gross tonnage of 8,000 tons, a capacity of 100 persons, an economic/maximum speed of 12.0/16.54 knots, an endurance of 15,000 nautical miles, and a self-sustaining period of 60 days. It is classed by the CCS.

The main laboratories on board: general wet laboratory, general dry laboratory, scientific research laboratory, general clean laboratory, temperature-controlled laboratory, and in-situ direct seawater analysis laboratory.

The vessel is equipped with deep, medium, and shallow-water (full-depth) multibeam systems, a sub-bottom profiler, ultra-short baseline system, an acoustic Doppler current profiler, a fish finder, and a TV grab.

The main scientific research support equipment on the vessel: a. fixed installation equipment such as two CTD winches (12,000 m/10,000 m), two optical cable winches (5,000 m/18,000 m), and one hydrobiological winch (3,000 m); b. mobile installation (container) equipment such as a ROV winch (6,000 m), a variable-depth towing winch, a digital seismic cable winch (3,000 m), and a 240-channel digital seismic system.

On the main deck stern section, the vessel is equipped with a A-frame (30t), a starboard-side telescopic arm crane (25t), a CTD water sampling and LARS (6t), a telescopic folding arm crane on the boarding deck (20t x 16 m), and a telescopic folding arm crane at the bow of the forecastle deck (1.5t x 16 m).

"Zhongshan Daxue" is currently a multipurpose oceanographic research vessel, with the largest displacement (excluding polar research vessels), strongest comprehensive scientific research capabilities and most innovative design features in China. Its main design features are as follows:

1) Outstanding Speed and Efficiency, Reaching International Advanced Levels

   Through trial voyage, the vessel reached a speed of 12.00 knots (above the designed estimate of 11.5 knots) when a single generator was operating, 14.97 knots (exceeding the designed value of 14.5 knots) with two generators running, and a maximum speed of 16.54 knots (surpassing the designed target of 16.00 knots) with all three generators in operation. These speed and efficiency performance indexes surpassed the requirements in the technical specifications.

   Equipped with German variable-speed diesel engines, the vessel has a total generator capacity of 3,050 kilowatts (three sets) and 1,090 kilowatts (one set). Through testing, it was confirmed that whether cruising at an economical speed (with one engine) or maintaining cruising speed (with two engines) under fixed engine speeds, the power load of the power station remained at 75% to 85%. This range is deemed optimal for both power station operation and diesel engine fuel consumption. It is calculated that if the vessel operates under economical speed at 11.5 knots using a single engine, the daily fuel consumption is 11 to 12 tons. If the diesel engine operates in a variable-speed mode, the fuel consumption is even lower.

2) Adoption of an Integrated Design of Anti-bubble Invisible Bulb Bow, Ensuring Overall Performance of Embedded Acoustic Transducer

   The bow of the research vessel contains lots of embedded acoustic equipment, and the bubbles generated in the bow in windy and wavy conditions can easily leak down to the location of the equipment, adversely affecting its measurement performance and accuracy. To address this technical challenge, the vessel adopted a novel design called anti-bubble invisible bulb bow integrated design (invention patent). This design guides the bubbles to a path that does not interfere with the transducer's operation while also ensuring the vessel's speed and economy, ensuring the overall performance of the vessel.

3) First-Ever Adoption of Rim Permanent Magnet Side Thruster, Significantly Reducing Vibration and Noise Throughout the Vessel

Research vessels equipped with conventional side thrusters face challenges in reducing noise levels further, especially under dynamic positioning conditions. The rim-driven permanent magnet side thruster effectively reduces vibration, lowers noise, and simplifies maintenance, optimizing scientific survey operations and living conditions under DP1 conditions.

The incorporation of rim permanent magnet side thrusters on this vessel is a first endeavor. Through meticulous design and construction and ship trials, it is confirmed that the vessel fulfills the CCS' requirements for a Level 2 comfort rating at a cruising speed of 14.5 knots and under DP1 dynamic positioning conditions. When cruising at 14.5 knots, over 90% of the living compartments meet the highest comfort standards set by the CCS (less than 49 decibels).

4) Channel-Type Side Thrust Cover, Enabling Hull Surface Integration

The implementation of channel-type side thruster cover ensures complete conformity with the vessel's 3D hull curve when closed, achieving a seamless integration with the surface. This technology effectively addresses the adverse effects of vortexes and bubbles within the thruster channel on underwater acoustic equipment. Meanwhile, it guarantees hull integrity, minimizes hull resistance, and reduces fuel consumption.

5) First-Ever Adoption of Active Anti-rolling Fin at All Speeds, Enhancing Operational Sea Condition Adaptability

Equipped with a pair of active anti-rolling fins at all speeds, the vessel incorporates the advantages of traditional low-speed passive controllable anti-rolling tanks and fixed anti-rolling fins at cruising speeds. This significantly improves adaptability to sea conditions across a full range of speeds, from zero to maneuvering speed, enhances scientific research efficiency, and reduces motion sickness among personnel.

6) Adoption of DC Busbar and Energy Storage Battery, Ensuring Ultra-Silent Scientific Research with Full Electric Propulsion

This vessel is the first to adopt the configuration of DC busbar + energy storage battery, effectively reducing diesel fuel consumption and exhaust emissions. It dynamically adjusts network energy and quickly suppresses voltage fluctuations, thereby improving energy efficiency and electrical power quality. The introduction of battery systems serves the purpose of peak shaving and providing separate power for ultra-silent operating conditions, ensuring the accuracy of scientific research data collection and processing. Through optimization of power quality, the energy quality of the electric propulsion system can be significantly improved. With comprehensive harmonic suppression measures and the use of clean network control technology, the power supply quality of the entire vessel's AC network is greatly improved. The trial voyage data of the vessel far exceeds the relevant requirements for total harmonic distortion of the CCS and the technical specifications. The power configuration of the power station is reasonable, and the generators operate near their economic points under typical operating conditions at a fixed engine speed. The operating costs are well controlled. If a variable speed mode is used, further cost reductions can be achieved.

7) Dedicated Container Storage Compartment Meeting the Urgent Needs of Scientific Research Surveys

The vessel is equipped with a dedicated container storage compartment (stern equipment compartment), capable of storing four standard 20-foot containers. Through specially designed container handling equipment, the vessel provides various modular storage solutions for different scientific research missions, meeting the urgent demand for storage space in scientific research.

8) First-Ever Adoption of L-type All-direction Low-noise Propellers, Meeting the Underwater Radiation Noise Requirements of the Det Norske Veritas. This is the first domestic research vessel to adopt L-type all-direction low-noise propellers. Through ship trials, the underwater radiation noise level meets the requirements of DNV Silent A+S at a speed of 14.5 knots and DNV Silent F at a speed of 10 knots. The underwater radiation noise of this vessel is at the highest level among all domestic and international all-direction propeller-type research vessels.

9) Establishment of Helicopter Landing Platform, Enhancing the Transfer Capability of Scientific Research Personnel and Supplies

The vessel has a helicopter landing platform at the bow, greatly improving the transfer capability of scientific research personnel and supplies. This also expands the scope of scientific research observation in three-dimensional space.

10) Dedicated UAV Hangar and USV Container

To adapt to the development trend of unmanned equipment in the future, the vessel is equipped with a dedicated unmanned aerial vehicle (UAV) hangar and an unmanned surface vessel (USV) container, which can meet diverse unmanned scientific research needs and realize the deployment of unmanned equipment clusters.

11) Adoption of Lift-type Stern Roller, Providing Cable Protection and Flexible Use of Deck Space

The vessel's lift-type stern roller can be adjusted vertically to serve as a protective roller for different towing equipment, thereby significantly reducing friction between the cable and the hull structure. When lowered, the roller can be completely concealed below the deck, greatly expanding the deck space for stern operations.

12) Arrangement of a Scientific Research Observation Room on the Compass Deck, Enhancing Research Observation Capabilities

Modern research vessels demand higher standards for top-mounted scientific research spaces and experiments. The vessel includes a dedicated scientific research observation room on the compass deck, positioned at a higher level and offering independent space. This provides scientists with excellent visibility and observation conditions at sea.

13) Weather Radar Installation Platform Meeting High-precision Requirements for Meteorological Surveys

To meet the high requirements of meteorological survey discipline, the vessel's compass deck is equipped with a specialized weather radar. The installation of the radar requires high precision and poses challenges, with strict requirements for deformation, vibration performance, harmonics, and electromagnetic interference. To address this, a dedicated door-shaped weather radar installation platform was designed to integrate seamlessly with the vessel's radar mast. This design ensures the vessel is able to conduct comprehensive and high-precision meteorological surveys while maintaining safe navigation.

From May 9 to 17, 2021, the vessel successfully passed a routine performance trial voyage with excellent results and completed underwater radiation noise measurements. The vessel's underwater radiation noise levels met the requirements for DNV Silent A+S certification at a speed of 14.5 knots and obtained the corresponding certificate, as well as the requirements for DNV Silent F certification at a speed of 10 knots and obtained the corresponding certificate. This achievement represents the highest level of underwater radiation noise among current domestic and international research vessels equipped with all-direction propulsion systems.

Chapter 6
Ocean Management and Offshore Survey

1 Overview

With China's growing awareness of marine resource exploration, utilization, and protection, and guided by the vision of building a strong maritime nation and safeguarding national maritime rights and interests, the research, design, and construction of research vessels, though starting relatively late, have rapidly advanced. China's research vessels have evolved from retrofitting old equipment to independently designing and developing world-class vessels.

Prior to the founding of New China, due to the lack of scientific research equipment, the country had little knowledge of marine resources in neighboring waters. Following the founding of the country, in line with the country's economic development and defense needs, several multipurpose specialized offshore research vessels were built. These included the refitted "Jinxing" oceanographic survey vessel and the self-designed "Qixiang 1" meteorological survey vessel, as well as the "Xingguang 1" and "Xingguang 2" geophysical vessels nominated by Li Siguang and refitted in 1963, and the "Haiyang 1" and "Haiyang 2" marine geological survey vessels built in 1973, among others. These research vessels are specifically introduced in Chapters 1–5 of this book. During this period, according to the actual needs, the central authorities allocated special funds to build various multipurpose survey vessels, such as the "Fendouzhe 4", "Fendouzhe 5", "Xingye" geological survey vessels, the "Runjiang 1", "Yanping 2", "Dongyuan 01", "Dongyuan 02", and surface catamaran survey vessels. Specialized vessels include the "Qingyan Haishi 1" experimental mother ship of Tsinghua University, the "Kan 407", "Ye Zhizheng" marine geological survey vessel, and the "Zhongguo Kaogu 01" marine archaeological vessel. These vessels have played important roles in the marine chemical, biological, meteorological, geological, physical, and geomorphic surveys in China's offshore areas.

L. Han et al., *History of the Research and Development of Chinese Marine Research Vessels*, https://doi.org/10.1007/978-981-96-8414-4_6

2 Classical Research Vessels

2.1 Marine Geological Survey Vessels "Fendou 4" and "Fendou 5"

The "Fendou 4" marine geological survey vessel (see Fig. 1) was constructed by Guangzhou Shipyard and delivered in 1978. It is under the management of the Guangzhou Marine Geological Survey. The vessel was designed with a length of 69 m, a molded breadth of 10 m, a draft of 3.5 m, and a full load displacement of 1,200 tons. It is equipped with dual/single engines with a power of 1,100 horsepower, along with adjustable pitch propellers. It can achieve a speed of 16 knots, with an endurance of 6,000 nautical miles, a self-sustaining period of 30 days, and a capacity of 70 persons.

This vessel is capable of conducting seabed soil sampling, vibrating piston sampling, seismic, gravity, magnetic, and shallow-to-deep water multi-beam surveys, as well as various hydrological and ocean engineering investigations. In March 1992, "Fendou 4" conducted a comprehensive marine geological and geophysical survey in the South China Sea, focusing on gas exploration and integrating gravity, magnetic, sounding, and multi-channel seismic methods. In 2014, "Fendou 4" completed six voyages and spent 147 days at sea, performing multiple geological survey missions in various areas. It conducted multi-beam surveys covering a total distance of 27,900 km, as well as comprehensive geophysical surveys spanning 2,700 km, and multi-channel seismic surveys covering 1,400 km.

Fig. 1 Marine geological survey vessel "Fendou 4"

In August 2015, a geological survey team composed of the "Tanbao" and "Fendou 4" survey vessels from the Guangzhou Marine Geological Survey of the China Geological Survey conducted a one-week dual-ship seismic exploration in relevant areas of the South China Sea. The advantages of dual-ship seismic exploration include precise positioning and flexible exploration.

The "Fendou 5" vessel (see Fig. 2) is a sister ship of the "Fendou 4." Both vessels were constructed by Guangzhou Shipyard. "Fendou 5" was delivered in 1980 and is also under the management of the Guangzhou Marine Geological Survey. This vessel is capable of conducting drilling, multi-beam surveys, geological sampling, multi-channel seismic surveys, engineering geophysical surveys, marine gravity and magnetic surveys, and other investigations. It can also perform geophysical, environmental, and engineering geological surveys. The vessel is equipped with a navigation positioning system, a vibrating piston sampler (primarily used to collect samples of bottom sediments), a box corer (for surface sample collection), a large piston sampler, and a trawl sampler. It can conduct deep-water projects with a maximum depth of 400 m.

"Fendou 5" has been involved in multiple major projects, including surveys of oil and gas in the Nansha area and the Pearl River estuary, as well as natural gas hydrate surveys. It has even ventured as far as the Matsu Archipelago for survey purposes. On October 11, 1999, "Fendou 5" set sail from the Whampoa Military Academy Pier in Changzhou, Guangzhou. Its mission was to determine the possibility of natural gas hydrates in the Xisha trough of the northern slope of the South China Sea through high-resolution multi-channel seismic surveys. After more than 20 days of work, the investigations for three planned 534.3-kilometer survey lines were completed. Following one month of data processing, this expedition discovered

Fig. 2 Marine geological survey vessel "Fendou 5"

important geophysical markers of the presence of natural gas hydrates in the surveyed area, which, combined with other relevant information, led to the inference of natural gas hydrates in this region. The findings were reported to the State Council by the Second Marine Brigade of the Guangzhou Marine Geological Survey. Wen Jiabao, who was then the Vice Premier of the State Council, remarked, "The discovery of bottom simulating reflectors indicating the presence of natural gas hydrates in the South China Sea represents a breakthrough in the exploration of this new resource for our country."

Since its departure on March 11, 2015, "Fendou 5" has completed five survey voyages, spending a total of 128 days at sea. The vessel has conducted multi-beam surveys covering a distance of 7,665 km in regions such as Ledong, Sansha, Xiamen, and Quanzhou, as well as comprehensive geophysical surveys, including single-channel survey, sub-bottom profiling, and sounding, totaling 3,988 km. It has also conducted navigational current measurements covering 481 km and carried out current observations, CTD measurements, and water collection at 76 stations. Additionally, the vessel has performed five shallow drillings and completed tasks such as cold seepage observations, seabed static sounding, and pressure sampling.

"Fendou 5" has achieved breakthroughs in safety management, quality control, and production progress.

2.2 *Engineering Geological Survey Vessel "Kan 407"*

In 1981, Shanghai Offshore Oil Bureau decided to transform the original triple-purpose working vessel "Kan 407" (see Fig. 3) into a marine engineering geological survey vessel to address the needs of oil drilling site surveys in the East China Sea and ensure the safety of drilling platforms and conduct offshore seas engineering geological surveys.

The vessel was redesigned by the Technical Equipment Research Institute of the Marine Geological Survey and refitted at the Shanghai East China Sea Shipyard. After the refit, the vessel was transformed into a shallow geophysical survey vessel for offshore engineering projects. It is equipped with shallow drilling and geophysical instruments, enabling operations such as seabed drilling, vibrating piston sampling, basic soil testing, sounding, sub-bottom profile survey, electric spark survey, and side-scan sonar scanning.

The vessel was designed with a length of 56.4 m, a molded breadth of 12.5 m, a draft of 6.2 m, a displacement of 1,600 tons, a speed of 9 knots, an endurance of 7,000 nautical miles. Its power system features a MET KRMB-9 diesel engine with a power capacity of 1,654 kilowatts and a rotation speed of 825 revolutions per minute. Scientific research equipment, including a scientific fish finder system, a radar wave tester, and a ship attitude tester, are provided.

Between 1983 and 1993, "Kan 407" conducted surveys at 50 oil drilling sites in the East China Sea, and also cooperated in completing an engineering geological

Fig. 3 Engineering geological survey vessel "Kan 407"

survey of the western region at a scale of 1:1,000,000 in the East China Sea. In addition, it also carried out systematic sub-bottom profiling surveys, sounding surveys, and side-scan sonar scanning. The vessel even assisted foreign companies, domestic oil corporations, and marine geological survey units in a wide range of site engineering geological investigations and marine engineering geological investigations, stretching from the Bohai Sea and the Yellow Sea to the South China Sea Pearl River Estuary and the Hong Kong region. The work content covered nuclear power plant site selection, harbor terminals, marine geological hazards, shipwrecks, submarine pipelines, hazardous substances, seabed sediments, and topographic and geomorphologic surveys. In 2012, the vessel achieved a national record by successfully drilling a full-section core sample to a depth of 240 m during an investigation and evaluation project of underwater freshwater resources in the Zhoushan Sea area. In 2013, as part of the East China Sea Environmental Effect Study Project, it once again surpassed the national record, with a full-section core sample depth reaching 300 m. This marked the first time that a scientific drilling operation on China's continental shelf exceeded a depth of 300 m.

2.3 Marine Geological Survey Vessel "Tanbao"

In 1993, the Guangzhou Marine Geological Survey purchased a marine geological survey vessel from WesternGeco, US. The vessel, originally built in September 1978 in Japan, was later refitted and renamed as the "Tanbao" (see Fig. 4).

This vessel, measuring 87 m in length, 14 m in molded breadth, and with a draft of 5 m, was refitted into a multi-function geological survey vessel capable of conducting high-resolution, conventional, and low-frequency deep-water oil and gas seismic surveys.

Fig. 4 Marine geological survey vessel "Tanbao"

In 2003, "Tanbao" successfully carried out the first single-vessel long-line high-capacity seismic acquisition, seabed seismometer operation, and synchronous operations combining other geophysical exploration methods. It obtained clear geological data of the underwater mid- to deep-bottom as well as deep reflections from the Earth in the South China Sea's deep-water area.

From March to November 2014, "Tanbao" completed seven voyages (segments) spanning a total of 209 days. During this period, it conducted comprehensive geophysical surveys, including multi-channel seismic surveys, and single-channel seismic surveys, in various locations such as Qiongdongnan, Xisha, Dongsha, the central South China Sea, and the Western Pacific. The vessel successfully completed 15,000 km of multi-channel seismic surveys (including gravity, magnetic, and sounding) and 1,350 km of single-channel seismic surveys.

2.4 Research Vessel "Haida Hao"

"Haida Hao" (see Fig. 5) is a fully welded, steel-hulled multipurpose research vessel with five decks and a 360-degree panoramic wheelhouse. The main tasks of "Haida Hao" include two-cable 3D geophysical surveys, deep towing in 4,000-m marine electrical surveys, underway investigation in sedimentary environment geophysical surveys, geological surveys, hydrobiological surveys, ROV surveys, underwater infrastructure inspection, deep-sea field monitoring, and shallow-water core drilling to depths of 300 m within a range of 150 m. This vessel filled the gap in China's 3D seismic monitoring and deep towing capabilities in marine geophysical surveys.

Fig. 5 "Haida Hao"

The vessel was designed with a total length of 68 m, a molded breadth of 15.6 m, a molded depth of 7.6 m, a draft of 2.25 m, a displacement of 2,650 tons, an endurance of 5,000 nautical miles, a self-sustaining period of 30 days, a maximum speed of 13 knots, and a capacity of 49 persons.

In May 2014, "Haida Hao" embarked on a 60-day scientific research voyage to the East China Sea. Carrying a number of scientists and technicians, the vessel was given the mission to conduct marine geological surveys and the analysis of geological samples.

2.5 *Multipurpose Survey Vessel "Yanping 2"*

"Yanping 2" (see Fig. 6) is an 800-ton satellite-guided multipurpose oceanographic survey vessel. It is equipped with special laboratories for physics, geology, plankton, benthos, microbiology, and marine chemistry. Other equipment onboard includes the mapping-type satellite positioning system, infrared rangefinder, sub-bottom profiler, echo sounder, hydraulic feed drill, self-recording current meter, wave meter, water level gauge, 3,000-m deep-water winch, and other marine engineering hydrological and geological survey equipment. Laboratory instruments mainly include atomic spectrometer, gas chromatograph, high-speed freezing centrifuge, and mineral identification instrument. This vessel can carry out various marine scientific survey activities in all coastal waters of China.

Fig. 6 Multipurpose survey vessel "Yanping 2"

The vessel was designed with a length of 52.6 m, a molded breadth of 8.68 m, a draft of 3.4 m, a displacement of 818 tons, a speed of 16.35 knots, an endurance of 2,500 nautical miles, and a self-sustaining period of 45 days.

In 2012, "Yanping 2" completed the marine survey mission of the "Fujian and Taiwan Strait Crustal Deep Structure Land-Sea Joint Survey" project for the shipborne seismic source system of Earthquake Administration of Fujian Province. In 2013, the Earthquake Administration of Fujian Province successfully designed and developed China's first shipborne air gun seismic source system, which can meet the requirements of both underway and stationary induced operations in different water areas such as rivers, lakes, and oceans. Multiple experimental verifications of this system have been conducted aboard the "Yanping 2".

2.6 Wave-Piercing Catamaran Survey Vessel "Dongyuan 01"

The design technology of wave-piercing catamaran-type ships in China started in 1997. With the support of the National Economic Mobilization Office, CSSC (Beijing) Dongyuan Technology Development Co., Ltd., organized related research institutions over a period of nearly five years to digest and absorb foreign ship form design technologies. Through performance research and testing, they mastered the key technologies of wave-piercing catamaran ship design. The wave-piercing catamaran survey vessel "Dongyuan 01" was constructed based on a comprehensive

Fig. 7 Wave-Piercing catamaran survey vessel "Dongyuan 01"

analysis on overall performance of "Haixia" vessel, while integrating its construction mission. The vessel began its construction in 2007 and was delivered for use in December 2009. The vessel was named "Dongyuan 01" (see Fig. 7), with the side number "Beidiao 990".

The vessel was designed with a length of 60 m, a molded breadth of 18 m, a demihull width of 3.5 m, a molded depth of 5.9 m, a maximum draft of 2.2 m, a designed displacement of 380 tons, a speed of 35 knots, an endurance of 1,000 nautical miles, a self-sustaining period of 7 days, and a capacity of 8 persons.

The vessel features a wave-piercing, all-aluminum alloy twin-hulled ship form and is powered by high-speed diesel engines with a four-engine, four-pump water jet propulsion system. It has the characteristics of high speed, low wave-making resistance, good seakeeping performance, large loading capacity, and good stability. It has excellent maneuverability and can perform stationary turning. "Yuandong 01" currently serves as the largest high-speed wave-piercing catamaran ship by tonnage in China, mainly used for oceanographic survey, marine transport, and maritime rescue in offshore areas. The vessel is equipped with a meeting room, a multi-function hall, and a special laboratory of 60 square meters. Other facilities include a trim adjustment system, a topography and geomorphology survey system, an internal and external communication system, a high-pressure water mist fire extinguishing system, hydrographic measurement devices, a recovery winch, and a 1-ton hydrological davit. A wet deck with a 1.6-m-diameter free flooding test moonpool is arranged at the stern, which is used for directly deploying measurement equipment to the seabed. It is equipped with low-noise power supplies of 380 V, 220 V, and 24 V, as well as UPS power supplies.

2.7 Small-Waterline Catamaran Survey Vessel "Dongyuan 02"

The vessel "Dongyuan 02" (side number: "Beidiao 991") (see Fig. 8) is a small-waterline catamaran survey vessel jointly designed by CSSC Dongyuan Technology Co., Ltd., China Ship Scientific Research Center, and China Ship Development and Design Center. The vessel was built by Wuchang Shipbuilding Industry Group Co., Ltd. and delivered in April 2008.

The vessel was designed with a length of 64 m, a molded breadth of 23 m, a submerged width of 5.1 m, a maximum draft of 5 m, a displacement of 1,441 tons, a maximum speed of 11 knots, a minimum speed of 3 knots, an endurance of 1,500 nautical miles, a self-sustaining period of 7 days, and a capacity of 32 persons.

The vessel is powered by an integrated electric propulsion system and is made of steel. The main supporting equipment, including the diesel engines, generators, and electric propulsion systems, are all integrated by subsidiaries of China State Shipbuilding Corporation. It was put into service in April 2008 at the Dalian Institute of Measuring & Control Technology. "Dongyuan 02" features excellent seakeeping performance and maneuverability, low fuel consumption, spacious working deck, and advanced special equipment, which enable it to carry out towing and survey experiments, marine engineering construction, and maritime rescue missions at low speeds. Due to tonnage limitations, the vessel's hull steel plate thickness is only 8 mm, and its fuel and water storage capacity is less than 100 tons. This restricts its ability for test modifications and results in insufficient endurance ability and needs of frequent resupply during sea trials. These factors have somewhat reduced the vessel's efficiency during its work in the South China Sea.

After delivery, "Dongyuan 02" underwent an adaptive refit to meet the requirements of offshore engineering support tasks in deeper waters beyond its designed navigation areas. This refit involves the installation of an A-frame and a joint

Fig. 8 Small-Waterline catamaran survey vessel "Dongyuan 02"

hydraulic station at the stern, the replacement of the original hydrological winch, strengthening of parts of the stern's structure, installation of a cross-well bridge over the midship moonpool, modification of the hydraulic crane on top of the moonpool, and strengthening of parts of midship structure.

"Dongyuan 02" is equipped with a spacious open moonpool located between RF 31# to 79# in the central section of the twin-hulled structure. There is one fixed lift-and-turn device with a load capacity of four tons and another one with a load capacity of two tons that can move both vertically and horizontally. These devices can descend up to six meters below the vessel's bottom with an accuracy of one degree, enabling underwater precise positioning of large and medium-sized tested equipment and underwater operations with a depth of up to 11 m at a maximum speed of three knots. There are also two sliding platform cars above the moonpool, along with a cross-moonpool bridge and a heavy-duty winch with a load capacity of three tons. To facilitate experiments, a four-ton crane is installed on the starboard side of the rescue deck above the moonpool, with a reach of 3.5 m beyond the ship's side.

The stern is equipped with an A-frame with a load capacity of five tons, as well as a three-ton towing winch and a five-ton towing winch. Additionally, there is a 20-ton folding arm crane and a 500-m hydrological winch. Other equipment includes a laser range finder, stabilized gyrocompass, underwater acoustic communication apparatus, integrated navigation console of C-band and F-band satellite communication stations, depth sounder, electronic chart plotter, and low-noise power supplies of 380 V, 220 V, and 24 V. The vessel also features UPS power supplies, a flotation condition adjustment system, and a ship-wide monitoring system.

The test deck at the stern of "Dongyuan 02" covers an area of 120 square meters and is equipped with five laboratories including two located on each side in the midship section and one in the stern, totaling an area of approximately 125 square meters. The vessel also features a large meeting room, a test command center, a dining room, and other functional compartments and supporting facilities. Despite the additional weight of 23.8 tons resulting from the refit, the overall performance of "Dongyuan 02" has not significantly changed through calculation, except for a reduction in additional draft. The adaptive refit for different tasks greatly expanded the scope of offshore support missions that the vessel can undertake, effectively utilizing the advantage of the catamaran vessel's spacious working deck for operations.

2.8 *Multipurpose Survey Vessel "Dongyuan 3"*

The multipurpose survey vessel "Dongyuan 3" (also known as "Beidiao 993") (see Fig. 9) was designed by the China Ship Development and Design Center. It is a steel monohull with a twin-engine and twin-propeller propulsion system propelled by medium-speed diesel engines, which was delivered for use in 2001. This vessel is characterized by its excellent maneuverability, high adaptability for modifications, stable propulsion system, spacious working deck, and advanced special equipment

Fig. 9 Multipurpose survey vessel "Dongyuan 3"

technology. In 2012, the vessel underwent a refit to its fuel and water tanks, resulting in improved technical specifications such as extended endurance and self-sustaining period. It is capable of conducting marine environment surveys, geological explorations, and other missions along the coastal and southern regions of the South China Sea.

The vessel was designed with a length of 86 m, a molded breadth of 14 m, a molded depth of 6.8 m, a maximum draft of 3.9 m, and a displacement of 2,325 tons, a speed of 16 knots, an endurance of 2,000 nautical miles, a self-sustaining period of 20 days, and a capacity of 20 persons.

The bulb bow compartment of the vessel is made of special materials with acoustic transmission capabilities, with a horizontal range exceeding $\pm$ 90 degrees. It is also equipped with an open water well with a diameter of 6.4 m, located between FR 81# and 91#. Additionally, there is a 20-ton lift-and-turn device with a maximum travel distance of 9 m underwater.

A 200-kilowatt bow thruster is installed for easy maneuvering during docking and static testing. It is also fitted with an 8-ton telescopic crane with a span of 12 m at the bow, and a 3-ton crane with a span of 13.5 m at the stern. Three manual rotary hydrological winches, with capacities of 100 kg, 200 kg, and 300 kg, are available at the stern starboard, and on the larboard and starboard sides amidships. Low-noise power sources of 380 V, 220 V, and 24 V, as well as a UPS power source are equipped. The stabilized gyrocompass is equipped with underwater communication devices, electronic charts, integrated navigation display consoles, ship-wide monitoring system, and ship communication and navigation equipment.

Oceanographic equipment includes the UCTD, ship-based boundary layer and wave monitoring system, 10,000-m test instrument, and ADCP. The bow is equipped with a 1,500-m hydrological crane, and the stern is equipped with a 3,000-m geological sampling winch, a hydraulic crane, a gravity cable winch, a linear array winch, a 5-ton gantry, and other test towing facilities.

The vessel has four laboratories at front, middle and rear parts of the ship, covering a total area of 120 square meters. It also has a dedicated meeting room, a dining room, and a test command center, among other facilities.

2.9 Geological Survey Vessel "Xingye"

The geological survey vessel "Xingye" (see Fig. 10) was completed and delivered in March 2001. The vessel is owned by Shanghai Offshore Oil Geophysical Exploration Co. Ltd.

This vessel was designed with a length of 43.5 m, a molded breadth of 7.6 m, a draft of 2.8 m, a displacement of 455 tons, a speed of 12.5 knots, a working water depth of approximately 400 m, an endurance of 3,500 nautical miles, a self-sustaining period of 30 days, and a capacity of 24 persons (including 10 researchers). It is powered by two 6190ZLC diesel engines with a total power of 330 kilowatts, and three 1 Fc6 generators with a combined power of 64 × 3 kilowatts. The scientific research equipment includes a multi-beam depth sounder system used for gravity and magnetic and multi-beam surveys, as well as sampling. It can also be used for shallow geophysical exploration. Currently, the vessel has joined the national fleet of oceanographic survey vessels. The "Xingye" vessel is mainly used for marine geological multipurpose surveys, marine engineering geological surveys, geophysical exploration, and

Fig. 10 Geological survey vessel "Xingye"

multi-beam sounding. It has participated in national marine projects such as multi-beam sounding surveys, underwater pipeline surveys in the Chunxiao gas field in the East China Sea, surveys of the Canxue three-well field in the East China Sea, multi-beam submarine topography surveys of the western continental shelf of the East China Sea, gas pipeline inspections in the Pinghu oil and gas field, and 1:1 million-scale sub-bottom surveys in Shanghai. "Xingye" has carried out multi-beam sounding and side-scan sonar scanning tasks in the Yellow Sea and East China Sea.

2.10 *Offshore Research Vessel "Chuangxin"*

The offshore research vessel "Chuangxin" (see Fig. 11) is a small research vessel funded by the Institute of Oceanography of the Chinese Academy of Sciences (IOCAS). It is used for scientific exploration in Jiaozhou Bay and offshore areas of Qingdao, as well as for transportation and supply missions. The construction and operation of this vessel provides strong support for the smooth progress of offshore scientific research.

This vessel was designed with a length of 17.52 m, a molded breadth of 4.5 m, a molded depth of 2 m, a draft of 1.2 m, a gross tonnage of 44, a maximum speed of 15 knots, a minimum speed of 4 knots, a self-sustaining period of 4 days, and a capacity of 33 persons. The vessel was built by Jiangsu Wuxi Hongsheng Shipyard. With the construction starting in October 2005, it was launched in May 2006 and delivered to the IOCAS in September 2006.

The vessel is equipped with a twin-engine and twin-propeller propulsion system propelled by two Styer engines, with a generator power of 140 kilowatts. It has two laboratories, equipped with a moonpool and a running water tank, which can be used for deploying various underwater instruments and conducting various tests and analyses. An electric sampling winch and a small boom are arranged on the fore and

Fig. 11 Offshore research vessel "Chuangxin"

aft decks, and a 0.5-ton telescopic small folding arm crane is arranged on the aft deck for deploying equipment. The fore and aft compartments are equipped with working benches, and the wheelhouse is equipped with an advanced navigation system, as well as navigation aids such as GPS, depth sounder, and radar, making the operation of the entire vessel smooth and convenient.

From October 17 to 19, 2006, the offshore research vessel "Chuangxin" conducted its first scientific survey in Jiaozhou Bay, involving a total of 21 crew members and monitoring personnel. During the survey, the vessel demonstrated good overall performance. In this maiden voyage survey, the vessel traveled over 140 km at sea, conducted sampling at 14 stations, collected samples from 30 water layers, including plankton, phytoplankton, chlorophyll, transparency, and 5 nutrients, obtaining a total of 400 samples. It also acquired continuous CTD data at depths of up to 160 m. The mission lasted three days and was successfully completed.

2.11 Marine Geological Research Vessel "Ye Zhizheng"

Mr. Ye Zhizheng, a renowned Chinese marine geologist and educator, is a senior academician of the Chinese Academy of Sciences and a pioneer in the fields of sedimentary geology and marine geology. He is also a leading advocate for the study of paleooceanography in China and has made significant contributions to the development of the marine geology. To honor his illustrious career and carry forward his fine style of work and scholarly spirit, the Qingdao Institute of Marine Geology named their newly constructed marine geological research vessel "Ye Zhizheng" (see Fig. 12).

This vessel was designed with a length of 57 m, a molded breadth of 8.8 m, a molded depth of 3.7 m, a draft of 2.6 m, a displacement of 705 tons, a maximum speed of 16 knots, an endurance of 2,000 nautical miles, a self-sustaining period of 25 days, and a capacity of 36 persons (including 16 scientific researchers). Constructed by Rushan Shipyard in Shandong, the vessel was delivered in November 2004.

The vessel is a steel single-decked marine geological survey vessel with a long forecastle, square stern, twin engines, twin propellers, and twin rudders driven by diesel engines and propelled by propellers. Its designated working area includes the offshore, Bohai Sea, Yellow Sea, and East China Sea regions of China. Its power system is composed of two LB62502C9 diesel engines, each with a rated power of 1103 kilowatts, and they are paired with a 2:1 reduction gearbox.

The vessel is equipped with a multi-beam sounding system, a marine gravity measurement system, and a marine magnetometry system. In addition, the vessel is also fitted with a single-channel seismic survey system and various scientific research instruments and equipment that meet international advanced standards. It is capable of conducting seismic surveys, geophysical studies, engineering exploration, and ocean zone environmental geological surveys in offshore areas. All this makes it a technologically advanced and well-equipped specialized vessel for offshore scientific research operations.

Fig. 12 Marine geological research vessel "Ye Zhizheng"

The stern of the vessel is fitted with a hydraulic A-frame that can be tilted forward or backward. With a lifting capacity of 2.5/8.5 tons, the A-frame ensures a distance of 2.5 m from the lifting line center to the rear of the vessel when tilted backward. Additionally, a 2.5/8.5-ton electric cable winch and electrical control box are positioned within the midship area for sampling operations of the A-frame.

The "Ye Zhizheng" is capable of carrying out various types of surveys, including multi-beam surveys, gravity surveys, magnetic surveys, single-channel seismic surveys, sub-bottom profiling, side-scan sonar scanning, and geological sampling activities. It is well-equipped to undertake national marine geological survey tasks in offshore and coastal regions.

The commissioning of "Ye Zhizheng" has not only raised the equipment standards for marine geological scientific surveys in China, but also enhanced the oceanographic survey capabilities of the Qingdao Institute of Marine Geology. Furthermore, it provides invaluable means and reliable support for the integration of marine geological surveys with scientific research. Moreover, it provides equipment for marine engineering geological surveys and marine environmental geological surveys.

On July 2, 2005, the "Ye Zhizheng" vessel departed from Qingdao Port, embarking on its inaugural mission for a 50-day national special marine geological survey. In June 2007, "Ye Zhizheng", along with three other vessels, departed from the dock and embarked on survey missions in the southern Yellow Sea and Chengshantou areas.

On November 5, 2009, following a 177-day, 30,000-km journey, "Ye Zhizheng" returned to Qingdao Port after successfully completing a survey mission in the Zhoushan area of the East China Sea. This voyage overcame complex sea conditions and achieved the objectives of a 1:1 million survey of freshwater resources in the Zhoushan area, surpassing the initial survey targets set in early 2009. On June 5, 2015, the vessel sailed into the waters adjacent to Huangwei Yu and the Equator for oceanographic surveys, followed by a successful completion of its scientific research mission.

2.12 *Multipurpose Marine Environment Survey Vessel "Xiangyanghong 08"*

The new generation vessel "Xiangyanghong 08" (see Fig. 13) was funded by the SOA for construction. It is managed by the North China Sea Branch of the SOA. It is a 600-ton offshore survey vessel. The vessel was completed and delivered on November 25, 2008.

This vessel was designed with a length of 54.8 m, a molded breadth of 8.8 m, a draft of 2.4 m, a displacement of 608.3 tons, a speed of 11 knots, an endurance of 4,800 nautical miles, a self-sustaining period of 30 days, and a capacity of 24 scientific researchers. It is equipped with a bow thruster and adjustable pitch propeller.

Onboard the "Xiangyanghong 08" are facilities such as an integrated instrument laboratory (wet laboratory), a clean laboratory, a low-temperature sample storage room, a dry laboratory, and an aft laboratory. The deck is equipped with geological and hydrographic winches, as well as a gantry crane. Advanced geophysical survey equipment, including multi-beam system, side-scan sonar system, and mid-bottom profiling system, are installed.

The main tasks undertaken by the vessel include multipurpose surveys of marine environment in offshore areas, bathytopographic surveys, engineering geophysical surveys, and sediment sampling. The vessel's shipborne equipment mainly

Fig. 13 Multipurpose marine environment survey vessel "Xiangyanghong 08"

includes automatic marine environment monitoring system, satellite-based differential GPS positioning and navigation system, multi-beam sounding system, sub-bottom profiling detection system, and moving ADCP[1] observation system.

The marine ecological and environmental monitoring technology system onboard "Xiangyanghong 08" is a significant achievement of the "863" Program, enabling automatic sample collection, automatic water sample distribution, automatic analysis, and automatic data transmission.

On September 22, 2015, the vessel departed from Qingdao and returned on November 30, covering a total distance of 3,350 nautical miles over a span of 70 days. "Xiangyanghong 08" successfully completed oceanographic survey tasks for five projects. On November 24, while anchored in Pulaodian Bay in the Bohai Sea, the vessel encountered an extraordinary strong windstorm. With wind speeds reaching 26 m per second equivalent to Force 10 winds, the left anchor chain of the vessel snapped. The captain immediately ordered the emergency deployment of the right anchor. However, due to the intensity of the wind, the vessel started dragging its anchor, posing a threat to its safety. After urgently hoisting the anchor, the captain discovered that the right anchor had become entangled in fishing nets and then ordered the crew to clear them. At the time, the temperature at sea had dropped below freezing, and the crew endured the biting cold wind that felt like knives. From 10:00 am until 3:00 pm, the crew members persevered on the deck, tirelessly removing the majority of the fishing nets entangled in the anchor. When they finally returned to their compartments, the crew's fingers were frozen and numb, their perspiration had frozen on their hair, and their gloves were stuck to their hands, making it impossible to remove them. Seeing this, everyone onboard was deeply moved.

The crew of "Xiangyanghong 08" provided them a valuable experience through their actions. This spirit of heroic struggle can only be deeply understood by those who have experienced it together.

Since the 1960s, the North China Sea Branch has continuously conducted surveys of the Yellow Sea-Bohai Sea section, constantly adjusting the frequency and number of section surveys based on actual needs. Currently, four annual voyages are dedicated to the Yellow Sea-Bohai Sea section surveys. So far, a wealth of precious oceanographic survey data has been accumulated from these surveys, which serve as essential data for marine hydrological and meteorological forecasts, marine climate change assessments, the supervision and management of marine resources, and marine research.

On February 18, 2020, "Xiangyanghong 52" and "Xiangyanghong 08" set sail from the Qingdao Marine Research Base wharf of the North China Sea Branch of the Ministry of Natural Resources, embarking on a marine section survey mission in the Yellow Sea-Bohai Sea and Bohai Sea areas.

Marine section surveys are a fundamental long-term marine observation conducted in China. The purpose of is to conduct long-term continuous investigations of marine meteorological, hydrological, and routine chemical items through vertical

[1] Acoustic Doppler.

observation profiles in the surveyed area. It aims to acquire periodic and continuous observational data from standard sections in the offshore areas, and further to reveal the impact of human activities on China's offshore marine ecosystem. The Yellow Sea-Bohai Sea region serves as a focal point for interactions between land and sea, exhibiting complex physical and ecological environmental evolution characteristics, as well as typical regional responses. It is also an area prone to ecological disasters. Therefore, regular Yellow Sea-Bohai Sea section surveys carry significant importance.

During this voyage, "Xiangyanghong 08" conducted routine surveys of sub-bottom geological environmental elements in the Bohai Sea area. Survey tasks included multi-beam depth measurement, sub-bottom profiling, side-scan sonar survey, and sediment sampling. In the survey area, a comprehensive geological survey line spanning 120 km was completed, along with sampling at four stations. This survey further enhanced China's understanding of sub-bottom geological environmental elements in specific areas of the Bohai Sea.

2.13 *Multipurpose Survey Vessel "Runjiang 1"*

The multipurpose survey vessel "Runjiang 1" (see Fig. 14) was designed by Zhejiang Seahead Ship Design and Research Institute and constructed by Zhoushan Runhe Marine Science and Technology Development Service Co., Ltd.

Construction of this vessel began in 2009, and it was launched in April 2010 and delivered in July 2010. The vessel was designed with a length of 49.56 m, a length between perpendiculars of 44.75 m, a molded breadth of 8.8 m, a molded depth of

Fig. 14 Multipurpose survey vessel "Runjiang 1"

3.7 m, a draft of 2.8 m, a full load displacement of 659.2 tons, an endurance of 7,000 nautical miles, a self-sustaining period of 30 days, and a capability to withstand Force 8 winds. With these features, it is suitable for operations in offshore areas at water depths ranging from 6 to 300 m. The vessel is also equipped with a specially designed geological laboratory, a comprehensive laboratory, a frozen sample repository, a hydrological winch, a geological winch, a gravity sampler, and other specialized research equipment. Other equipment include the automatic identification system, GPS, depth sounder, navigation radar, satellite phone, and work boat. The vessel can accommodate 18 researchers to carry out continuous operations for up to 30 days. Since its operation, the vessel has completed over 20 research missions and received recognition from numerous research institutes and marine engineering companies.

In September 2020, the College of Optical Science and Engineering of Zhejiang University and its branch in Ningbo Innovation Center of Zhejiang University successfully developed the first domestic marine high spectral resolution laser radar. Together with the self-developed marine elastic laser radar system, it participated in a sail experiment organized by the Second Institute of Oceanography, Ministry of Natural Resources. The equipment was installed on the "Runjiang 1" vessel and underwent sea trials in the East China Sea and South China Sea. Researchers from institutes such as the Shanghai Institute of Technical Physics of the Chinese Academy of Sciences, Xiamen University, and Ocean University of China participated in the experiment.

2.14 Marine Archaeological Vessel "China Kaogu 01"

In order to investigate, excavate, and protect underwater cultural heritage, the Underwater Cultural Heritage Conservation Center of the National Cultural Heritage Administration entrusted the China Ship Development and Design Center to design the "China Kaogu 01" vessel (see Fig. 15). In October 2012, the National Cultural Heritage Administration signed a construction contract with Chongqing Changhang Dongfang Shipbuilding Industry Co., Ltd. The construction commenced in April 2013, and the vessel was launched in January 2014 and delivered in May 2014. The vessel is managed by the North China Sea Branch of the SOA, commissioned by the Underwater Cultural Heritage Conservation Center of the National Cultural Heritage Administration.

The vessel was designed with a length of 57.91 m, a molded breadth of 10.8 m, a molded depth of 4.8 m, a draft of 2.6 m, a displacement of 980 tons, a speed of 12 knots, a self-sustaining period of 30 days, an endurance of 1,000 nautical miles, and a capacity of 30 persons (including 8 crew members, as well as 22 specialized personnel comprising divers, salvagers, and archaeological scientists). "China Kaogu 01" features a three-floor superstructure. Apart from facilities like an underwater heritage conservation laboratory, diving workspace, pantry, kitchen, and dining room, it also provides nearly 20 compartments for accommodation, each equipped with a separate bathroom. The vessel adopts a fully electric Z-drive propulsion system,

Fig. 15 Marine archaeological vessel "China Kaogu 01"

which is both environmentally friendly and energy-efficient. It is equipped with a wide range of archaeological equipment, enabling the mapping, recording, photography, videography, and site clearance of underwater relics. Additionally, the vessel is fitted with a dedicated hydraulic folding arm crane capable of lifting cultural relics weighing up to three tons from the water. The entire deck is covered with teak wood flooring, which is waterproof and moisture-resistant. After these damp cultural items are brought on board, they are first placed on the first deck for drying and cleaning. The deck area available for drying cultural relics is 260 square meters.

The "China Kaogu 01" vessel is equipped with exploration equipment. When underwater cultural relics are detected during marine surveys, the vessel stops at the surface, and archaeologists dive underwater to conduct their work. The onboard hydraulic folding arm crane is designed specifically for cultural relics recovery. It resembles an arm that extends outwards the side of the vessel, enabling the lifting of underwater relics onto the deck. Moreover, there are two specialized compartments on board, namely the air-sealed compartment and the decompression chamber. The air-sealed compartment is intended for better handling of the relics by isolating them from the surrounding environment, and the decompression chamber assists archaeological divers in managing decompression during their submerged activities.

The vessel has three major features. Firstly, it is well equipped. The onboard equipment is double that of a general-purpose civilian ship, featuring facilities such as kitchen, dining room, underwater heritage conservation laboratory, and more. With two food stores, this vessel is capable of providing a 30-day supply of food for personnel on board. This enables continuous research work at sea for up to 30 days without worrying about food and drinking water shortages. Secondly, it exhibits exceptional stability. Despite weighing only 950 tons, the vessel can withstand winds and waves up to Force 8. When traveling at a speed of 12 knots, it can operate up to 200 km away from the coastline. The vessel can return to harbor within nine hours when sailing at full speed. To enhance stability, the center of gravity is lowered, and an anti-rolling tank is specially fitted to ensure a smoother sailing. While the vessel is underway, the anti-rolling tank can automatically adjust based on the magnitude and period of the vessel's rolling motion, providing a steadier sailing in rough seas and waves. Lastly, it features an air-sealed compartment for better protecting the cultural relics recovered.

The vessel is equipped with an archaeological equipment room which includes a submersible capable of accommodating six persons. This submersible has a top speed of 20 knots, allowing archaeologists to dive into the water and search for treasures.

"China Kaogu 01" has five major functions:

Firstly, it provides a support platform for underwater archaeology. In the past, underwater archaeological work mainly relied on rented fishing vessels. Compared to fishing vessels, the vessel offers a larger space and provides more comprehensive support for underwater archaeological activities.

Secondly, it is equipped with a decompression chamber. Diving is a high-risk operation, and in the event of a diving accident, the sooner the archaeological team members can enter the decompression chamber, the better the effect of relieving the pressure caused by the underwater environment on their blood pressure. The decompression chamber serves as an emergency platform for the safety of underwater operations.

Thirdly, it functions as an on-site platform for protecting the relics recovered. The vessel features an underwater relics protection laboratory, which is equipped with various chemical agents and processing equipment, providing strong support for the preliminary on-site protection of recovered relics.

Fourthly, it provides a work and rest platform for personnel at sea. Underwater archaeological work is physically demanding and typically has a short operating time. Archaeologists can only conduct underwater work during low tide, with a working period of about two to three hours per day. Therefore, efficiency is crucial for underwater archaeological work. The vessel can anchor near the work site, allowing archaeologists to return to the vessel for rest and wait for the next low tide, reducing the time required for the vessel to depart and return to port.

Fifthly, it provides a logistical support platform. The working hours of archaeologists underwater can vary each day. The vessel adjusts meal times to accommodate the work schedule of the archaeological team and pays attention to the variety of food options, striving to improve the quality of meals. It plays a role in providing logistical support.

On October 10, 2014, the "China Kaogu 01" vessel officially launched a one-month underwater archaeological project in the Dongkengtuo sea area of Tangshan Haigang Economic Development Zone in Hebei Province. Two ancient sunken ships were discovered by the archaeological department in the Dongkengtuo sea area, including the "Dongkengtuo 1" ship, which is the earliest dated and best-preserved copper-clad wooden ship from the late Qing Dynasty (1636–1912) to the Republican era in China.

In 2015, the Department of Tourism, Culture, Radio Television and Sports of Hainan Province, the Underwater Cultural Heritage Conservation Center of the

National Cultural Heritage Administration, and the People's Government of Sansha City jointly organized and implemented an underwater archaeological mission in the Xisha Islands. The archaeological work was conducted in the Yongle Atoll area. A team of 25 members, on board the fleet with "China Kaogu 01" as the mother ship, set sail from Qingshan Port in Wenchang and carried out a 45-day archaeological excavation. This expedition discovered a large quantity of underwater cultural heritage from various historical periods, including the "Shanhu Island No. 1" shipwreck and the "Jinyin Island No. 1" shipwreck, which contained a wealth of architectural components and stone carvings from the mid-Qing Dynasty.

2.15 *Research Vessel "Xiangyanghong 18"*

Research vessel "Xiangyanghong 18" (see Fig. 16) is one of China's technologically advanced research vessels among the 1,500-ton vessels. It is currently used for marine research in the Bohai Sea, Yellow Sea, East China Sea, South China Sea, and other marine regions.

Designed by the China Ship Development and Design Center and built by the Wuhan Shipyard, the construction began in 2010. It was launched in May 2015 and delivered in December the same year. Since then, it is managed and operated by the First Institute of Oceanography, Ministry of Natural Resources.

The vessel was designed with a length of 86.4 m, a length between perpendiculars of 77.8 m, a molded depth of 6.4 m, a molded breadth of 12.5 m, and a displacement of 1,900 tons, a cruising speed of 12 knots, a maximum speed of 15 knots, an endurance of 8,000 nautical miles, a self-sustaining period of 50 days, meeting the requirements for non-restricted waters navigation.

Fig. 16 Research vessel "Xiangyanghong 18"

"Xiangyanghong 18" has a total laboratory area of 264 square meters, including a general wet laboratory, a general dry laboratory, a general laboratory, a sample processing laboratory, and an acoustic instrument control room. The general laboratory also serves as the network center, the instrument room, the data processing center, and the wheelhouse for air-sea experiments. The vessel is also equipped with facilities for deck survey operations, suitable for scientific surveys in physical oceanography, atmospheric science, air-sea interaction, geology, geophysics, marine ecology, environmental protection, remote sensing, telemetering, coastal zones, and marine engineering.

The scientific objectives that can be achieved by "Xiangyanghong 18" include observing ocean dynamic processes and catastrophic events, investigating ocean circulation and climate change, exploring marine geology and geophysics, developing marine mineral, biological, and genetic resources, studying marine ecosystems and carbon cycling, protecting the marine environment and promoting sustainable development of the marine economy. It can also conduct marine engineering surveys, verifying the authenticity of marine remote sensing information, testing and calibrating high-tech marine research devices.

"Xiangyanghong 18" is capable of carrying out multipurpose deep-sea explorations in all major ocean regions, as well as major maritime missions related to ocean dynamics, geological environments, ecological environments, underwater resources, comprehensive energy exploration, and military oceanography. It provides technical support and guarantees for major marine scientific and technological issues such as marine resources, energy, national defense security, and mitigating the impact of natural disasters.

The specialized survey equipment aboard "Xiangyanghong 18" includes deep-water and shallow-water multi-beam systems, ultra-short baseline system, and automatic surface multi-element measurement system.

In addition, "Xiangyanghong 18" is equipped with an advanced all-electric propulsion system. At the time, this system had only been used on the research vessel "Kexue" in Qingdao. Since then, it has become a standard feature on newly built research vessels. After completion, the vessel was delivered to the China Oceanic Survey Fleet, playing an important role in the development of China's marine industry.

"Xiangyanghong 18" and "Xiangyanghong 01" formed a joint fleet that set sail from Qingdao to conduct sea trials for survey equipment in the waters of the Yellow Sea and other offshore areas in China. This sea trial marked the completion of the management and operation system for the research vessel by the First Institute of Oceanography. Since then, the Institute's research vessels have officially entered the operational phase, which is of great significance for advancing the Institute's scientific research work. As planned, "Xiangyanghong 01" would primarily carry out sea trials for the 480-channel seismic survey system. This system is the most advanced equipment among similar devices in domestic marine research institutions. It features complex technology and high integration difficulty, with important scientific value for future study on deep seabed geological structure as the core survey system of scientific research vessels. The marine multi-channel digital seismic system for this sea trial includes an integrated navigation subsystem, a data collection subsystem, an air gun seismic source subsystem, an underwater positioning subsystem, and a winch deployment subsystem. This voyage would require a three-kilometer-long data collection cable, which called for the escort of another vessel.

Therefore, "Xiangyanghong 18" was chosen for providing towing escort service for "Xiangyanghong 01". Then, it would also conduct comprehensive sea trials of specialized survey equipment, including the multi-beam system, ultra-short baseline system, surface multi-element continuous collection system, and control support system.

On October 7, 2018, the vessel "Xiangyanghong 18", under management of the First Institute of Oceanography of the Ministry of Natural Resources, set sail from Xiamen Cruise Wharf, marking the commencement of the "2018 Comprehensive Open Voyage" scientific expedition. Meanwhile, it conducted multidisciplinary surveys and comprehensive research in hydrology, geology, chemistry, biology, and remote sensing in the waters of eastern China. This expedition lasted for 35 days and covered a distance of over 4,200 nautical miles.

This comprehensive observation and systematic research in China's offshore waters provided valuable first-hand data for further exploration of the current state of natural resources and environmental changes, in-depth analysis of land-sea interactions, and the study of key marine environmental issues.

On January 20, 2021, "Xiangyanghong 18" set sail from the outfitted quay of Fujian Mawei Shipyard to the Northwest Pacific Ocean. After completing a trial voyage for the newly installed deep-sea steel cable winch, it docked at the deep-sea base terminal. This was the first voyage of "Xiangyanghong 18" in 2021, lasting for 15 days. The overall performance of the vessel as well as the performance of the newly installed 8,000-m steel cable geological winch were tested in the Northwest Pacific Ocean. Through sea trials, the overall performance and survey capabilities of the vessel were enhanced. The survey range was expanded to deep-sea oceans, including the East Indian Ocean and the Northwest Pacific Ocean, along with enhanced capabilities for conducting scientific research in deep and far sea areas.

2.16 *Test Mother Ship "Qingyan Haishi 1"*

"Qingyan Haishi 1" (see Fig. 17) is the first test mother ship serving for scientific research and technological development, which is jointly owned by the city of Shenzhen and Tsinghua University. Designed by the China Ship Development and Design Center and constructed by Tongfang Jiangxin Shipbuilding Co., Ltd., the vessel's design began in August 2016 and construction officially started in November 2016.

The vessel was designed with a length of 49.8 m, a molded breadth of 13 m, a molded depth of 4.4 m, a designed draft of 2.9 m, and a designed displacement of 1020 tons. It is classed by CCS, with a designated navigation area in offshore waters. The vessel can hold 23 persons, including 11 crew members and 12 researchers.

Fig. 17 Test mother ship "Qingyan Haishi 1"

Under the conditions of deep calm water design draft and smooth hull, with sea conditions ranging from grade 0 to 2 and winds not exceeding Force 3, the vessel can achieve a designed speed of 11 knots when the main engines operate at 95% MCR. When operating at economical speed, the vessel has an endurance of 2,000 nautical miles and a self-sustaining period of 30 days. Its wind resistance capability is Grade 12, meeting the requirements specified in the *Special Wind Resistance Requirements for Marine Survey Ships* (CB/T3526-1994).

The vessel is equipped with a three-meter by two-meter rectangular moonpool on its main deck. In addition, it features four-anchor positioning capabilities, satisfying the anchor moored positioning requirements at a water depth of 40 m and in sea conditions of grade 4. With a DP-1 dynamic positioning system, the vessel can achieve a positioning accuracy of 1 m in sea conditions of grade 4. A passive anti-rolling tank and bilge keel are installed to reduce the vessel's rolling amplitude and improve its seakeeping capability. The vessel's vibration and noise limit levels meet the requirements specified in the *General Requirements for Vibration and Noise of Laboratory and Survey Decks of Marine Survey Ships* (CB/T3527-1993). Besides, the vessel's main deck is equipped with a hydraulic folding arm crane with a 10-ton working load and a reach of 13 m. The vessel adopts a round bilge hull form, with an inclined and square stem, twin engines, and twin adjustable pitch propellers, along with a bow thruster.

The primary mission of "Qingyan Haishi 1" is to serve the Public Service Platform of the Shenzhen Marine Equipment Offshore Test Field. The commissioning ceremony for the mother ship signified not only the activation of "Qingyan Haishi 1" but also the activation of the internationally advanced marine engineering test and research platform, integrating the offshore test field, indoor engineering water tank, and supercomputer.

On October 29, 2018, "Qingyan Haishi 1" completed partial navigation test projects in the Jiujiang segment of the Yangtze River Basin. The vessel's test speed, equipment performance, vibration, noise, and other indexes met the requirements of the design assignment. From November 12 to 13, 2018, the vessel underwent sea trials in Shanwei, Guangdong. From April 28 to 30, 2019, it successfully completed its first scientific research expedition in the Pearl River Estuary, conducting a multipurpose survey of the Pearl River for its physical oceanography, as well as marine biological and ecological environment. The vessel completed surveys at a total of 21 observation stations, with all performance indexes in line with the design requirements.

2.17 *Oceanographic Survey Vessel "Xiangyanghong 21"*

"Xiangyanghong 21" (see Fig. 18) is a multipurpose oceanographic survey vessel built by the East China Sea Branch of the Ministry of Natural Resources. It was built to better meet the demands of modern oceanographic survey, expand the scope of surveying areas, and enhance oceanographic survey capabilities.

Designed by the China Ship Scientific Research Center and built by the Wuchang Shipyard, the vessel is operated by the East China Sea Branch of the Ministry of Natural Resources. During the design process, the design team initiated the concept of "comprehensive, information-based, user-friendly, and integrated design". This approach was meticulously implemented, ensuring the vessel's overall performance and technological superiority. "Xiangyanghong 21" is another outstanding product resulting from the Center's long-term dedication to the field of research vessels. More remarkably, the vessel has become the mainstay of China's 1,000-ton research vessels, distinguished by its excellent economic efficiency and overall navigability.

Fig. 18 Oceanographic survey vessel "Xiangyanghong 21"

The vessel was designed with a total length of 79.8 m, a molded breadth of 11.6 m, a molded depth of 5.3 m, a draft of 4 m, a displacement of 1925 tons, an endurance of 8,000 nautical miles, a self-sustaining period of 50 days, and a speed of 15 knots. Its main task is to conduct multipurpose surveys in marine topography, geomorphology, geophysics, geology, hydrology, and pipelines.

"Xiangyanghong 21" features several unique designs to enhance the efficiency of oceanographic surveys:

(1) It is equipped with full-speed anti-rolling fins, providing enhanced adaptability to adverse sea conditions and ensuring the reliability of instrument testing data.
(2) The living compartments and working compartments onboard meet the vibration and noise indexes required by CCS for optimum comfort.
(3) The vessel is fitted with scientific survey system and research equipment, including a 6,000-m geological winch, a 2,000-m hydrological winch, a 1,000-m magnetometer winch, a 2-ton folding arm crane, and an A-frame, enabling it to fulfill the requirements of conventional oceanographic survey missions.

In order to adapt to the evolving needs for oceanographic surveys, the East China Sea Branch of the Ministry of Natural Resources once again entrusted the original design unit, the China Ship Scientific Research Center, to carry out a refitting design for the "Xiangyanghong 21" vessel in 2019. The refit project was then contracted to the Huanghai Shipyard. The refit commenced in January 2019 and was completed by July 2020.

The refitted "Xiangyanghong 21" vessel incorporates the advanced ABB pod electric propulsion system, improving the energy conversion efficiency, fuel efficiency, and maneuverability. Moreover, it is equipped with a DP1 dynamic positioning system. It enables the vessel to automatically maintain its latitude and longitude coordinates using GPS, providing invaluable support for instrument deployment and sample collection during deep-sea research. The overall performance and hull structure of the vessel meet the requirements of international conventions, current CCS specifications, and relevant standards. The refit enhanced the vessel's performance and reliability, elevated the level of automation and capabilities for marine monitoring and surveys, and strengthened the vessel's multipurpose survey capabilities. It also improved the working and living environment for the crew and researchers, and further extended the vessel's service life.

2.18 Research Vessel "Xiangyanghong 51"

The research vessel "Xiangyanghong 51", formerly known as the "China Haijing 1111" and "Shuguang 05", was completed in December 1972 and is now managed by the North China Sea Branch of the Ministry of Natural Resources. The vessel has always been at the forefront of China's marine monitoring, surveillance, and scientific research efforts. In June 28, 2019, the vessel underwent a large-scale technical refit by

Fig. 19 Research vessel "Xiangyanghong 51"

Tianjin Xingang Shipbuilding Heavy Industry Co., Ltd. The project involved significant technological renovations while retaining some of the original vessel's steel structure and equipment. The overall refit principles focused on economic feasibility, advanced rationality, safety and reliability, and ease of maintenance, while ensuring compliance with relevant regulations, standards, and the needs of conducting offshore marine resource surveys.

In August 2021, the refitted research vessel "Xiangyanghong 51" (see Fig. 19) was successfully delivered, marking a significant breakthrough for Tianjin Xingang Shipbuilding Heavy Industry Co., Ltd. in the field of ship reconstruction. The vessel is classed by CCS and is primarily used for specialized survey and research in marine geology, hydrology, sound field, and other fields.

As a research vessel, "Xiangyanghong 51" is equipped with a variety of equipment, including complex electrical and piping systems, and scientific research equipment with high requirements for installation precision. The project team implemented meticulous planning and refined management to ensure the smooth progress of the refit preparations. They strictly followed the production schedule, successfully completing the refit. This project provides valuable experience for future construction of similar scientific research vessels and contributes to the development of China's marine survey and research endeavors. The vessel was designed with a length of 74 m, a molded breadth of 10 m, a molded depth of 4.7 m, a capacity of 50 persons, a full-load displacement of 1,200 tons, and a maximum speed of 14 knots.

As a research vessel, "Xiangyanghong 51" meets the needs of multipurpose surveys of marine resources and the environment in its jurisdictional waters. It supports operations of ROVs at depths of up to 1,000 m, large-scale underwater towed body moving and towing surveys, as well as 1,000-m bottom sampling and in-situ rock-soil testing operations. In addition, it is fitted with the capability of dynamic positioning DP-I. In the future, "Xiangyanghong 51" will mainly be engaged in meteorological observations, hydrological surveys, marine gravity field research, marine sound field research, marine geology surveys, and other specialized investigations in China's waters. It is primarily used for cruising and monitoring the exclusive economic zones, continental shelves, and other jurisdictional waters, playing an important role in safeguarding China's national maritime rights and interests.

2.19 Oceanographic Survey Vessels "Xiangyanghong 02" and "Xiangyanghong 03"

The first-generation oceanographic survey vessel, "Xiangyanghong 02" (see Fig. 20), was designed in the 1960s as a marine hydrological and meteorological survey vessel. Guangzhou Shipyard began construction of the two vessels in 1968. In May 1972, they were launched and delivered to the South China Sea Branch of the SOA. The two vessels were named "Xiangyanghong 02" and "Xiangyanghong 03", respectively.

The new-generation oceanographic survey vessel, "Xiangyanghong 03", was designed by the MARIC and built by the Wuchang Shipyard. With construction starting in June 2013, the vessel was delivered for use in March 2016 and is now managed by the Third Institute of Oceanography, SOA. The vessel was designed with a length of 99.6 m, a molded depth of 8.9 m, a molded breadth of 17.8 m, a designed draft of 5.6 m, a designed displacement of 4,800 tons, a designed maximum speed of 15 knots, and an endurance of 15,000 nautical miles. The design and construction of the vessel fully considered the features of the scientific research equipment and the hull structure, such as embedded installation for the large multi-beam sounding equipment. It adopts an all-direction electric propulsion system with strong maneuverability. In addition, the vessel has dynamic positioning DP1 capability and can withstand winds of up to Force 12, allowing for high-precision positioning in any sea area. It is equipped with shipborne control devices, testing and detection systems,

Fig. 20 Oceanographic survey vessel "Xiangyanghong 02" (First Generation)

a ship computer network system, and other systems. Remarkably, it serves as a modern multipurpose oceanographic research vessel that meets the requirements of non-restricted waters navigation. Integrating multiple disciplines, functions, and technological means, it caters to the demands for interdisciplinary marine research. The vessel also features scientific survey capabilities covering marine dynamic environment, geological environment, ecological environment, seabed resources, and energy exploration and experiment.

"Xiangyanghong 03" (see Fig. 21) stood as one of the most advanced multipurpose oceanographic research vessels in China at the time with its world-class observation capabilities. Since being put into service, it has undertaken nearly one hundred voyages for major ocean investigations in China. On July 14, 2018, "Xiangyanghong 03" departed from Xiamen and conducted approximately a 150-day ocean scientific expedition in the Western and Eastern Pacific regions, covering a distance of approximately 15,000 nautical miles. In December 2018, it successfully completed the mission and returned to Xiamen.

During the research, "Xiangyanghong 03", with its unique "thousand-mile eye" equipment, could explore deep into the mysterious underwater world and acquire important marine monitoring data. The "6,000-m integrated optical towed body" (referred to as the "optical towed body") (see Fig. 22) resembles an iron cage in appearance, but it houses a HD camera, an attitude sensor, a pressure sensor, an altimeter, an illumination light, and a laser ruler, making it a highly integrated underwater observation platform.

The optical towed body is primarily used for rapid towing operations in deep-sea environments, which adapts well to various complex seabed terrains and possesses multi-functional submarine hydrothermal investigation capabilities. It integrates techniques of submarine hydrothermal anomaly detection technique and seabed high-definition camera technique, allowing for underwater filming and photography. It can

Fig. 21 Oceanographic survey vessel "Xiangyanghong 03" (New Generation)

Fig. 22 6,000-m Integrated optical towed body

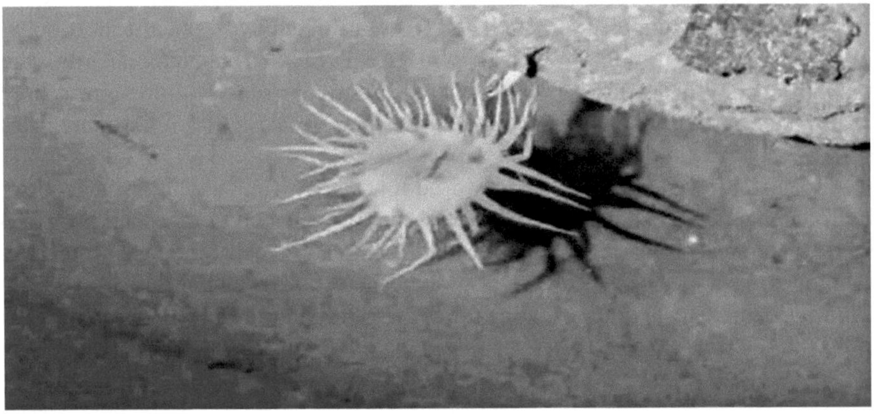

Fig. 23 Deep-Sea organisms detected by the optical towed body

also collect, transmit, display, and record GPS and other relevant data from the underwater electronics compartment and the onboard laboratory in real time. With the optical towed body deployed, scientists aboard "Xiangyanghong 03" can observe the vast underwater plains, rugged oceanic ridge mountains, and a wide range of deep-sea organisms in the ocean (see Fig. 23).

2.20 *Large Buoy Operation Vessel "Xiangyanghong 22"*

In December 2019, China's first independently designed and built 3,000-ton large buoy operation vessel, "Xiangyanghong 22" (see Fig. 24), constructed by CSSC

Fig. 24 Large buoy operation vessel "Xiangyanghong 22"

Wuchang Shipbuilding Industry Group, was delivered to the East China Branch of the SOA. This vessel is the first large buoy operation vessel in China, with the capability to handle large ocean monitoring buoys with a diameter of 10 m and a weight of 60 tons. It is also one of the vessels with the strongest lifting and hoisting capabilities internationally.

The vessel was designed with a length of 89 m, a molded breadth of 18 m, a molded depth of 7.2 m, and a maximum speed of not less than 16 knots. It has an endurance of 10,000 nautical miles and is capable of global navigation, with the ability to withstand winds up to Force 12. In addition to fulfilling marine monitoring and support missions, the vessel itself also possesses the capability to conduct section surveys and other research tasks. Its peer, "Xiangyanghong 31", completed a trial voyage in 2022.

2.21 Multipurpose Oceanographic Survey Vessel "Xiangyanghong 28"

The first generation of China's multipurpose oceanographic survey vessel, "Xiangyanghong 28" (see Fig. 25), designed by the MARIC and constructed by Wuchang Shipyard, was launched in June 1986 and delivered in October of the same year.

The-new generation "Xiangyanghong 28" multipurpose oceanographic survey vessel is under the management of the East China Branch of the SOA. It is an advanced vessel that underwent a five-month adaptive refit after serving as a hydrographic survey vessel for 32 years. After the refit, the vessel was officially put into service

Fig. 25 Multipurpose oceanographic survey vessel "Xiangyanghong 28"

in November 2014, becoming a new force within the East China Sea regional fleet for marine survey and exploration.

The vessel was designed with a total length of 74.55 m, a molded breadth of 10 m, a maximum draft of 3.5 m, a displacement of 1,216.7 tons, an endurance of 4,000 nautical miles, and a speed of 16 knots.

"Xiangyanghong 28" is propelled by dual main engines, dual propellers, and dual rudders, providing excellent maneuverability, as well as advanced and reliable communication and navigation capabilities. The living facilities and working environment on board are also significantly improved after the refit. The vessel is equipped with several laboratories for biological, chemical, geological, and hydrological purposes, as well as a comprehensive data center. It is also equipped with the most advanced survey equipment in China at the time, including the mid- and deep-water multi-beam sounding systems, geophysical exploration equipment, depth sounder (10,000 m), and water sampling and distribution system, which greatly improved the efficiency of oceanographic survey operations, and significantly enhanced the SOA East China Branch's capabilities for oceanographic survey in the adjacent and offshore waters.

On May 9, 2016, during the execution of marine ecological environmental monitoring tasks in the East China Sea, the East Sea Environmental Monitoring Center of the SOA discovered a red tide in the waters east of Huaniao Mountain in Zhoushan, Zhejiang at 30 degrees 50 min north latitude and 123 degrees east longitude. The area affected by the red tide was approximately 470 square kilometers. The vessel initiated emergency monitoring of the red tide and reported it immediately. On May 10, 2016, "Xiangyanghong 28" sailed past the East Sea Bridge. While conducting marine ecological environmental monitoring in the East China Sea, the vessel returned from the waters of Shengsi, Zhejiang, back to the Yangtze River Estuary. The research team focused on monitoring the health conditions of the Shanghai Yangtze River Estuary and the areas near the "trumpet mouth" of Hangzhou Bay.

Chapter 7
Enterprising Design Teams

Since the establishment of New China, the country's research vessels have gone from nonexistence to being at the forefront of the world, from being small to becoming large, and from being backward to making great strides. This achievement is attributed to the leadership of the Party and the advantages of mobilizing resources under the socialist system with Chinese characteristics. It is the result of self-reliance, hard work, and perseverance, as well as the efforts of the research vessel design teams. Below is the introduction to two of these teams, aiming to showcase their contributions to the development of Chinese research vessels.

1 Design Team for the "Xiangyanghong 10" Oceangoing Survey Vessel

"Xiangyanghong 10" (see Fig. 1). The development of this oceangoing survey vessel was a systematic project. Eight years was spent from 1971 to 1979 before the vessel was successfully delivered. It was the result of collaboration between various organizations including the user, the designer, and the constructor,

as well as the collective intelligence and efforts of them. Leaders and scientific personnel from the MARIC formed a team of nearly a hundred people, who have made significant contributions to the birth of the research vessel "Xiangyanghong 10".

Among all the ships of its kind in the world at the time, "Xiangyanghong 10" stood out for its impressive tonnage and most complete facilities. The design team overcame numerous technical challenges in designing such a high-performance oceanographic research vessel.

In April 1976, "Xiangyanghong 10" was launched, as shown in Fig. 2.

The first challenge was to address the problem of electromagnetic interference and communication. The powerful RF (radio frequency) magnetic field generated by the

L. Han et al., *History of the Research and Development of Chinese Marine Research Vessels*, https://doi.org/10.1007/978-981-96-8414-4_7

Fig. 1 Oceangoing survey vessel "Xiangyanghong 10"

Fig. 2 "Xiangyanghong 10" in launching in April 1976

vessel's 30-kilowatt high-power shortwave transmitter caused non-linear wideband RF interference, presenting a significant issue of electromagnetic compatibility that would affect the functioning of communication equipment. Through joint efforts, the design team, along with collaborating factories, successfully ensured the harmonious coordination of operations such as short- and medium-term weather forecasting, severe weather alerts, ship-to-shore communication relay, information transmission, hydrographic surveys, and gravity investigations. During the "580" test, the team achieved accurate meteorological support and faultless signal communication with 100% accuracy.

The second challenge was to ensure ship stability. As the survey vessel would frequently operate in adverse sea conditions, it had to boast excellent stability, capable of withstanding winds up to Force 12. The design team employed various methods for calculation and comparison, studied relevant foreign regulations, calculation methods, and their evolution and trends, ultimately working out applicable wind pressure calculation formulas, values, as well as correction methods for initial heel angle.

The last challenge was to solve the problem of ship vibrations. Given the vessel's mission requirements, a significant number of precision instruments, large antennas, and tall masts were equipped onboard. Vibration of the ship could cause severe damage to these instruments and equipment. Through collaboration and strenuous efforts, the design team successfully solved numerous vibration-related challenges.

In April 1980, the oceangoing survey vessel "Xiangyanghong 10" embarked on its inaugural long-distance voyage, participating in a long-range carrier rocket full-range test. The vessel achieved high-altitude and surface meteorological observations, accurate weather forecasting, unimpeded communication, precise navigation operations, normal functioning of electromechanical equipment, and safe maneuverability.

During China's first-ever geostationary communications satellite launch test, "Xiangyanghong 10" ventured into the Pacific Ocean three times, ensuring smooth, accurate, and timely communication transmission and information relay. It also performed gravity surveys, hydrographic measurements, and provided weather forecasts for the experimental sea area for future 12, 48, and 72 h, achieving an accuracy rate of over 83%. The team aboard the vessel received two collective second class merits.

To fulfill the ambitious strategic objective proposed by the SOA, which emphasized the need to "fully investigate China's seas, venture into the three oceans, and reach Antarctica", "Xiangyanghong 10" undertook numerous expeditions to Antarctica. In December 1984, the vessel embarked on its maiden voyage to Antarctica, playing a crucial role in scientific exploration and station establishment in the region. Over a span of 142 days, it traversed 26,000 nautical miles in the Southern Ocean. The ship safely returned to Shanghai Port and the team received a collective first class merit.

On January 26, 1985, while conducting ocean surveys near the Antarctic Circle, "Xiangyanghong 10" encountered a severe Antarctic storm with winds surpassing Force 12, with a maximum wind speed recorded as 34 m per second. Towering waves, resembling mountains, surged on the sea surface one after another, with wave heights reaching 13 m and wave lengths of over 100 m. Foam and spray whizzed past the bridge, which stood over 30 m tall, and the fore and aft decks of the vessel were periodically engulfed by surging waves. At times, the vessel was lifted by the waves, exposing the propellers above the water surface, causing the vessel to tilt at a maximum angle of 31 degrees, accompanied by substantial vibrations and noise.

Braving the ferocious wind and waves, members of the expedition team courageously fought to salvage and protect survey equipment, instruments, and specimens. Some were even swept off their feet by surging waves over a meter high, and luckily enough, they were rescued by their colleagues. After more than 20 h of struggling, the entire expedition team and crew managed to escape the perilous situation and return safely. The vessel's resilience in enduring a storm with winds exceeding Force 12 set a precedent in the history of marine research in China.

Following seven days of meticulous inspection, it was confirmed that the vessel's control system, power system, fire protection system, hull structure, and overall performance remained in excellent condition.

In 1979, the vessel was awarded the first prize of the Major Scientific Research Achievement Award by the State Administration of Science, Technology and Industry for National Defense (SASTIND).

In April 1980, China's first long-range carrier rocket full-range test in the Pacific Ocean, participated by the vessel, made a hit.

In April 1984, the vessel participated in China's first launch of a geostationary experimental communications satellite and achieved success.

In November 1984, the vessel embarked on its first expedition to the Southern Ocean. The team members established the Great Wall Station, China's Antarctic research station, on George Island. It marked the first time the five-starred red flag flew above Antarctica.

In May 1985, as assessed by the National Science and Technology Progress Award Evaluation Committee, the design unit of the "Xiangyanghong 10" ocean-going survey vessel was awarded the National Science and Technology Progress Award Special Prize (refer to Fig. 3).

In 2006, "Xiangyanghong 10" was recognized as one of China's top ten famous vessels.

2 Design Team for the Multipurpose Oceanographic Research Vessel "Kexue"

In April 2014, a magnificent research vessel set sail, fully loaded with scientists, embarking on a scientific expedition to the deep blue ocean. This vessel, named "Kexue", was a multipurpose oceanographic research vessel designed and developed by the design team from the MARIC. Over the years, "Kexue" has successfully completed a number of national scientific research missions, such as exploring hydrothermal fluids, cold seeps, and seamounts in the Western Pacific, leading to a series of important research findings. It became the first vessel to autonomously observe active hydrothermal vents in the Okinawa Trough, producing high-resolution underwater topographic maps. Additionally, the vessel accomplished the groundbreaking feat of deploying 18 sets of large-scale deep-sea subsurface buoy arrays in the main currents of the Western Pacific, which marks a global first. It also conducted

Fig. 3 Certificate of the special prize of National Science and Technology Progress Award

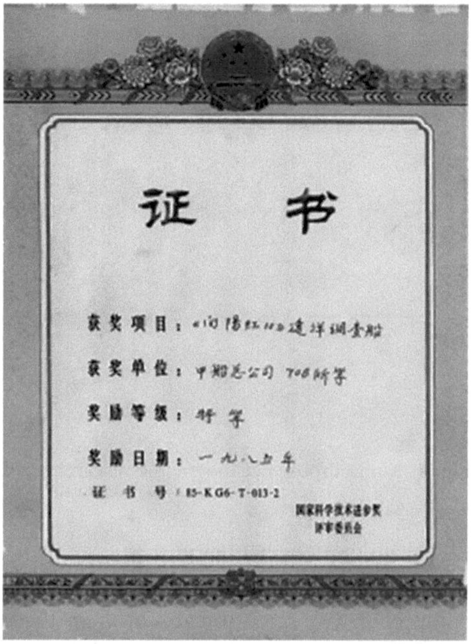

a pioneering large-scale geophysical survey in the south central regions of the South China Sea. These achievements have propelled China's deep and far-sea scientific research capabilities and level into the international forefront. As a research vessel with independent intellectual property rights, "Kexue" (Fig. 4) has broken numerous domestic and international records. It integrates multiple disciplines, functions, and technological means, allowing for real-time and synchronized deep-sea detection, truth-preserving sampling, on-site analysis, as well as information transmission and joint processing with the land-based team. It significantly enhanced China's overall capabilities and research standards in overall marine explorations, providing robust support for multipurpose oceanographic research and the advancement of the country's marine strategic layout. This signifies a substantial leap in China's marine research capabilities.

Conducting research on marine environments, marine biological resources, seabed geology, and resource exploration is vital for China's strategic goal of becoming a strong maritime nation. As a crucial platform for ocean exploration and research, multipurpose oceanographic research vessels play a key role in strengthening China's marine research capabilities and serve as a testament to a nation's comprehensive strength. However, most of China's existing research vessels were built in the late 1970s and early 1980s, with relatively outdated overall performance and detection equipment. Most notably, they lacked the capability to undertake complex deep-sea comprehensive scientific research, which hindered the development of China's marine science.

Fig. 4 Multipurpose oceanographic research vessel "Kexue"

To implement China's strategy of becoming a strong maritime nation and strengthen the construction of marine research equipment, as well as to deepen comprehensive scientific research in the deep oceans, the National Development and Reform Commission approved the major project of the National Major Science and Technology Infrastructure—the research vessel in 2007.

Upon receiving the news, the MARIC, led by Academician Zhang Bingyan of the Chinese Academy of Engineering as the chief designer, actively carried out preliminary demonstrations and research work based on the national future development strategy. In order to meet the requirements of modern multipurpose oceanographic research, the team aimed to ensure that the vessel's overall performance would reach international advanced standards. Leveraging their years of experience in research vessel design, they targeted newly delivered or under-construction research vessels worldwide, digesting and absorbing technical knowledge from domestic and foreign counterparts. On this basis, they tackled key technologies, such as the design of an anti-bubble bulb bow for multi-beam system embedded installation on research vessels, an economically efficient hull design with an optimal length-to-breadth ratio, as well as vibration and noise reduction and electromagnetic compatibility. In May 2009, through an open bidding process, the MARIC won the project with its advanced technology, detailed and precise biding documents, and high-level objectives.

For a complex and specialized research vessel like this, any change in the technical state of a system or equipment would involve multiple disciplines. To design a high-quality vessel, it is essential for different specialties to unite, cooperate closely, and work towards a common goal. Academician Zhang Bingyan, as the chief designer, led the team with a spirit of meticulousness and exploration. He excelled in researching and venturing into new fields, fearlessly taking risks and charting undiscovered territories. His commitment to the deep blue ocean was unwavering, making him an expert with innovative theories and extensive practical experience in the field of

Fig. 5 Academician zhang bingyan in speech at a design exchange meeting

oceanographic survey vessel design. Figure 5 shows Academician Zhang Bingyan in speech at a design exchange meeting.

At the first regular meeting of "Kexue" project held by the design team, Academician Zhang Bingyan emphasized that in addition to familiarizing themselves with the design specification, design personnel of all specialties should also pay attention to two aspects throughout the design process: weight and center of gravity, as well as vibration and noise reduction. During the meeting, he elaborated on the specific issues that each specialty should consider and focus on during the design phase of the vessel. For the overall specialty, emphasis should be placed on overall layout, the matching of rudders and stern lines, the matching of bow lines with acoustic equipment, as well as the control of weight and center of gravity. The structure specialty should pay special attention to reinforcing the structure of locations where large equipment would be installed, the structural openings for equipment such as ultra-short baseline system, multi-beam system, sub-bottom profiler, and lifting fin, as well as the control of weight, center of gravity, overall vibration, and local vibration. The outfitting specialty should arrange conventional outfitting equipment and scientific research equipment reasonably, and pay attention to the influence of the position of large equipment on the weight and center of gravity of the vessel. For the compartment specialty, attention should be given to the rational layout of living compartments and laboratories, the coordination of compartment height with needs of other specialties, the suitable selection and arrangement of vibration and noise reduction materials, as well as the application of lightweight materials. The marine engine specialty should ensure the reasonable arrangement of engine room facilities and the proper routing of pipelines, select suitable vibration and noise reduction solutions for these facilities, and control weight and center of gravity. The electrical specialty should focus on estimating the electrical load and configuring the power station, addressing electromagnetic interference and harmonic waves, as well as controlling and coordinating the selection of equipment. The HVAC (heating, ventilation, and air conditioning) specialty should concentrate on selecting the appropriate air ducts, determining their routing and layout, implementing measures for vibration and noise reduction, as well as coordinating with the ceiling height of the compartment. Academician Zhang Bingyan's forward-thinking mindset and rigorous approach during the design process, akin to that of a commander in preparation for a great battle, instilled a sense of confidence and meticulousness among the design team. Despite facing numerous challenges, he had a clear plan and confidently strategized to overcome them. The team members were greatly inspired by him. Figure 6 shows the scene of technical communication between the design team and the user unit.

Wu Gang, Deputy Chief Designer of the "Kexue" project, was not only responsible for the overall specialty but also oversaw the general layout of the scientific research system, the performance of podded electric propulsion system, the workflow of scientific research operations, and multiple other key technologies. He coordinated and stayed informed about the implementation of tasks in each specialty, assisted Academician Zhang Bingyan in solving technical challenges, and wholeheartedly devoted himself to ensuring the successful delivery of the vessel.

The functional diagram of the research vessel "Kexue" is shown in Fig. 7.

Due to the numerous equipment loaded on the research vessel, the user unit had high requirements for vibration and noise reduction. They specified that the measured noise values during the actual sea trial should be lower than the limit

Fig. 6 Scene of technical communication between the design team and the user unit

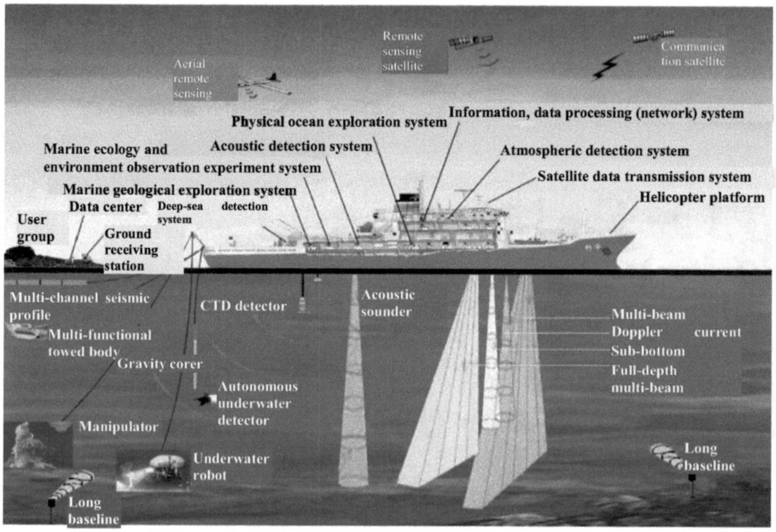

Fig. 7 Functional diagram of the research vessel "Kexue"

set by the national standard Regulation on *Noise Levels on Board Sea-Going Ships* (GB5979-1986). Academician Zhang Bingyan requested that each specialties take measures to address this issue from the beginning of the design process. All specialties cooperated and collaborated, implementing multiple measures to resolve this problem. For the overall specialty, consideration was given to the general arrangement of noise-generating equipment away from living and survey compartments to minimize noise transmission. At the beginning of design, the structure specialty made efforts to maintain the continuity of the hull structure, align upper and lower bulkheads, and strengthen the bases of large equipment to reduce vibrations. The interior outfitting specialty selected high-quality high-damping materials for the main engine bases, auxiliary engine bases, important compartments' decks, and bulkheads to reduce vibration and prevent noise propagation. Sound insulation materials were used between living compartments, public areas, laboratories, and corridors to reduce interference between spaces. The marine engine specialty employed double-layer vibration isolation materials or floating rafts for main diesel generator sets and harbor generator sets. In addition, the emergency generator sets were equipped with elastic vibration absorbers, and hatches leading to the outside of the engine rooms were capped to prevent noise propagation. The electrical specialty implemented a variety of measures for electromagnetic compatibility (EMC) and harmonic mitigation and control, ensuring the functionality of the vessel's experimental systems. The HVAC specialty adopted medium-pressure low-noise centrifugal fans with variable frequency control for air conditioner ventilation fans. Elastic connectors were used to connect the inlet and outlet pipelines of ventilation fans and water pumps. The airspeed in the ventilation system was controlled to reduce noise generation. Through joint efforts of all specialties, the vessel's vibration reduction and noise control

Fig. 8 Signing ceremony of the construction contract for the "Kexue" research vessel

measures passed the inspection of the constructor and user. During its trial voyage, the noise levels at all measurement points were found to be below the limits specified in the Regulation on *Noise Levels on Board Sea-Going Ships* (GB5979-1986) when the vessel was sailing at full speed. This achievement met the highest comfort standards specified for passenger ships, receiving satisfying feedback from the user. Figure 8 shows signing ceremony of the construction contract for the "Kexue" research vessel. Figure 9 shows the return of design team members after trial voyage.

The weight and center of gravity are critical factors that not only affect the stability, load capacity, and longitudinal strength of a research vessel but also significantly influence its seakeeping performance. This further impacts the effectiveness of the anti-rolling system and the helicopter takeoff and landing capabilities. From design to construction, the design team bore the concept of weight and center of gravity control in the whole process. Weight and center of gravity were estimated twice during the design phase, as required by the chief designer, and detailed calculations were performed after the detailed design phase. During the shipyard's production design stage, the design team paid continuous attention to the changes in weight and center of gravity, and elaborated to the constructor on the importance to strictly control them. Through the collective efforts of the design team and the shipyard, weight and center of gravity were effectively controlled as tested by the actual ship verification, which matched the design estimates very closely without using any fixed ballasts.

Through repeated analysis and calculations of various technical indicators, the design team conducted an analysis on the vessel's main dimensions and economic feasibility. They compared key indicators such as displacement tonnage, speed and resistance, endurance, and scientific research operating area, ultimately identifying

Fig. 9 Return of design team members after trial voyage

the optimal main dimensions and hull form coefficients. In the profile design of research vessels, apart from focusing on speed, wave resistance, and seaworthiness, specific attention should be paid to the optimization of the bow shape and its compatibility with the acoustic equipment.

The design team worked closely together, focusing on addressing key issues such as optimizing the layout of the scientific research system and the ship system, analyzing the economic feasibility of hull form, optimizing profile design, and reducing vibration and noise. At the design verification stage, the design team carefully classified, arranged, and demonstrated the future tasks and missions that the vessel would undertake, the requirements for shipborne equipment and its installation, the experimental space, modular functionality, and the need for expandable modifications. They mastered the interconnection and optimized arrangement of systems, ensuring the functional effectiveness of scientific research operations on the research vessel.

In September 2012, the "Kexue" vessel was officially delivered for use in Qingdao, marking a new breakthrough in China's marine research capabilities. This was of great significance for safeguarding maritime rights and interests, developing and utilizing marine resources, protecting the marine environment, and facilitating sustainable utilization of marine resources and the environment.

The design and manufacturing units of the vessel were awarded the first prize of Science and Technology Progress Award by the China State Shipbuilding Corporation, the first prize of Shanghai Science and Technology Award, the first prize of

Fig. 10 Awards awarded to "Kexue"

Science and Technology Progress Award by the Chinese Society of Naval Architects and Marine Engineers, and the special prize of Marine Engineering Science and Technology Award by the China Association of Oceanic Engineering (see Fig. 10).

Chapter 8
Outlook for the Development of Research Vessels

As one of the essential tools for conducting marine research, vessels are indispensable for exploring the oceans. As a platform carrying marine research equipment, research vessels are the prerequisite and fundamental requirement for performing functions related to marine research. Future research vessels will undergo continuous upgrades driven by advancements in science and technology, addressing current limitations such as poor adaptability to extreme environments, low efficiency in scientific operations, insufficient precision in navigation and positioning, and particularly issues related to structural material pressure resistance and corrosion durability in polar regions. Innovations in hull design, propulsion systems, and automated controls will emerge to continuously meet the evolving demands of scientific research and exploration. Many experts suggest the timely establishment of a national-level policy research institution or research center dedicated to top-level research in marine science. The focus should be put on linking policies with the development of engineering equipment through scientific medium and long-term planning. It is necessary to establish a unified high-level management center or association for marine research equipment. This center is intended to classify and grade research vessels and carriers, promote standardization and series production, implement open sharing and unified management, and maximize overall efficiency. Another suggestion is to promote the application of large-scale and multi-type unmanned equipment for networked observations and explorations, and to foster the coordinated development of ocean satellites, research vessels, subsurface and surface buoys, and unmanned equipment. Additionally, the development of a stereoscopic observation system involving air, space, sea, and underwater domains should be prioritized. It is also recommended to expedite major projects such as deep-sea and polar investigations, emphasizing their leading role in spurring the need for new technologies and new equipment. Furthermore, policies supporting the localization of research and detection equipment and laboratory instrumentation should be introduced. They also proposed the establishment of a new generation of marine research platforms, focusing on unmanned equipment for extreme or special environments and promoting their cluster-based

L. Han et al., *History of the Research and Development of Chinese Marine Research Vessels*, https://doi.org/10.1007/978-981-96-8414-4_8

intelligent collaborative operation. The future holds the following characteristics for research vessels.

1 Diversified Ship Types

Current research vessels, whether monohulls or small-waterline catamarans, predominantly operate on the water surface. However, future research vessels will introduce semi-submersible types for "settling" in specific areas. There will be no strict distinction between the above-water and underwater parts of such research vessels. Under favorable sea conditions, the majority of the vessel's hull will remain above the water surface. In the event of severe weather, such as storms and high waves, the hull will submerge to adapt to the harsh environment. These research vessels will incorporate advanced wind and wave energy generating sets, utilizing natural energy sources from the marine environment. This approach allows for self-sustainability, energy conservation, and environmental friendliness.

2 Modular Integration

As the requirements for multidisciplinary and 3D integrated observation in marine research increase, modular design and construction will play a significant role in enhancing the economic efficiency and effectiveness of research vessels. Future research vessels will feature movable equipment on desks and modular laboratories, and enhanced capabilities for expanding station capacity. This will not only simplify the complexity of ship design and construction but will also enhance ship utilization efficiency. There will be greater flexibility in the use of research vessels and longer service life of instruments and equipment. All this will significantly improve the efficiency of oceanographic surveys. Currently, modularization is primarily achieved through the utilization of containers fixed on the open decks of vessels' bow and stern areas. These containers can be customized to cater to various scientific research missions, such as radiation research and clean sampling. In the future, modular design will become more refined and diverse, enabling the collection of data from multiple disciplines and fields during a single voyage.

3 Higher Energy and Environment Efficiency

Currently, the majority of research vessels are propelled by diesel engines. While this choice is cost-effective and convenient for operation and maintenance, it causes pollution and the efficiency is not very high. Increasingly stringent regulations by the International Maritime Organization have necessitated the use of ultra-low sulfur fuel, additional scrubbers, selective catalytic reduction (SCR) systems, exhaust gas recirculation (EGR) systems, and other new green environmental protection technology systems to control emissions of sulfur oxides and nitrogen oxides. In the future, research vessels will adopt more hybrid power systems and even single non-fossil fuel propulsion methods. This could include using clean energy sources such as liquefied natural gas, high-capacity fuel cells, solar power, and other clean energy technologies. Research vessels are poised for a power revolution, utilizing environmentally friendly and clean fuels to significantly reduce emissions of sulfur oxides and nitrogen oxides into the atmosphere.

4 Intelligentization

Research vessels, like other ship types, will also evolve towards intelligence. Future research vessels will not be limited to general automation features such as dynamic positioning and autonomous navigation but will evolve into intelligent ships. Utilizing various technologies, such as sensors, communications, the Internet, and the Internet of Things (IoT), research vessels will be able to automatically perceive and acquire data and information from various internal equipment, systems, and external environments including marine environments, weather, logistics, and ports. With automatic control technologies and big data processing, intelligent navigation, management, and maintenance of the vessel can be realized.

High-level intelligentization of the ship system will reduce the number of crew members, improve their quality as well as their living and working conditions, and significantly enhance the precision of surveys in time and space. It will enable automatic time–space synchronized observations, automatic track tracing for mapping and analysis, and even the automatic generation of initial scientific research reports.

Future research vessels will incorporate novel computer and artificial intelligence technologies to gradually automate engine rooms and navigation, and further to enhance navigation safety and improve the reliability of research operations. In addition, numerous unmanned devices and unmanned deep-diving submersibles will be equipped to execute more complex scientific research tasks.

Unmanned technology will be widely applied, with unmanned aerial vehicles and unmanned deep-diving submersibles becoming standard features on research vessels. The new-generation unmanned devices will significantly reduce the need for communication and monitoring by research personnel. Once deployed, these devices can independently carry out detection, target identification, and sampling tasks, thus undertaking data collection in various water surface and underwater environments that are beyond human capabilities. In the event of unpredictable changes in the environment, they are also be capable of self-adjusting, overcoming obstacles, and collaborating with other unmanned detection instruments through the application of navigation assistance and communication relays. Consequently, the vessel's ability to handle complex scientific research tasks will be significantly enhanced.

Intelligent unmanned research vessels will break the conventional mode of oceanographic surveys and pave the way for utilizing intelligent research vessel-borne equipment to support mass unmanned equipment to coordinate marine observations across different regions.

Intelligent research vessels will enable collaboration among aerial, surface, submersible, and deep-sea unmanned systems by intelligent remote control and autonomous navigation. They will obtain and assess real-time information on marine environments, predict temporal and spatial changes in the ocean, environment, and climate, conduct research and innovation related to multi-scale oceanic changes and their impacts on climate resources, and maintain offshore wind fields, among other capabilities.

It is foreseeable that unmanned systems (AUV/ROV/ARV/Glider/small USVs, etc.) will be widely used in marine research in the near future. Advanced research vessels will inevitably become mother ships or flagships for unmanned systems. The next generation of research vessels should be able to accommodate clustered unmanned scientific research equipment, and act as an equipment platform that requires minimal human intervention or even operates unmanned, allowing for both intelligent navigation and unmanned scientific research operations. This platform will highly integrate new technologies, such as ship adaptability technology for harsh sea conditions, intelligent navigation and remote control technology, unmanned/clustered robotic technology, remote sensing, and high-speed communication. Prominent features include the ability to: a. carry out continuous scientific surveys and intelligent operations in extreme sea conditions (which are typically 2–3 levels higher than current research vessels); b. rapidly deploy and retrieve batches of unmanned scientific research equipment at multiple points; c. employ highly automated analysis and experimental systems; and d. collaborate and intelligently network with airborne and underwater unmanned systems. This new type of intelligent scientific research platform represents a revolutionary shift from the traditional "human-operated" model and will be a focal point for future research vessel development.

5 Networked Transmission

With advances in computer and communication technologies, future research vessels will engage in real-time information exchange with onshore laboratories, enabling seamless integration of onboard and onshore data. Onshore control rooms, as well as onboard laboratories and working decks, will be equipped with video interaction devices such as cameras and displays that facilitate remote guidance on oceanographic surveys from onshore scientists. Each research vessel at sea will serve as an "eye" for the onshore laboratory whereby data from onboard scientific research equipment will be synchronized with the onshore laboratory in real time, significantly improving the timeliness of scientific research data. Informatization technology will be integrated into scientific research management, enabling network connections between different onboard systems, ship-to-shore systems, and ship-to-ship systems. This integration will facilitate the sharing of hardware and software resources, as well as database information, to enable real-time analysis, storage, and exchange of survey data.

6 Stereoscopic Observation

Stereoscopic observation will serve as a crucial operation method for future marine observations. It will involve the integration of marine radar station network, surface and subsurface buoy network, seabed observation network, surface drifting buoy network, marine ecological monitoring point network, satellite-based marine observation (monitoring) system, research vessel fleet, profile drift buoy network, and marine mobile observation (monitoring) system. Stereoscopic observation will combine various types of data related to marine space, environment, ecology, and resources, while integrating advanced marine observation technologies and methodologies. This integration will enable multi-factor, high-density, all-weather, and fully automated global marine stereoscopic observation, providing essential technical support for marine ecological preservation, disaster prevention and reduction, response to changes in oceanic climate, safeguarding of maritime economic security, and participation in global marine governance.

Future research vessels will be equipped with devices that can interact with other stereoscopic observation systems, providing support for research vessel operations through the exchange of information with other detection systems.

7 Development of Scientific Research Equipment

7.1 Ocean Carriers

The advancement of ocean carrier technology has brought about revolutionary changes in marine scientific exploration, particularly in the field of unmanned carriers, also known as marine robots. These ocean carriers possess remarkable adaptability to complex environments, diverse modular carrying capabilities, and high levels of unmanned intelligence. They are capable of enduring and performing long-term work in highly dangerous, polluted, and harsh sea conditions. China has initially established a system composed of manned submersibles, remotely operated vehicles, autonomous underwater vehicles, autonomous remotely operated vehicles, underwater gliders, wave gliders, unmanned surface vehicles, hydraulic work-class ROVs, and AUVs. The country has also successfully developed a comprehensive range of full-depth equipment, ranging from portable and lightweight to medium-sized and heavy-duty equipment, which shares similar performance indexes with international counterparts. In the field of ARVs, China has kept pace with the global development trend of hybrid robots. Key achievements include the "Haidou" series of 11,000-meter-depth capability and the "Wenhai 1" with a 6,000-meter-depth capability. China has become the third nation, following Japan and the US, with the ability to independently develop autonomous remotely operated robots capable of operating at depths exceeding 10,000 m. In the field of surface and underwater gliders, China has made remarkable progress and has developed exemplary equipment such as the "Haiyan" series, "Haiyi" series, and "Black Pearl". In 2017, the "Haiyan" underwater glider reached a dive depth of over 6,000 m, maintaining 91 days of continuous operation, and traveling 1,884 km. Notably, these records broke new ground in terms of underwater glider dive depth, endurance, and number of observation profiles. It also experienced its first application in Arctic scientific exploration.

The development of these key equipment has provided robust technical support for China's seabed resource explorations, geological and deep-sea biological surveys, among other areas.

7.2 Ocean Subsurface and Surface Buoys

The research on subsurface and surface buoy technology began in the 1960s. Since then, leading marine nations in the world have achieved long-term operational capabilities in this field. These technologies have matured in terms of reliability, accuracy, stability, and other aspects. While starting the development of ocean subsurface and surface buoys at a later stage, China has achieved breakthroughs in certain areas. The first buoy was developed in the 1960s, and in the 1980s, the first generation of self-contained subsurface buoys were deployed successfully in the South China Sea. In 2012, China successfully developed and deployed the "Bailong" 7000-m buoy, which entered the global ocean observation system. With this achievement, China became the third country, after the US and Japan, to possess deep-sea climate buoy observation capabilities. In 2016, the Chinese Academy of Sciences overcame the challenge of long-term stable real-time transmission of subsurface buoy data in the western Pacific, significantly expanding the depth of real-time transmission of observational data from 1,000 m to 6,000 m. In 2017, China completed the foundation construction of "South China Sea Subsurface Buoy Observation Network" with over 40 sets of self-developed subsurface buoys for real-time monitoring of marine environments, enabling comprehensive observation coverage of the deep-sea basins in the South China Sea. Through years of exploration and development, China has established subsurface and surface buoy operational observation systems in coastal areas, as well as regional networks such as the South China Sea Stereoscopic Observation Network and the Western Pacific Ocean Scientific Observation Network in high seas. The future development of ocean subsurface and surface buoy technology will provide invaluable data to support the country's long-term continuous observation of marine environments and climate forecasting, as well as the study of marine dynamic environmental systems.

7.3 Ocean Exploration Equipment

Ocean exploration systems include water detection, atmospheric detection, seabed detection, deep-sea extreme environment detection, and on-site verification of remote sensing information. The quality of oceanographic data and the level of autonomous control in marine research are determined by the effectiveness of deep-sea detection systems represented by acoustic positioning technology, seabed detection systems represented by the multi-beam detection system, and water detection systems represented by the CTD detection system and Doppler current profilers.

Mapping of marine topography and geomorphology is a fundamental requirement for marine research, and multi-beam sounding systems are considered an advanced mapping solution highly favored by global marine research institutions. China has successfully developed various shallow-water multi-beam sounding systems. The first deep-sea ultra-short baseline positioning system product, which provides positioning detection for the "Jiaolong" HOV, demonstrates stable and reliable performance. In 2017, China achieved a positioning accuracy of 0.3 m and a positioning efficiency of over 90% using an independently developed high-precision acoustic comprehensive positioning system on the "Deep Sea Warrior" HOV, signifying that China's deep-sea comprehensive positioning capability has reached a world-leading level. Future developments in China's ocean exploration equipment will focus on advancements in long-range and high-precision positioning.

7.4 Marine Sensors

Marine sensors play a critical role in ocean observation and monitoring. These sensors encompass a range of technologies that are considered core technologies in the field of marine research equipment. With national support, China has made breakthroughs in the field of marine environmental monitoring sensors. However, there still exists a certain gap compared to foreign counterparts in terms of performance, stability, and accuracy, especially in high-end CTD profiler sensors and meteorological sensing equipment.

Future marine research will undoubtedly be interdisciplinary in nature, and sensors will serve as the core foundation of the marine science Internet of Things. China must continue to advance in the research, application, and promotion of basic component technologies and methods related to marine sensors to achieve new breakthroughs.

7.5 *Miniaturization, Automation, and Diversification of Equipment*

As scientific and technological advancements continue, scientific research equipment will inevitably evolve towards miniaturization, automation, and diversification.

Miniaturization is an inevitable trend in the development of marine research equipment. Reducing the size of any equipment without compromising its functionality presents a challenging task. If cranes, winches, and detection devices on research vessels can be made smaller, more space can be allocated for other scientific research equipment onboard. This can significantly expand the capabilities of research vessels. Achieving higher precision and functionality within a smaller form factor signifies an important index of the design level of scientific research equipment systems.

Automation has long been a pursuit of humanity. Future scientific research equipment systems will primarily rely on "self-action" based on databases, powerful computing and analysis capabilities, with the operator playing a secondary role. Scientific research operations can be completed through remote control and appropriate intervention from shore.

Diversification primarily refers to the application of new technologies on research vessels. With rapid advancements in deep-sea technology, microelectronic technology, automatic control technology, materials, and processing techniques, various technologies previously limited to areas such as offshore oil exploration, cable-less autonomous submersibles, underwater gliders and unmanned underwater robots, deep-water multi-beam sounding systems, deep-water stratigraphic profiling systems, and seismic acquisition systems of long spread, high capacity, and high resolution can now be applied across various marine research activities. It is conceivable that technologies, such as quantum communication and nanotechnology, could become key components driving major breakthroughs in specific scientific research equipment and systems, and bring about groundbreaking changes in ocean exploration technology.

Epilogue

In 1950, China's annual shipbuilding capacity was merely over 10,000 tons. At that time, there were no 10,000-ton vessels sailing on the rivers and seas of China that were designed and built domestically. Over the past 70 years, under the leadership of the Communist Party of China, scientists and shipbuilding workers have increased China's annual shipbuilding capacity to over 60 million tons by 2018. Today, we can not only design and build conventional ships but also high-tech, high-value-added ships that are considered the "jewels" of the shipbuilding industry, making China the world's leading shipbuilding nation.

In commemoration of the 100th anniversary of the founding of the Communist Party of China in 2021, the MARIC, Shanghai Society of Naval Architects and Marine Engineers, Jiangnan Shipyard Group, Hudong-Zhonghua Shipbuilding Group, Shanghai Waigaoqiao Shipbuilding Co. Ltd., Shanghai Merchant Ship Design and Research Institute, and Shanghai Jiao Tong University Press have collaborated to publish a series of books entitled "History of Chinese Ships Research and Development", as a tribute to the 100th anniversary of the Party's founding. The series consists of 10 volumes: *History of Chinese Tanker Research, History of Chinese Container Vessel Research and Development, History of the Research and Development of Chinese Marine Research Vessels, History of Chinese Dredger Research and Development, Research History of Chinese Liquefied Gas Carrier, History of Chinese Working Ship Research and Development, History of Chinese Bulk Carrier Research and Development, History of Chinese Passenger Ship Research and Development, History of Chinese Hovercraft Research and Development, Research History of Chinese Offshore Oil & Gas Development Units.*

The compilation of this series of books has been supported and actively participated by academician Zeng Hengyi from the Chinese Academy of Engineering, as well as senior and young experts, designers, and technicians in the field of ship research and development. Their concerted efforts have contributed to the successful completion of this series.

The series of books provides comprehensive and reliable information and is characterized by its originality. More than 70 experts and technicians engaged in ship and marine engineering research, design, and construction have contributed to its

L. Han et al., *History of the Research and Development of Chinese Marine Research Vessels*, https://doi.org/10.1007/978-981-96-8414-4

compilation. These individuals are witnesses and contributors of the development and achievements of China's shipbuilding industry. From the moment they undertook the task of compilation, they devoted themselves tirelessly to this endeavor, transforming their passion for ship science and their love for the homeland into the pages of these books. Each volume has undergone extensive discussions and multiple revisions, from the initial outline to the first draft and final draft, all aimed at achieving the highest standards in publication.

Moreover, the showcased products in this series are not only a reflection of the era's accomplishments but also serve as precious historical materials and lessons learned from China's ship research and development, which provide inspiration and valuable insights for young professionals involved in ship design.

During the compilation process, the series has received support and care from various organizations and leaders. Special thanks go to Dr. Wu Gang, the Chief Designer of multipurpose research vessels at the MARIC, and Dr. Zhang Fumin, who participated in the writing and review of the volume on research vessels. Additionally, heartfelt thanks are extended to the diligent editors for their remarkable contributions to this series. Appreciation is also given to the numerous anonymous ship designers who participated in the reviewing process and provided valuable suggestions. This series references books, newspapers, viewpoints, and images, for which we express our gratitude. Due to the extensive time span and the challenges involved in data collection, some ship types may have been inadvertently excluded. Whenever possible, Chinese pinyin have been used for ship names, personal names, and place names. English names have been adopted when industry conventions default to them.

This series may contain inaccuracies or shortcomings, and we sincerely welcome criticism and corrections from experts and readers.

Zhang Yi

Bibliography

1. Qu N, Han L, Wu G (2019) Research vessels. Shanghai Scientific & Technical Publishers, Shanghai
2. Zhang Y (2017) Wholehearted dedication to Xiangyanghong: A story of shipbuilding expert Zhang Bingyan. China Science and Technology Press, Beijing
3. Zhang Y (2016) Writing blue dreams with life: biography of Zhang Bingyan. Shanghai Jiao Tong University Press, Shanghai
4. Wu G, Huang W (2018) Design of multipurpose oceanographic research vessels. Shanghai Jiao Tong University Press, Shanghai
5. Sun S (2016) Into the deep ocean: multipurpose oceanographic research vessel. Zhejiang Education Publishing House, Zhejiang
6. Feng S (1998) An introduction to marine science. Higher Education Press, Beijing
7. Shi M, Gao G, Bao X et al (2000) Methods of oceanographic survey. China Ocean University Press, Qingdao
8. Yang S (2012) Haiyang 6" in 2011: record of the 23rd scientific expedition of China Ocean. China Ocean Press, Beijing
9. Li W, Ning J (2017) Introduction to ship types. Dalian Maritime University Press, Dalian
10. Cui J (2012) Ocean chronicles. China Youth Press, Beijing
11. Zhang B (2011) Humanized design of "Haiyang 6". Chin Eng Sci 13(4):24–28, 36
12. Liang Q (2017) Compilation of Chinese ship types (2012–2016). Shanghai Jiao Tong University Press, Shanghai
13. Zhang F, Zhang Y (2014) Analysis of several issues in general design of recent marine multipurpose survey vessels. Ship Sci Technol 36(S1):26–31
14. Chen L, Su Q, Dong L et al (2014) Comparative analysis of domestic and foreign oceanographic survey vessels. Ship Sci Technol 36(S1):2–7
15. Tang Q (2017) Analysis of the coordination between anchor bolts and transverse side operation equipment. Ship Boat 28(2):71–75
16. Zhou N (2014) Current status and future vision of China's oceangoing survey vessels. Ship Sci Technol 36(S1):15–20

L. Han et al., *History of the Research and Development of Chinese Marine Research Vessels*, https://doi.org/10.1007/978-981-96-8414-4

Index

L. Han et al., *History of the Research and Development of Chinese Marine Research Vessels*, https://doi.org/10.1007/978-981-96-8414-4

T

U

MIX
Papier aus verantwortungsvollen Quellen
Paper from responsible sources
FSC® C105338

If you have any concerns about our products,
you can contact us on
ProductSafety@springernature.com

In case Publisher is established outside the EU,
the EU authorized representative is:
Springer Nature Customer Service Center GmbH
Europaplatz 3, 69115 Heidelberg, Germany

Printed by Libri Plureos GmbH
in Hamburg, Germany